Thermal Cracking
in Concrete
at Early Ages

Thermal Cracking in Concrete at Early Ages

Proceedings of the International Symposium held by RILEM (The International Union of Testing and Research Laboratories for Materials and Structures) at the Technical University of Munich and organized by RILEM Technical Committee 119 and the Institute of Building Materials of the Technical University of Munich, Germany. The Symposium is supported by the Japan Concrete Institute (JCI), the German Concrete Association (DBV), the Association of the Bavarian Building Industry (BBIV), and the American Concrete Institute (ACI).

Munich
October 10–12, 1994

EDITED BY

R. Springenschmid

Institute of Building Materials,
Technical University of Munich, Germany

CRC Press
Taylor & Francis Group
Boca Raton London New York

CRC Press is an imprint of the
Taylor & Francis Group, an **informa** business

A CHAPMAN & HALL BOOK

CRC Press
Taylor & Francis Group
6000 Broken Sound Parkway NW, Suite 300
Boca Raton, FL 33487-2742

First issued in paperback 2019

© 1995 RILEM
CRC Press is an imprint of Taylor & Francis Group, an Informa business

No claim to original U.S. Government works

ISBN-13: 978-0-419-18710-3 (hbk)
ISBN-13: 978-0-367-44930-8 (pbk)

A catalogue record for this book is available from the British Library

Publisher's Note
This book has been produced from camera ready copy provided
by the individual contributors in order to make the book available
for the Symposium.

Visit the Taylor & Francis Web site at
http://www.taylorandfrancis.com

and the CRC Press Web site at
http://www.crcpress.com

Contents

Preface   xi
Préface   xiii
Scientific Council   xvii
RILEM Technical Committee TC 119 TCE   xvii
Organising Committee   xviii

PART ONE   HEAT OF HYDRATION   1

1   **Numerical and experimental adiabatic hydration curve determination**   3
E.A.B. KOENDERS and K. van BREUGEL

2   **Thermal and mechanical modelling of young concrete based on hydration process of multi-component cement minerals**   11
T. KISHI and K. MAEKAWA

PART TWO   PREDICTION OF TEMPERATURE DEVELOPMENT   19

3   **Prediction of temperature distribution in hardening concrete**   21
Ch. WANG and W.H. DILGER

4   **Low-heat Portland cement used for silo foundation mat - temperatures and stresses measured and analyzed**   29
M. YAMAZAKI, H. HARADA and T. TOCHIGI

5   **Modelling of heat and moisture transport in hardening concrete**   37
P.E. ROELFSTRA and T.A.M. SALET

6   **Modelling of temperature and moisture field in concrete to study early age movements as a basis for stress analysis**   45
J.-E. JONASSON, P. GROTH and H. HEDLUND

7   **Influence of the geometry of hardening concrete elements on the early age thermal crack formation**   53
G. De SCHUTTER and L. TAERWE

PART THREE DETERMINATION AND MODELLING OF MECHANICAL PROPERTIES   61

8   **Stiffness formation of early age concrete**   63
P. PAULINI and N. GRATL

9   **From microstructural development towards prediction of macro stresses in hardening concrete**   71
S.J. LOKHORST and K. van BREUGEL

10   **Effect of creep in concrete at early ages on thermal stress**   79
H. UMEHARA, T. UEHARA, T. IISAKA and A. SUGIYAMA

11   **Basic creep and relaxation of young concrete**   87
G. WESTMAN

12   **Estimation of stress relaxation in concrete at early ages**   95
H. MORIMOTO and W. KOYANAGI

13   **Stresses in concrete at early ages: comparison of different creep models**   103
I. GUENOT, J.M. TORRENTI and P. LAPLANTE

14   **Young concrete under high tensile stresses - creep, relaxation and cracking**   111
A. GUTSCH and F.S. ROSTÁSY

15   **The thermal stress behaviour of concrete based on the micromechanical approach**   119
H. OHSHITA and T. TANABE

16   **Numerical simulation of the effect of curing temperature on the maximum strength of cement-based materials**   127
K. van BREUGEL

PART FOUR MEASUREMENT OF THERMAL STRESSES IN THE LABORATORY   135

17   **Development of the cracking frame and the temperature-stress testing machine**   133
R. SPRINGENSCHMID, R. BREITENBÜCHER and M. MANGOLD

18   **Investigation of concrete behaviour under restraint with a temperature-stress test machine**   145
G. THIELEN and W. HINTZEN

19 Determination of restraint stresses and of material properties during
    hydration of concrete with the temperature-stress testing machine   153
    K. SCHÖPPEL, M. PLANNERER and R. SPRINGENSCHMID

20 Material characterization of young concrete to predict thermal
    stresses   161
    A. NAGY and S. THELANDERSSON

PART FIVE MEASUREMENT OF THERMAL STRESSES
IN SITU   169

21 Thermal stress in full size RC box culvert   171
    T. MISHIMA, H. UMEHARA, M. YAMADA and M. NAKAMURA

22 Thermal cracking in wall of prestressed concrete egg-shaped
    digester   179
    T. YOSHIOKA, S. OHTANI and R. SATO

23 Study of external restraint of mass concrete   187
    M. ISHIKAWA and T. TANABE

PART SIX INFLUENCE OF CONSTITUENTS AND
COMPOSITION OF CONCRETE ON CRACKING
SENSITIVITY   195

24 The effect of slag on thermal cracking in concrete   197
    M.D.A. THOMAS and P.K. MUKHERJEE

25 Minimization of thermal cracking in concrete members at early
    ages   205
    R. BREITENBÜCHER and M. MANGOLD

26 The effect of thermal deformation, chemical shrinkage and
    swelling on restraint stresses in concrete at early ages   213
    K. SCHÖPPEL and R. SPRINGENSCHMID

27 Effect of autogenous shrinkage on self stress in hardening concrete 221
    E. TAZAWA, Y. MATSUOKA, S. MIYAZAWA and S. OKAMOTO

28 High performance concrete: early volume change and cracking tendency 229
    E. SELLEVOLD, Ø. BJØNTEGAARD, H. JUSTNES and P.A. DAHL

29 Factors influencing early cracking of high strength concrete   237
    I. SCHRAGE and Th. SUMMER

PART SEVEN   COMPUTATIONAL ASSESSMENT OF STRESSES AND CRACKING

PART SEVEN   COMPUTATIONAL ASSESSMENT OF STRESSES AND CRACKING   245

30   Experience in controlled concrete behaviour   247
E. MAATJES, J.J.M. SCHILLINGS and R. De JONG

31   Numerical simulation of crack-avoiding measures   255
J. HUCKFELDT, H. DUDDECK and H. AHRENS

32   Thermal prestress of concrete by surface cooling   265
M. MANGOLD

33   Defining and application of stress-analysis-based temperature
difference limits to prevent early-age cracking in concrete structures   273
P.E. ROELFSTRA, T.A.M. SALET and J.E. KUIKS

34   Numerical simulation of temperatures and stresses in concrete at
early ages: the French experience   281
J.M. TORRENTI, F. de LARRARD, F. GUERRIER, P. ACKER and
G. GRENIER

35   A practical planning tool for the simulation of thermal stresses
and for the prediction of early thermal cracks in massive concrete
structures   289
P. ONKEN and F.S. ROSTÁSY

36   Prediction of temperature and stress development in concrete structures   297
E.S. PEDERSEN

37   Sensitivity analysis and reliability evaluation of thermal cracking
in mass concrete   305
K. MATSUI, N. NISHIDA, Y. DOBASHI and K. USHIODA

38   Deformations and thermal stresses of concrete beams constructed
in two stages   313
R. SATO, W.H. DILGER and I. UJIKE

39   Thermal stresses computed by a method for manual calculations   321
M. EMBORG and S. BERNANDER

40   Thermal effects, cracking and damage in young massive concrete   329
J.P. BOURNAZEL and M. MORANVILLE-REGOURD

41   Temperature field and concrete stresses in a foundation plate   337
P. PAULINI and D. BILEWICZ

PART EIGHT PRACTICAL MEASURES FOR AVOIDANCE OF CRACKING – CASE RECORDS

PART EIGHT PRACTICAL MEASURES FOR AVOIDANCE
OF CRACKING – CASE RECORDS   345

42 Report on construction of water-impermeable concrete structures
    with high-level ground-water ("Weisse wannen" - "White troughs")
    in Bavaria   347
    J.-St. KREUTZ

43 On the reliability of temperature differentials as a criterion for
    the risk of early-age thermal cracking   353
    M. EBERHARDT, S.J. LOKHORST and K. van BREUGEL

44 Why are temperature-related criteria so unreliable for predicting
    thermal cracking at early ages?   361
    M. MANGOLD and R. SPRINGENSCHMID

45 Inherent thermal stress distributions in concrete structures and
    method for their control   369
    A.R. SOLOVYANCHIK, B.A. KRYLOV and E.N. MALINSKY

46 Practical experience with concrete technological measures to avoid
    cracking   377
    R. SPRINGENSCHMID, R. BREITENBÜCHER and M. MANGOLD

47 Risk of cracking in massive concrete structures - new developments
    and experiences   385
    S. BERNANDER and M. EMBORG

48 Thermal cracking in the diaphragm-wall concrete of Kawasaki Island 393
    D.D. LIOU

49 Measures to avoid temperature cracks in concrete for a bridge deck 401
    W. FLEISCHER and R. SPRINGENSCHMID

50 Avoidance of early age thermal cracking in concrete structures -
    predesign, measures, follow-up   409
    M. EMBORG and S. BERNANDER

51 Water-tight design, artificial cooling or extra reinforcement 417
    W.G.L. WAGENAARS, K. van BREUGEL

52 A large beam cooled with water shower to prevent cracking 425
    M. YAMAZAKI

53 Reduction of thermal stresses in structures with air-cooling 433
    H. HEDLUND, P. GROTH and J.E. JONASSON

54 **Countermeasure for thermal cracking of box culvert**   441
    M. IWATA, K. SAITO, K. IKUTA and T. KAWAUCHI

55 **High performance concrete: field observations of cracking tendency
    at early age**   449
    R. KOMPEN

56 **Establishment of a new crack prevention method for dams by
    RCD method**   457
    N. SUZUKI, T. IISAKA, S. SHIRAMURA and A. SUGIYAMA

Author index   465
Subject index   467

Preface

Cracks in mass concrete structures caused by the hydration heat of the cement have been a well-known phenomenon since the beginning of this century. Methods of avoiding such cracks have been developed mainly for large concrete dams and other massive hydraulic engineering structures. In order to reduce the heat development, pozzolanas and since 1932, low-heat cements have been used. Further progress aimed at reducing the high maximum temperature caused by the hydration heat has been made by the use of very low cement contents, coarse aggregates, cooling the concrete materials, limitation of lift-heights and by pipe cooling.

Although the processes of heat generation and dissipation were familiar to engineers, the specification of the maximum permissible temperature difference between the concrete mass and its base and between the inside and the surface of a concrete block was based solely on experience. The essential properties of a specific concrete such as its tensile strength or its coefficient of thermal expansion were not considered.

In the past decades, cracks in the structural concrete of foundations, bridges, tunnel linings and other medium-sized concrete elements have become an increasing problem. It has been found that drying shrinkage is often of minor importance. The heat of hydration as well as other temperature changes were established as the main causes of restraint stresses and cracks in unreinforced as well as in reinforced concrete.

In the late sixties the first attempts were made to estimate the stresses caused by restrained thermal deformations and to compare them with the increasing tensile strength of the young concrete. Two points turned out to be extremely difficult:

- The results of thermal stress calculations depend strongly on the evaluation of the increasing stiffness of the concrete during its transition from a semi-liquid to a solid state. However, the stiffness is difficult to measure and predict.
- Restraint stresses cannot be determined with conventional methods and therefore no data were available to verify the results of stress calculations.

In 1969 the first laboratory equipment, the cracking frame, was developed which allowed model tests. By measuring the stress response of young concrete to changing temperature we gained a deeper understanding of the changes which occur when the expansion or contraction of a concrete element is prevented and consequently converted into stresses. The Munich Temperature Stress Testing Machine (1984) and similar machines in several other research institutes now allow stress measurement for any degree of restraint.

In recent years much research work has been devoted to the calculation of the early age restraint stresses and to the determination of the risk of cracking. Computer programs have been based on the properties of materials, the hydration heat development, the increase of stiffness and the decrease of relaxation capacity, the increasing tensile strength, the coefficient of thermal expansion and the influence of chemical reactions on the deformation. All these factors depend largely on age, temperature, cement type and concrete mix composition. Realistically speaking, it is only possible to assess roughly the effects of these factors. However, much progress has been made with models for the approximation of materials properties. Such models require assumptions about the restraint conditions and the expected temperature conditions on site.

Promising new methods have been developed in Japan and France to measure restraint stresses in situ. The comparison of test results both in the laboratory and in the field with the results of calculations is a source of further progress in this area.

In recent years high-strength concrete has proved to be an extremely sensitive material regarding cracking at an early age. This is not only a consequence of the hydration heat: Autogenous shrinkage due to self-desiccation and chemical reactions of the sulphate phase can also be important.

Unexpected cracking of structural or mass concrete can not always be attributed to the inexperience of the field engineer, and the limited knowledge of many problems in this area has strongly encouraged research all over the world. In 1989 RILEM, the International Union of Testing and Research Laboratories for Materials and Structures, established a Technical Committee, TC 119, on "The Avoidance of Thermal Cracking in Concrete at Early Ages". As well as the exchange of opinions and experience, the tasks of this committee are to prepare a State of the Art Report and make recommendations for the following test methods:

- Determination of the semi-adiabatic hydration heat
- Determination of the adiabatic hydration heat
- Restraint stress measurements in the laboratory with the cracking frame
- Restraint stress measurements in situ with the stress-gauge.

The International Symposium in Munich between October 10th and 12th, 1994 thus takes place in a dynamic period of development. A number of questions regarding the avoidance of early age cracking have been solved and the results are ready for practical application. To answer other questions we have many suggestions based on test results or from theoretical considerations.

It still has to be determined how these methods can be applied most effectively in practice, whether they need further development or whether they can only serve as a stimulation for further research.

In order to limit the scope of the work, drying shrinkage is not taken into consideration. Furthermore the role of steel reinforcement to avoid wide crack opening and the advantageous use of prestressing are not treated in this symposium.

The avoidance of early age cracking is a task which requires theoretical knowledge, sound engineering judgement and extensive experience. Furthermore, dedicated engineers must ensure that all the necessary considerations are carried out in practice.

I hope this symposium becomes a milestone of progress of our field.

Rupert Springenschmid
Munich, July 1994

Préface

La fissuration du béton dans les constructions massives, due à la chaleur d'hydratation du ciment est un phénomène bien connu depuis le début du siècle. Des méthodes permettant d'éviter les fissures de ce type ont été développées surtout pour les grands barrages en béton ou pour les structures massives du génie hydraulique. Les pouzzolanes et, depuis 1932, les ciments à faible chaleur d'hydratation sont utilisés pour limiter la chaleur libérée. D'autres progrès tendant à réduire le pic de température dû à la chaleur d'hydratation ont été réalisés par la mise en oeuvre de dosages en ciment fortement réduits, par l'emploi de granulats plus gros, par réfrigération des constituants du béton, par la limitation de la hauteur des levées, et par le refroidissement par circulation d'eau dans des serpentins.

Bien que les processus de production et de dissipation de la chaleur sont familiers aux ingénieurs, la spécification d'une différence de température admissible entre la masse de béton et la fondation et entre le coeur et la surface d'un massif de béton faisait exclusivement appel à l'expérience. Les propriétés fondamentales d'un béton donné, telles que sa résistance à la traction ou son coefficient de dilatation, n'étaient pas prises en compte.

Au cours des dernières décennies, les fissures dans le béton de structure de fondations, de ponts, de revêtements de tunnel, ou d'autres constructions de taille moyenne sont devenues un problème croissant. Il a été montré que le retrait de dessiccation est souvent d'importance mineure. La chaleur d'hydratation, ainsi que les changements de température, sont reconnus comme étant la principale cause des contraintes dues aux déformations empêchées et des fissures, tant dans le béton non armé que dans le béton armé.

A la fin des années soixante, les premières tentatives pour évaluer les contraintes dues aux déformations thermiques empêchées et pour comparer celles-ci à la résistance à la traction du béton jeune ont été faites. A cette occasion, deux points se sont révélés extrêmement difficiles.

- Les résultats des calculs de contrainte thermique dépendent fortement de l'évaluation de la raideur du béton, raideur qui croît lors de la transition de l'état semi-liquide à l'état solide. Cependant, il est difficile de mesurer et de prévoir la raideur du béton.
- Les contraintes dues aux déformations empêchées ne peuvent être mesurées par des méthodes conventionnelles et, de ce fait, on ne disposait d'aucune donnée pour vérifier les résultats des calculs de contrainte.

En 1969 a été développé le banc de fissuration, premier équipement de laboratoire permettant des essais sur éprouvettes reproduisant les conditions de chantier. La mesure de la réponse en contrainte du béton jeune aux changements de température a rendu possible une compré-hension plus approfondie des changements qui se produisent lorsque l'allongement ou le raccourcissement d'une structure en béton est empêché et provoque l'apparition de contraintes. L'appareil d'étude des contraintes thermiques construit à Munich en 1984 ainsi que d'autres dispositifs similaires développés dans de nombreux autres laboratoires de recherche permettent désormais une évaluation des contraintes sous n'importe quelles conditions de déformation imposée.

Un effort de recherche important a été consacré ces dernières années au calcul des contraintes dues aux déformations empêchées au jeune âge, ainsi qu'à la détermination du risque de fissuration. Les programmes de calcul doivent prendre en compte les propriétés des matériaux, le développement de la chaleur d'hydratation, l'augmentation de la rigidité et la diminution du coefficient de relaxation, la résistance à la traction et son évolution dans le temps, le coefficient de dilatation thermique, ainsi que l'influence des réactions chimiques sur les déformations. Tous ces facteurs dépendent fortement de l'âge, de la température, du type de ciment, et de la composition du béton. De façon réaliste, il n'est que grossièrement possible d'évaluer ces influences. Un progrès important a été accompli avec des modèles permettant de mieux évaluer les propriétés des matériaux. Ces modèles requièrent également des hypothèses sur les conditions aux limites mécaniques ainsi que la prévision des températures in situ.

De nouvelles méthodes fort prometteuses ont été développées en France et au Japon afin de mesurer in situ les contraintes dues à ces déformations empêchées ou gênées. La comparaison de résultats d'essais de laboratoire ou sur le terrain avec les résultats de calculs est la source de plus amples progrès dans ce domaine.

Plus récemment, le béton à haute résistance s'est révélé être un matériau souvent plus sensible à la fissuration au jeune âge. Ce phénomène n'est pas uniquement une conséquence de la chaleur d'hydratation. Le retrait endogène dû à l'auto-dessiccation et aux réactions chimiques de la phase sulfatique peut également jouer un rôle important.

La fissuration inattendue d'un béton de structure ou d'un béton massif ne peut pas toujours être attribuée à l'inexpérience de l'ingénieur de chantier. De ce fait, la connaissance limitée de nombreux problèmes dans ce domaine a fortement suscité un effort de recherche dans le monde entier. En 1989, la Réunion Internationale des Laboratoires d'Essais et de Recherches sur les Matériaux et les Constructions a créé la Commission Technique 119 "Prévention de la fissuration d'origine thermique dans le béton au jeune âge". Outre l'échange de points de vue et d'expérience, le rôle de cette commission est de préparer un rapport sur l'état de l'art et d'élaborer des recommandations pour les méthodes d'essai suivantes:

- Détermination de la chaleur d'hydratation par calorimétrie semi-adiabatique.
- Détermination de la chaleur d'hydratation par calorimétrie adiabatique.
- Mesure en laboratoire des contraintes dues aux déformations empêchées à l'aide du banc de fissuration.
- Mesure in situ des contraintes dues aux déformataions empêchées.

Le colloque international ayant lieu à Munich du 10 au 12 octobre 1994 s'inscrit dans une période dynamique de développement. Un grand nombre de questions ayant trait à la prévention de la fissuration précoce ont été résolues et les résultats sont disponibles pour la mise en pratique. Nos suggestions de réponse à d'autres questions sont fondées sur des résultats d'essais ou dérivent de considérations théoriques. Il convient encore de déterminer comment ces méthodes peuvent être employées le plus efficacement possible, si celles-ci nécessitent un développement plus poussé ou si elles ne sont que destinées à stimuler des recherches plus approfondies.

Afin de limiter l'étendue du domaine, le retrait de dessiccation, le rôle de l'armature en acier pour limiter l'ouverture des fissures, ainsi que l'emploi avantageux de la précontrainte ne sont pas traités dans ce colloque.

La prévention de la fissuration au jeune âge est un exercice qui requiert des connaissances théoriques, un sens aigu de l'ingéniérie et une vaste expérience. En outre, les ingénieurs consciencieux doivent s'assurer que toutes les mesures nécessaires sont mises en oeuvre.

J'espère que ce colloque marquera une étape des progrès de notre domaine.

Rupert Springenschmid
Munich, Juillet 1994

Scientific Council

RILEM Technical Committee TC 119 TCE

Organizing Committee

M. Plannerer (Secretary), Technische Universität München, Germany
R.E. Beddoe, Technische Universität München, Germany
M. Bierschneider, Technische Universität München, Germany
J.-L. Bostvironnois, Technische Universität München, Germany
M. Mangold, Technische Universität München, Germany
K. Schuhmann, Deutscher Betonverein, Germany
W. Stoermer, Bayerischer Bauindustrieverband, Germany
R. Springenschmid, Technische Universität München, Germany

Conference Secretariat

Baustoffinstitut, Technische Universität München
Baumbachstr.7, 81245 München, Germany

Note: It is hoped that a further volume will be produced after the Symposium containing supplemantary papers, reports on the discussion sessions, the conclusions of the Symposium and other material. Readers interested in this volume should contact the Conference Secretariat at the address above.

PART ONE

HEAT OF HYDRATION

(La chaleur d'hydratation)

1 NUMERICAL AND EXPERIMENTAL ADIABATIC HYDRATION CURVE DETERMINATION

E.A.B. KOENDERS and K. van BREUGEL
Delft University of Technology, Delft, The Netherlands

Abstract
Adiabatic hydration curves of concrete mixes are used as input for computer programs utilized for the calculation of
the temperature distribution in hardening concrete. At present no standardised test equipment has been specified with
which adiabatic curves should be determined. A comparison of
test results obtained with different adiabatic calorimeters
has revealed a substantial scatter and has evidenced the
need for a thorough analysis of the cause of this scatter.
Results of such an analysis are presented in this paper.
Problems encountered in adiabatic testing are dealt with. A
computer-based numerical model is presented, with which hydration curves can be calculated as a function of the clinker composition of the cement and the mix composition of the
concrete.
Keywords: Adiabatic Calorimetry, Heat of Hydration,
Modelling of Heat of Hydration, Numerical Simulation.

1 Introduction

For the evaluation of thermal problems in hardening concrete
several computer programs have been developed. For most of
these programs the adiabatic hydration curve of the concrete
mix is an essential part of the input. Until recently the
determination of these hydration curves had to be done experimentally with adiabatic calorimeters. A comparison of
multi-laboratory test results has revealed a substantial
scatter. This scatter originates from differences in the
steering algorithms used for adjusting the temperature of
the control medium, the size of the sample and the type of
the mould. A practical drawback of experimental testing is
that each change of the concrete mix requires an other test.
This is time consuming and does often not fit in the tight
time schedule of a project.
    A new trend in materials science is to simulate the har-
dening process in cement-based materials numerically as a
function of the dominant rate controlling parameters. The
advantage of such numerical models would be, apart from
saving time, that pure adiabatic condition can be simulated.

Thermal Cracking in Concrete at Early Ages. Edited by R. Springenschmid. Published 1994
by E & FN Spon, 2–6 Boundary Row, London SE1 8HN, UK. ISBN: 0 419 18710 3.

In the following some of the basic features of a numerical simulation program will be presented. In the model called HYMOSTRUC, the acronym for HYdration, MOrphology and STRUCtural development, hydration curves can be predicted as a function of the characteristics of the cement and the concrete mix. Due attention is given to the accuracy with which numerical predictions are possible and this as compared to the accuracy achievable with experimental tests.

2   Numerical simulation of hydration and microstructural development in cement-based materials

2.1 Basic features of the simulation model
In the computer-based simulation model HYMOSTRUC hydration curves are calculated as a function of the particle size distribution and the chemical composition of the cement, the water/cement ratio ω_0 and the reaction temperature. Unlike most previously proposed models the effect of physical interactions between hydrating cement particles on the rate of hydration of individual cement particles is modelled explicitly. For numerical evaluation of the interaction between hydrating particles due attention had to be given to the stereological aspect, i.e. the spatial distribution of the cement particles in the paste.

2.2 Stereological aspects
In HYMOSTRUC cement particles are considered to be distributed homogeneously in the paste. An arbitrary particle is considered to be located in the centre of a cell "I_x^c" (Fig. 1). A cell "I_x^c" is defined as a cubic space in which the central particle has a diameter x and further consists of $1/N_x$ times the original water volume and $1/N_x$ times the volume of all particles with a diameter smaller than x μm. N_x is the number of particles with diameter x μm in a certain paste volume.

For the assumed homogeneous spatial distribution it is relatively easy to determine the amount of cement found in a fictitious shell with thickness d surrounding a cement particle with diameter x μm.

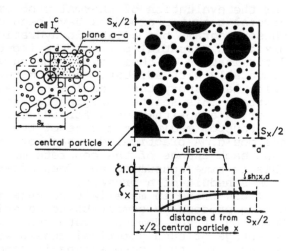

Fig. 1. Cell concept and shell density factor $\zeta_{sh;x,d}$.

For this purpose a shell density factor $\zeta_{sh;x,d}$ has been defined (Fig. 1, bottom part). Going from the periphery of particle x in outward direction the shell density gradually increases from zero at the outer periphery of the anhydrous particle to the cell density ζ_x.

2.3 Particle expansion and particle interaction mechanism

In order to obtain a workable algorithm for the determination of the interaction between hydrating and expanding cement particles the following assumptions were made:

- a. Particles of the same size hydrate at the same rate.
- b. The ratio ν between the volumes of the reaction products and the reactant decreases with increasing temperatures.
- c. Reaction products precipitate in the close vicinity of the cement particles from which they are formed.
- d. Dissolution and expansion of the hydrating cement particles occur concentrically.

The simplifications introduced here are, to a large extent, inherent to the statistical approach that was adopted. For a more extensive justification of these simplifications reference is made to (van Breugel).

The expansion and interaction mechanisms are schematically shown in Fig. 2. On contact with water an arbitrary cement particle x starts to dissolve under formation of reaction products. These product are formed partly inside and partly outside the original surface of the particle. In this way cement particles exhibit an outward expansion, thereby making contacts with neighbouring particles. At a certain time, say t_j, the depth of penetration of the reaction front is $\delta_{in;x,j}$ with a corresponding degree of hydration of this particle $\alpha_{x,j}$. Further hydration of this particle goes along with further expansion and embedding of neighbouring particles. The encapsulation of other particles causes and extra

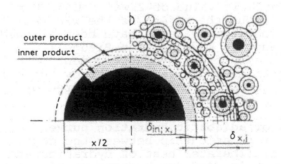

Fig. 2. Interaction mechanism for expanding particles.
Left part: free expansion, formation of inner and outer product. Right part: embedding of particles, several stages.

expansion and, as a consequence of this, the encapsulation of even more particles. For this mechanism of continuous expansion and embedding of particles an algorithm has been developed with which the stereological aspect of structural development can be quantified.

2.4 Rate of penetration of the reaction front

The rate of penetration of the reaction front in an individual cement particle x at time t_j is computed with the "basic rate formula" viz. (in a reduced form):

$$\frac{\Delta\delta_{in;x,j+1}}{\Delta t_{j+1}} = K_0(.) * [(\frac{\delta_{tr}(.)}{\delta_{x,j}})^{\beta_1}]^{\lambda} * F\{\omega_0, T(\alpha)\} \qquad (1)$$

with $\Delta\delta_{in;x,j+1}$ the increase of the penetration depth in time step Δt_{j+1}, $K_0(.)$ the basic rate factor in $\mu m/h$, $\delta_{tr}(.)$ the transition thickness in μm, being the total thickness of the product layer $\delta_{x,j}$ at which the reaction changes from a phase-boundary reaction ($\lambda = 0$) into a diffusion controlled reaction ($\lambda = 1$). The factor $F\{\omega_0, T(\alpha(t))\}$ is a function of the w/c-ratio and the actual reaction temperature.

In an extensive evaluation program, in which a relationship between the two model parameters $K_0(.)$ and $\delta_{tr}(.)$ and the clinker composition of the cement was investigated, the following expressions have been established for the default values of afore mentioned parameters:

$$K_0(C_3S) = 0.02 + 6.6 \ 10^{-6} * [C_3S\%]^2 \qquad [\mu m/h] \qquad (2)$$

and:

$$\delta_{tr}(C_2S) = -0.02 * [C_2S\%] + 4 \qquad [\mu m] \qquad (3)$$

For $K_0(C_3S)$ a standard deviation would hold of 0.008 $\mu m/h$. Upper and lower bound values for the transition thickness $\delta_{tr}(C_2S)$ can be obtained by increasing or decreasing the default values according to eq. (3) by about 0.5 μm, respectively. For β_1 a default value of 2 was found. For an extensive discussion about the effect of the w/c ratio and the way in which temperature effects have been allowed for in the model reference is made to (van Breugel).

By adding the amounts of hydrated cement of the individual particle fractions and dividing them by the original amount of anhydrous cement the overall degree of hydration $\alpha(t)$ at time t is obtained.

2.5 Prediction of adiabatic hydration curves

Assuming a linear relationship between the degree of hydration $\alpha(t)$ and the liberated heat of hydration and that all the liberated heat is used for heating up of the hydrating sample, adiabatic hydration curves can be computed. In these calculations the specific heat of the concrete is considered to be a function of the degree of hydration.

3   Experimental determination of adiabatic hydration curves

3.1   Adiabatic calorimetry: Factors affecting the accuracy

Factors which are of paramount importance in view of the accuracy of adiabatic testing are:

a. Accuracy of the temperature measurements.
b. The steering algorithm (implicit or explicit)
c. The type of the moulding
d. The size of the sample
e. Heat dissipation to the environment and the way this heat loss is compensated.

The effect of these parameters can be investigated experimentally by varying them one by one or with the help of computer programs with which the adiabatic test conditions can be simulated. The latter method has the advantage of being more flexible and quicker than experimental testing. Numerical analyses of the effects of the control precision, type of steering algorithm, type of moulding and size of the sample were carried out with a 3-D computer program called ASA3D (Adiabatic Sensitivity Analysis - 3 Dimensional).

3.2   Adiabatic test device specifications

Up till now several adiabatic calorimeters have been developed. An inventory of adiabatic and semi-adiabatic calorimeters is being prepared by RILEM (Wainwright). Pending the work of RILEM and still in the absence of standardised test equipment, a new experimental set-up has been developed.

A schematic view of this test set-up is shown in Fig. 3. The test specimen is a cube with a volume of 3.375 litre. The material of the moulding is 20 mm thick polyethylene (PE) with a thermal conductivity coefficient of $\lambda = 0.17$ W/mK. Instead of a PE also a steel mould can be used with $\lambda = 50$ W/mK. The control medium is water.

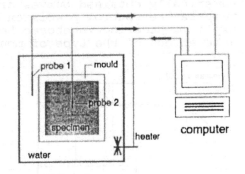

Fig. 3 Adiabatic test device - Schematic.

For adjusting the temperature of the control medium a heater is used with a capacity of 3000 Watt. Maximum acceleration 60°C/h. With this heater a control precision of ±0.05°C can be reached. This relatively high capacity is necessary for controlling the boundary condition if a rapid hardening material is tested, for example high strength concrete.

Temperature measurements are carried out with PT100 elements with a precision of, for the time being, ±0.05°C (probe 1) and ±0.01°C (probe 2).

3.3 Steering algorithms
Adjustment of the temperature of the control medium can be done by explicit or implicit steering.

In the explicit steering method the temperature of the control medium is adjusted step-wise to the temperature in the core of the sample at the end of each time step. During this step some heat may dissipate from the sample.

In case of implicit steering temperature adjustment is done taking into account the rate of reaction in the preceding time step by using extrapolation techniques. During each time step the temperature is adjusted gradually. Dissipation of heat to the environment is prevented almost completely in this way, which will result in a higher accuracy.

4 Results

4.1 Evaluation of variations in steering algorithm
The effects of variations in the type and accuracy of the steering algorithm on adiabatic hydration curves have been evaluated numerically and experimentally. Fig. 4 shows the results of numerical simulations. An adiabatic curve obtained with implicit steering is compared with curves obtained with explicit steering. In the latter case the control precision was adjusted at 0.05°C and 0.2°C, respectively.

4.2 Numerical evaluation of the effect of type of mould
In Fig. 5 the effect is shown of the type of mould, i.e. steel and polyethylene, on adiabatic hydration curves. The numerically obtained curves are computed for a simulated temperature control precision of ±0.2°C (explicit steering). Explicit steering is chosen here so as to enforce a noticeable effect of the type of mould. Experimental results are

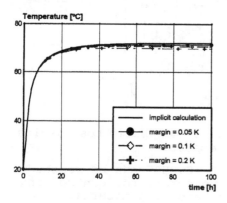

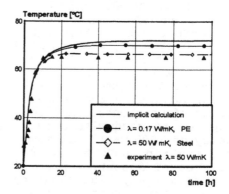

Fig. 4. Influence of steering accuracy of the adiabatic test device.

Fig. 5. Effect of type of mould on adiabatic hydration curves (explicit steering).

inserted in the figure as well. The two computed curves are
compared with an adiabatic curve obtained by implicit steering. It appears that in case of explicit steering the type
of mould can have a substantial influence. When an explicit
steering algorithm is used reliable results can only be a-
chieved if a mould with a high insulation capacity is used,
by increasing the control precision from 0.2°C to, for example, 0.01°C, or by increasing the size of the sample.

4.3 RILEM Round Robin test
With the aim to get a better judgement of the multi-laboratory scatter of adiabatic hydration curves a Round Robin
test has been carried out. Cement, sand and aggregate were
provided by one of the participating laboratories. Tests
were carried out with mixes with w/c ratio = 0.6. Cement,
sand and aggregate were mixed in proportion 1 : 2.5 : 3.5.
The calculated clinker composition of the cement was: C_3S =
53% and C_2S = 16%. Specific surface (Blaine) = 372 m^2/kg.
The nominal initial temperature of the mix was T_0 = 20°C.

In Fig. 6 the temperatures reached after 200 hrs are presented. The mean value of the temperature at that time was
66.8°C. The difference between maximum and minimum temperature is 8.8°C. The multi-laboratory standard deviation was
computed at 3.8°C. Assuming a normal distribution of the
test results the 5% upper and lower bound temperatures at
200 hrs are 73.0 and 60.6°C, respectively.

4.4 Adiabatic hydration curves predicted with HYMOSTRUC
Theoretical adiabatic hydration curves calculated with HYMO-
STRUC are inserted in Fig. 6 as well. The solid curve, which
has been calculated with the default values of the model parameters $K_0(C_3S)$, $\delta_{tr}(C_2S)$ and β_1, represents the mean value

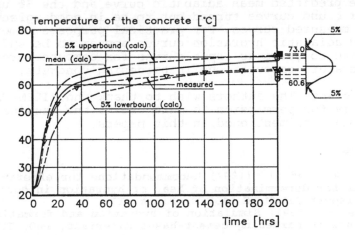

Fig. 6 Adiabatic test data obtained in round robin test
compared with numerically predicted hydration curves.

of the adiabatic curve. The predicted temperature at 200 hrs
is 68.5°C, which is 1.7°C higher than the mean value obtained in the Round Robin test. The predicted upper and lower bound curves are presented with dashed lines. The calculated 5% upper and lower bound temperatures at 200 hrs were
70.5 and 65.5°C. This range of 5°C, in which 90% of the adiabatic results are expected, is of the same order of magnitude as the corresponding range in the Round Robin test.

A comparison of the predicted solid curve with one of the
measured adiabatic curves (dotted line) reveals that in the
early stage of hydration these curves coincide quite well.
After some time they start to deviate. One of the reasons
for this deviation must most probably be attributed to
inadequate correction of heat losses in the test device.

5 Conclusions

The results of adiabatic tests depend on, among other
things, the control precision of the test device, the size
of the sample and the type of the mould. Numerical simulations of adiabatic tests with the computer program ASA3D
evidenced that substantial differences, up to 8°C after 100
hrs hydration, are to be expected due to differences in a-
fore mentioned parameters. In a Round Robin test, in which
seven laboratories participated, a difference between maximum and minimum temperature of 8.8°C after 200 hrs was found
with a standard deviation of 3.8°C. This result was in good
agreement with the results of the numerical simulations.

With the computer program HYMOSTRUC adiabatic hydration
curves of the same mix as used in the Round Robin test were
predicted as a function of the cement type and mix composition. The predicted mean adiabatic curve and the 5% upper
and lower bound curves turned out to be in good agreement
with the measured curves. For practical purposes numerically
predicted adiabatic hydration curves appear to be sufficiently accurate to serve as input curves for computer programs
for temperature calculations in hardening concrete.

Acknowledgment The authors wish to thank the RILEM Task
Group TC119 for permission to make use to the results of the
Round Robin test mentioned in this papers.

References

Wainwright P.J. et.al. (1993) **Recommendations for experimental
methods for determination of heat of hydration in concrete**,
Draft report RILEM TC 119TCE, Paris, 17p.
Van Breugel K. (1991) **Simulation of hydration and formation of
structure in hardening cement-based materials**, PhD, TU-Delft,
295p.
Koenders, E.A.B. (1994) Numerische Empfindlichkeitsanalyse von
adiabatischen Hydratationskurven, in Proc. **Forschungskolloquium des DAfSt**, Delft University of Technology, Delft.

2 THERMAL AND MECHANICAL MODELLING OF YOUNG CONCRETE BASED ON HYDRATION PROCESS OF MULTI-COMPONENT CEMENT MINERALS

T. KISHI and K. MAEKAWA
Engineering Research Institute, Civil Engineering, The University of Tokyo, Japan

Abstract
This paper aims at developing a predictive method on both heat generation and associated evolution of strength for young concrete. Mineral compounds of cement clinker and pozzolans are focused and the hydration degree of them are computed step by step with modified Arrhenius's law of chemical reaction. The specific free water and calcium hydro-oxide, that is an activator for pozzolans, are assigned as state variables representing chemical environment of pore solution. The effect of fly ash on cement and slag hydration retarded by the adsorption of calcium ion is taken into account. The strength and instantaneous stiffness of hardening concrete are related to the accumulated heat of each mineral compound and versatility of the mechanical model proposed is verified under varying temperature environments.
Keywords: Clinker Minerals, Slag, Fly ash, Hydration, Heat Generation

1 Introduction

For making thermally induced cracks avoidable, evaluation of thermal crack risk is required at design stage. Here, the hydration heat of cement in concrete has to be modeled as a source of temperature rise. Meanwhile, the strength and stiffness evolution of concrete should be also predicted for examining thermal crack occurrence. The heat generation and varying mechanica properties of concrete at early ages are strongly related to the hydration degree of each mineral compound consisting of cement. Thus, it is desired to consistently predict them with a unified concept concerning hydration progress of cement. This approach brings an engineering advantage to enable sensitivity analysis with respect to mix proportion of concrete and sorts of cement and pozzolans with different chemical compositions and blended ratios.

To meet the engineering challenge stated above, the model has to be applicable to wider variety of clinker compositions of Portland cement and replacement by pozzolans under varyin; temperature. On this line, this paper proposes "multi-component model" for hydration heat an the evolution of strength based on the Arrhenius's law of chemical reaction. Within this frame, interacting heat generation among constituent minerals is coherently treated in terms of free water and calcium hydro-oxide solution.

2 Interaction between Portland cement clinkers and pozzolans

Pozzolans react with calcium hydro-oxide as an activator which is produced by hydration of clinker minerals. Then, it is crucial to accurately appraise interaction between cement clinker

Thermal Cracking in Concrete at Early Ages. Edited by R. Springenschmid. Published 1994 by E & FN Spon, 2–6 Boundary Row, London SE1 8HN, UK. ISBN: 0 419 18710 3.

and pozzolans for versatile modeling. For this purpose, conduction calorimetry test was conducted for taking some knowledge on effects of pozzolans replacement. The conduction calorimetry of cement paste in various mix proportions of blended materials was at 50% water cement ratio and 20°C constant curing temperature. The physical properties and chemical compositions of powder materials are given in Table 1.

Table 1.   Chemical components and physical properties of cement and pozzolans used.

	specific gravity	Blaine (cm²/g)	Ing Los (%)	SiO$_2$	CaO	Al$_2$O$_3$	Fe$_2$O$_3$	MgO	SO$_3$	FeO	K$_2$O	Na$_2$O
OPC	3.15	3290	0.6	22.1	64.1	6.2	2.6	1.4	2.1	-	0.5	0.31
slag	2.89	4000	0.0	31.3	43.3	13.2	-	6.0	2.0	0.3	-	-
fly ash	2.33	3440	0.4	48.1	8.8	27.6	5.3	-	-	-	-	-

note) chemical components : weight percentage

2.1 Reaction of slag in mixed cement

Typical heat generation rates of cement with slag are shown in Figure 1a. In general, two peaks of heat rate are seen in time domain. The authors attempted to extract heat released from slag existing in the mixed cement paste by subtracting heat of cement clinker (pure OPC test). Here, let us assume independency of cement reaction which is not affected by slag. With this hypothesis, heat generation rates of slag in the mixture normalized by the content of slag can be obtained as shown in Figure 1b.

The heat generation rates are similar regardless of the blended contents of slag till the peak. While, there is a tendency that heat generation is quickly descending after the peak of hydration rate when higher replacement is performed. It can be simply presumed that slag can react independently under a condition where calcium hydro-oxide is sufficiently released from cement, but at the higher replacement of slag, the reaction of slag is stagnant because of shortage of calcium hydro-oxide in pore solution. In other words, the assumption that cement clinker minerals react with less influence brought by slag is acceptable to the hydration heat model. On the contrary, the influence of cement hydration on the reaction of slag is crucial for constructing multi-component modeling.

2.2 Fly ash in mixed cement

Heat generation of cement with fly ash at various mixing ratio is shown in Figure 2a. It is clear that the maximum heat generation rate is reduced with overall delay of heat in accordance with replacement of fly ash. Provided the assumption stated in slag, computed heat from fly ash was found to get negative. This indicates that mutual interaction between cement and fly ash is predominant unlike the "one way" interaction between cement and slag.

As generally known, fly ash retards the hydration of Portland cement especially at early stage of hydration and makes dormant period longer. When cement is mixed with fly ash, Ca^{2+} ion concentration in pore solution is depressed due to the removal by aluminum ions on the fly ash surface. This phenomena is explained such that fly ash surface acts like calcium sink. The depression of concentration retards the formation of calcium rich surface layer on clinker minerals which are precursor of reactivity. This will involve in delaying the formation of Ca(OH)$_2$ and C-S-H gel nucleation.

Heat generation of pure slag and binary mixture of slag and fly ash were measured with addition of calcium hydro-oxide as shown in Fig.2b. This experiment aims at clarifying interaction between slag and fly ash without cement clinker since the case of triple mixture is required to be modeled from an engineering view point.

The binary pozzolans release heat as shown in Figure 2b. Pure slag with calcium hydro-oxide has the peak of heat generation around 20 hours after mixing. This is nearly the same as the case of slag-OPC composite. It is observed that fly ash delays heat generation of slag as well as cement. The retardation effect of fly ash on the slag hydration can be concluded.

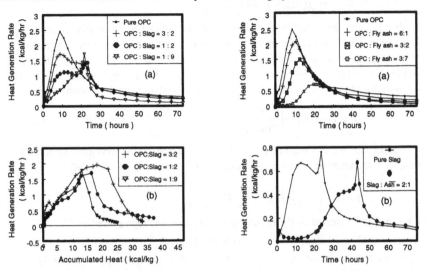

Fig.1 Conduction calorimetry of mixture and pure slag extracted.

Fig.2 Effect of fly ash on conduction calorimetry.

3 Multi-component model of cement hydration heat

3.1 Modeling

The specific heat rate of cement consists of constituent mineral based heat rates. The authors take up four clinker minerals (aluminate C_3A, alite C_3S, felite C_4AF, belite C_2S) and five patterns of reactions for Portland cement (Kishi et al.). In the case of mixed cement, reactions of slag and fly ash are combined with cement clinker minerals as,

$$\overline{H} = p_{mono}\overline{H}_{mono} + p_{C_3A}\overline{H}_{C_3A} + p_{C_2S}\overline{H}_{C_2S} + p_{C_4AF}\overline{H}_{C_4AF} + p_{C_3S}\overline{H}_{C_3S}$$
$$+ p_{FA}\overline{H}_{FA} + p_{SG}\overline{H}_{SG} \tag{1}$$

where, p_i = mass ratio of i-th component, $\overline{H}_i$ = specific heat rate of i-th component, subscript 'mono' represents the transformation of ettringite to monosulfate.

Suzuki et al. reported that Arrhenius's law can be extended to the composite with different chemical reactions of clinker minerals by adopting variable mean activation energy uniquely specified in terms of the accumulated heat of cement. As the hydration of cement are classified into mineral compounds, the activation energy of their reactions are assumed constant one by one. According to Arrhenius's law, the temperature dependent heat rate of reaction yields,

$$\overline{H}_i = \overline{H}_{i,To} \cdot \exp\left\{-\frac{E_i}{R}\left(\frac{1}{T} - \frac{1}{T_o}\right)\right\} \tag{2}$$

where, E_i = activation energy of i-th component reaction, R = gas constant and $\overline{H}_{i,To}$ = referential heat rate when temperature is To (=293 K).

The interaction among constitutive compounds is expressed as,

$$\overline{H}_{i,To} = \beta_i \cdot \Omega_i \cdot \gamma \cdot F\left(\int_0^t \overline{H}_i dt\right) \tag{3}$$

where, β_i represents the reduction of probability of the contact between unhydrated compound and free water. The factor Ω_i represents retardation of cement and slag reaction caused by fly ash. The factor γ represents reduction of hydration concerning pozzolans due to shortage of calcium hydro-oxide. The parameter β_i is associated with increasing thickness of cluster around unhydrated compounds and the decreasing free water during hydration as,

$$\beta_i = 1 - \exp\left\{-r\left(\frac{\omega_{free}}{100 \cdot \eta_i}\right)^s\right\} \tag{4}$$

where, r and s = material constants, ω_{free} = free water which can be consumed by further reaction, η_i = thickness of cluster around unhydrated compound (Kishi, et al) as,

$$\eta_i = 1 - \left\{1 - \left(\int_0^t \overline{H}_i dt\right) \cdot \left(\int_0^\infty \overline{H}_i dt\right)^{-1}\right\}^{\frac{1}{3}} \tag{5}$$

The retarding effect caused by fly ash on other compounds is represented as,

$$\Omega_i = 1 + \exp\left(-h \cdot p_{FA}\right) - \exp\left(-h \cdot p_{FA} \cdot \phi_{FA}\right) \tag{6}$$

where, h = material constant, p_{FA} = mass ratio of fly ash, ϕ_{FA} = hydration degree of fly ash. The reduction of pozzolans reaction caused by shortage of calcium hydro-oxide is formulated as,

$$\gamma = \left[1 - 0.06\exp\left\{-5.5\left(\frac{F_{CH} + \vartheta}{R_{CH}} - 0.5\right)\right\}\right]^k \tag{7}$$

where, k = material constant, F_{CH} = available Ca(OH)$_2$ in the solution, R_{CH} = Ca(OH)$_2$ necessary for reaction of pozzolans. The factor ϑ expresses that consumption of Ca(OH)$_2$ by fly ash is reduced when concentration of Ca(OH)$_2$ in pore solution gets less as follows.

$$\vartheta = u\left\{\exp\left(p_{FA} \cdot \phi_{FA}\right) - 1\right\}^v \tag{8}$$

where u and v = material constants. The multi-component model on heat generation is formulated. By solving Eq.1- Eq.8 with the thermodynamic energy conservation, temperature transition of concrete can be obtained in time and space domains (Harada, et al).

3.2 Verification of heat generation model

The material constants and referential heat rate of each compound are shown in Table 2 and Figure 3. These values were identified through adiabatic temperature rise tests and data back analysis processing. Theoretical specific heat is given as final heat generation of each clinker

Table 2 material parameters k=5, u=0.35, v=0.25 for pozzolans only, E/R : K, $\int H$: cal/g

	C$_3$A	C$_3$S	C$_2$S	C$_4$AF	ettringite	slag	fly ash
E/R	6500	6000	3000	3000	6000	5000	5000
r	2	2	2	2	2	3	2
s	2.5	2.5	2.5	2.5	2.5	-	2.5
$\int H$	207	120	62	100	330	110	50
h	1.5	1.3	0.9	1.2	-	0.95	-

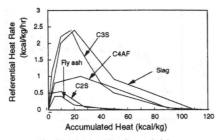 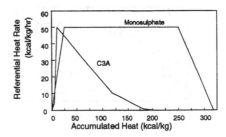

Fig.3 Assumed referential heat generation rate of cement clinkers and pozzolans.

Table 3 Concrete mix proportion, cement mineralogical analysis.

	C	W	S	G	PC	SG	FA	C_3A	C_3S	C_2S	C_4AF	SO_3
mix1	400	157	658	1129	100	-	-	10.4	47.2	27.0	9.4	1.85
mix2	400	157	663	1129	100	-	-	3.7	44.4	33.7	12.5	1.85
mix3	260	140	740	1125	75	-	25	4	26	53	11	1.7
mix4	260	140	738	1120	35	45	20	2	24	53	12	1.7
mix5	260	140	740	1122	40	30	30	5	15	65	11	1.6
mix6	260	140	774	1125	25	75	-	3	44	34	12	2.5
mix7	260	140	748	1133	70	30	-	3	7	75	9	2.5
mix8	260	140	738	1117	30	50	20	4	22	57	11	2.0
mix9	260	140	738	1120	25	55	20	5	36	40	12	1.4
mix10	260	140	718	1137	32	48	20	2	24	53	12	1.3

note) C : cement = cementitious powder, W : water, S : sand, G : gravel (kg/m^3)

PC : Portland cement (ordinary : OPC and moderate heat : MC), SG : slag, FA : fly ash

mineral. The adiabatic temperature rises of two Portland cement mixtures and eight mixtures with slag and fly ash as low heat type are adopted for verification as shown in Table 3. Application of the proposed model as shown in Figure 4 are ensured for different chemical components and temperature rises of mixed cements adopted are fairly predicted.

4 Strength development - generalized mineral water ratio law

4.1 Experiment of strength development

For making strength evolution model of young concrete with temperature rise, compressive loading tests of mortars were conducted under two patterns of temperature histories. Figure 5 shows temperature history patterns under which test specimens were cured and tested. Specimen was 5cm × 10cm cylinder and cured with sealing condition. Pattern 1 is supposed the analogy of temperature history of concrete in a massive structure at early ages. For comparison in pattern 2 the curing temperature was kept at room condition for about a week and then temperature was elevated for acceleration of hydration. The beginning of temperature rise in pattern 2 is slightly different in each mixture (MC : 7.4days, SG+MC : 6.8days, FA+MC : 9.1days). Mix proportions are shown in Table 4. The hydration degree of each mineral constituents are computed by the multi-component model of heat generation.

The evolution of strength at pattern 1 and pattern 2 with respect to time are remarkably different though the strength at final stage are almost the same as shown in Figure 7.

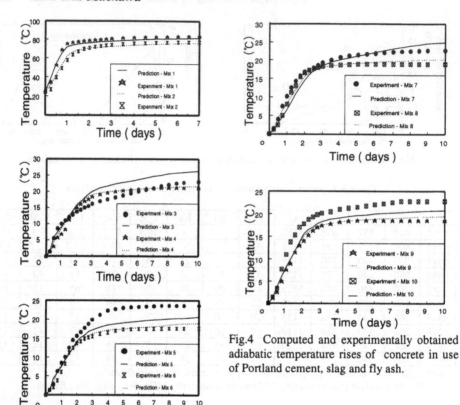

Fig.4 Computed and experimentally obtained adiabatic temperature rises of concrete in use of Portland cement, slag and fly ash.

The approach based on the hydration heat model is essential for modeling the evolution of strength. The hydration degree of each mineral constituents is not common at the nearly same strength or accumulated heat between two patterns because of the difference in thermal activity of them. This indicates that mineral compounds consisting of cement should be formulated with multi-component concept in the modeling of the evolution of strength.

4.2 Modeling

The authors (1993) reported that clear bi-linear relation was seen between the strength and the entire hydration level of cement which was computed by multi-component model and indicated the possibility to be able to estimate the evolution of strength with the hydration degree of cement. But no unique relation between strength and entire hydration level of whole cement is not obtained. The relation of them should be formulated according to mix proportion and used powder materials. As the model has to be applicable to any combination of materials, multi-component concept based on the mineral constituents was adopted for the strength evolution model. Four constitutive minerals(C_3S, C_2S, slag, fly ash) are taken as the effective components for the evolution of strength, and C_3A and C_4AF are assumed negligible in the proposed model. Then, the compressive strength is expressed in terms of the total differential equations as,

$$f_c' = \int df_c', \qquad df_c' = 25dQ_{3S} + 40dQ_{2S} + 27dQ_{SG} + 40dQ_{FA}$$

$$dQ_i = \frac{p_i}{W} d\phi_i, \qquad \phi_i = \int_0^t \overline{H}_i dt \bullet \left(\int_0^\infty \overline{H}_i dt \right)^{-1}$$

(9)

where, fc' = compressive strength (MPa), ϕ_i = hydration degree indicated by the accumulated heat normalized by final heat generation. The contribution of constitutive minerals are individually formulated and the concept of cement water ratio in the compressive strength is extended to each component in the above equation.

4.3 Verification
The experimental and analytical results in terms of accumulated heat of cement (See Figure 6) fairly coincide with each other. The relations of the strength evolution and the accumulated

Table 4 Mix proportion of concrete used

mix	MC	SG	FA	W	G	S
A	761	-	-	181	808	720
B	360	360	-	181	808	720
C	478	-	192	181	808	720

note) unit : kg/m³ MC : moderate heat Portland cement

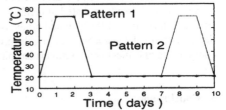

Fig.5 Induced temperature histories

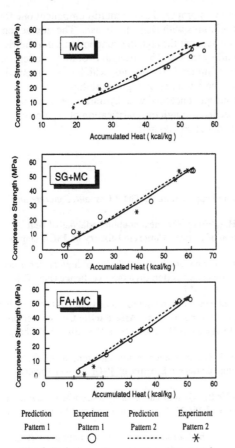

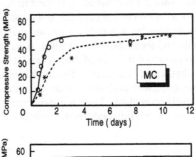

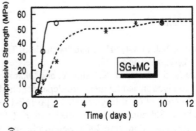

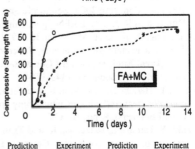

Prediction	Experiment	Prediction	Experiment
Pattern 1	Pattern 1	Pattern 2	Pattern 2
——	O	-------	✳

Fig.6 Relation of accumulated heat and compressive strength of concrete.

Prediction	Experiment	Prediction	Experiment
Pattern 1	Pattern 1	Pattern 2	Pattern 2
——	O	-------	✳

Fig.7 Strength development of concrete

heat at the temperature histories of both pattern 1 and pattern 2 are similar. In mathematical view of modeling, these relations do not necessarily match. Figure 7 shows the results with respect to the curing time, in which fair agreement can be seen between experimental and analytical results as well. It is emphatic that the proposed model can rationally deal with the evolution of strength with respect to time at any temperature history.

In general, the concrete strength varies in accordance with the curing temperature and it is reduced at longer age under the elevated temperature especially experienced at the early stage. Within this study, however, the evolution of strength can be computed only by dealing with the composition of already hydrated minerals as the hydration degree though it is not clear how the structural formation of the hydrated products affect the evolution of strength. Compared with the maturity model available recently, the proposed strength model based on chemical components of cement and pozzolans has physical generality and equivalence.

5 Conclusions

The hydration heat model of cement in concrete was proposed as the multi-component one for clinker based Portland cement as well as their mixtures with slag and fly ash. The interaction among powder materials was suitably taken into account through the conduction calorimetric study and the proposed model was experimentally verified. The strength evolution model was also proposed in terms of the accumulated heat of mineral constituents, which was computed by the hydration heat model. The models of heat generation and strength evolution were associated with each other under the unified concept concerned with hydration progress of mineral compounds. Further improvement is needed through the clarification of regulative factors having influence on the hydration of chemical compounds.

References

Carino, N.J. (1984) :The maturity method, Theory and application, **ASTM J.Cement**,*Concrete and Agregates*, pp.61-73.
Harada, S. Maekawa, K. Tsuji, Y. and Okamura, H. (1991) : Nonlinear coupling analysis of heat conduction and temperature-dependent hydration of cement, **Concrete Library of JSCE** (Japan Society of Civil Engineers), 14, pp.167-176.
Kishi, T. Shimomura, T. and Maekawa, K. (1993) : Thermal crack control design of high performance concrete, **Concrete 2000, Economic and durable construction through excellence** (Eds. R.K. Dhir and M.R. Jones), E&FN Spon, Dundee, pp. 447-457.
Santhikumar, S. Kishi, T. Maekawa, K. (1993) : Heat generation model for mixed Portland cement, blast furnace slag and fly-ash concrete, **Proceedings of the Fourth East Asia-Pacific Conference on Structural Engineering and Construction : Progress in Harmony** (eds. Y.K. Shin, S.P.Chang and H.M.Koh), Seoul, pp.1449-1454.
Suzuki, Y. Harada, S. Maekawa, K. and Tsuji, Y. (1988) : Evaluation of adiabatic temperature rise of concrete measured with the new testing apparatus, **Concrete Library of JSCE** (Japan Society of Civil Engineers), 9, pp.109-117.
Suzuki, Y. Harada, S. Maekawa, K. and Tsuji, Y. (1990) : Quantification of hydration-heat generation process of cement in concrete, **Concrete Library of JSCE** (Japan Society of Civil Engineers), 12, pp.155-164.
Uchikawa, H. (1986) : Effect of blending component on hydration and structure formation, **Principal report of the 8th international congress on cement chemistry**, Brazil.

PART TWO

PREDICTION OF TEMPERATURE DEVELOPMENT

(Calcul des champs de températures)

3 PREDICTION OF TEMPERATURE DISTRIBUTION IN HARDENING CONCRETE

Ch. WANG and W.H. DILGER
Department of Civil Engineering, The University of Calgary, Calgary, Canada

Abstract
This paper presents the development of a computer model to predict the temperature distribution in hardening concrete. A two-dimensional finite element thermal analysis is used to model the transient heat transfer between the concrete and the environment by taking into account the cement type and content, boundary and environmental conditions including solar radiation and artificial heating, if any. The output is the temperature distribution in the member at any moment. This is the key information for concrete maturity estimate and thermal stress assessment. The model can also be used in construction planning, such as designing the necessary curing measures to achieve the desired strength at a specific age or to control the temperature differentials in the structure to eliminate thermal cracking. Comparisons with data from various field temperature measurements confirm the analytical work.
Keywords: Temperature, Hardening Concrete, Hydration Heat, Heat Transfer, Thermal Analysis, Temperature Prediction.

1 Introduction

The prediction of the temperature distribution in hardening concrete is of great interest to designers and contractors for several reasons: thermal cracking and deformation control or prevention, and evaluation of the concrete strength development. The importance of temperature effects in hardening concrete has inspired extensive research worldwide in recent years (Tsukayama 1974, Thurston et al 1980, Breitenbücher 1989, Laube 1990, Wang 1994).

Temperature change is the cause of any thermal stress and thermal cracking. In concrete the hydration of portland cement is an exothermal process, releasing up to 500 Joules of heat per gram of cement (Neville 1981). The relatively low thermal conductivity of the concrete mass delays the heat dissipation into the surroundings, resulting in a substantial temperature rise in large members at the early ages. The concrete may also gain heat from the environment such as solar radiation, and from heat curing. When concrete has developed some apparent strength, any temperature change will cause stress and deformation, and often, cracking in structures. The only situation when temperature variation does not generate stress is in a statically determinate structure where the temperature varies linearly across the

Thermal Cracking in Concrete at Early Ages. Edited by R. Springenschmid. Published 1994 by E & FN Spon, 2–6 Boundary Row, London SE1 8HN, UK. ISBN: 0 419 18710 3.

section. Therefore, the prediction of temperature distribution and
history in hardening concrete is essential in order to estimate the
thermal stress and strain as well as to prevent thermal cracking.

Like all chemical reactions, cement hydrates faster at higher
temperature. After the concrete mix is cast, the only factor in its
strength development is temperature. In hardening concrete, the
temperatures are normally different at different locations of a
structure at any time. For a typical thick member, the concrete
strength gain at the core is much faster than near the surface during
the early ages because of the temperature differences. An accurate
estimate of concrete strength development, which is critical in
determining the time of form stripping and/or prestressing, requires
the prediction of the history of temperature distribution in the
concrete. Failure to determine the concrete strength accurately in the
field may either delay the construction, cause local damage or even
collapse of the structure.

This paper presents a computer model to predict the temperature
distribution in hardening concrete at any time. A two-dimensional
finite element thermal analysis is employed to model the transient heat
transfer between the concrete and the surroundings as affected by the
concrete mix, thermal boundary and environmental conditions. Because
of the time and temperature dependent nature of the hydration heat rate
and boundary heat transfer conditions, the solution of the problem
requires the step-by-step integration in the time domain.

The details of estimating concrete maturity and strength as well as
thermal stress from the temperature distribution and history are
presented elsewhere (Wang 1994, Wang and Dilger 1994).

2 Heat Transfer between Concrete and the Environment

At any time, the temperature distribution in a concrete cross section
is the dynamic heat balance between the heat generated inside the
concrete and the heat loss to, or gain from, the surroundings. In
hardening concrete, the heat generated inside is hydration of the
cement. For most of the actual structures whose length is much larger
than the width or thickness, thermal analysis can be treated as a two-
dimensional problem by assuming that the temperature distribution does
not vary along the length.

The temperature distribution within a two-dimensional body or the
transient heat flow within the boundaries of the body is governed by
the well-known Fourier's Law (Holman 1986):

$$k\left[\frac{\partial^2 T}{\partial x^2} + \frac{\partial T^2}{\partial y^2}\right] + q = \rho c \frac{\partial T}{\partial t} \qquad (1)$$

where, T = temperature, °C,
  t = time,
  k = thermal conductivity, W/(m°C)
  q = rate of heat generated inside the body, W/m^3
  ρ = density of the material, kg/m^3
  c = specific heat of the material, kJ/(kg°C)
  x,y = Cartesian coordinates in x,y directions.

There exist basically two types of boundary conditions for Eq.1. The first one is that the temperature along the boundary or a portion of the boundary is known, and the second one is that the energy transfer through the boundary is known. For ordinary engineering structures, the second type of boundary condition normally occurs. That is, the heat exchange between the concrete body and the environment in the forms of solar radiation, thermal radiation, convection, etc. needs to be evaluated in order to solve Eq.1 to obtain the temperature distribution in the cross section. The mathematical expression of the boundary condition is,

$$k\left[\frac{\partial T}{\partial x}n_x + \frac{\partial T}{\partial y}n_y\right] + q_b = 0 \tag{2}$$

where, $n_{x,y}$ = direction cosines of the unit outward normal to the boundary surfaces,

q_b = total boundary heat gain or loss, W/m^2, including solar radiation, thermal radiation and convection, etc.

3 Boundary Heat Transfer Conditions

It is obvious that the main difficulty in solving Eq.1 is to establish the boundary heat transfer conditions which are time and/or temperature dependent. The heat transfer through the boundaries of a concrete body takes place basically in five forms: solar radiation, convection, thermal radiation, evaporation and condensation. Of these, evaporation and condensation heat transfers are the least significant for concrete structure in normal conditions, and are neglected here in the modelling (Wang 1994).

The amount of solar radiation that reaches the concrete surface can be estimated from the structure's geographical location, orientation, altitude, atmospheric conditions, time of the day, day of the year (Duffie and Beckman 1974, Dilger and Ghali 1980). The amount of solar energy absorbed by the concrete depends on the solar radiation absorptivity of the surface which is affected by the colour and texture of the surface material.

The convection heat transfer is the heat loss to or gain from the surrounding air as a result of the air movement. It depends on the wind speed and the temperature difference between the concrete surface and the bulk air (Holman 1986).

Thermal radiation is the heat radiation emission by the concrete body. It is a function of the surface temperature and the environment temperature as well as the thermal emissivity of concrete.

When more accurate data is not available, the diurnal variation of the ambient air temperature can be assumed to follow a sinusoidal cycle between the maximum and minimum values (Hulsey 1976, Dilger and Ghali 1980).

4 Cement Hydration Heat Development

It is apparent that the total cement hydration heat of a concrete mix depends on the cement type and content. For a particular cement, the hydration rate (or the heat release rate) at any moment depends only on the total heat already released (i.e., the concrete maturity) and the current temperature (Neville 1981, Mindess and Young 1982). The temperature effects on cement hydration rate is sometimes called the temperature function. After a careful study of different functions (Wang 1994), the Arrhenius Function was chosen for this study.

When test results on adiabatic temperature rise with time for the concrete mix are available, the cement hydration heat rate at standard temperature (20°C) and any other temperature can be derived and applied to the computer model. More often, test results are, however, not available. In this case, the present study suggests, for example, the following relationships between hydration heat rate and concrete maturity at 20°C for ordinary portland cement (Wang 1994):

$$q = 0.5 + 0.54M^{0.5} \qquad \text{for M} \leq 10.0 \text{ hours}$$

$$\text{(3)}$$

$$q = 2.2 \exp[-0.0286(M - 10.0)] \qquad \text{for M} > 10.0 \text{ hours}$$

where M is the maturity of concrete in hours.

When the temperature is constantly 20°C, the concrete maturity is equal to its clock age, and when it is different from 20°C,

$$M = \int_0^\infty H(T)\, dt \qquad \text{(4)}$$

where H(T) is the temperature function and t is clock time.

With the introduction of the temperature function, the hydration heat rate at any temperature can be obtained for the analysis. It should be pointed out that the concrete maturity and temperature at a different location of the cross section is normally different. Therefore, the hydration heat rates at different points are normally different at a particular time as well.

5 Finite Element Transient Thermal Analysis

Following the work of Dilger and Ghali (1980) and Elbadry and Ghali (1983), three types of elements are used to model the concrete cross section including any formwork and/or insulation, i.e., 4-node bilinear rectangular elements and 3-node linear triangular elements for the body, and 2-node linear one-dimensional elements for the boundary.

The temperature field of the cross section at each time step is obtained by the variational finite element method combined with the Galerkin weighted residual method for the time domain solution. The non-linearity problem of the thermal radiation heat transfer through the boundary is bypassed by converting the non-linear radiation heat transfer into the quasi-linear "radiant convection". Since the fictitious radiant-convection-heat-transfer coefficient is only slightly temperature dependent (Maes 1980), the iteration process of finding the coefficient is avoided by the approximate extrapolation from the values of the two previous time steps.

Most of the time, concrete structures undergo changes in the early ages, such as the removal of formwork and/or insulation materials, and multi-lift casting. The thermal analysis of hardening concrete should be able to accommodate these changes. This is achieved in this study by automatically transferring the element nodal temperature and concrete maturity data from the end of the previous stage to the beginning of the current stage.

The whole analysis is carried out by computer programm FETAB coded in FORTRAN 77. It was originally developed by Dilger and Ghali (1980) to calculate the temperature distribution within the cross section of a steel-concrete composite box-girder bridge for boundary conditions of constant heat flux, convection and heat generation. Refinement and extension were done by Elbadry and Ghali (1982) to account for the time-varying boundary heat transfer conditions.

For the present study, the program is extended to include the hydration heat rate of various types of cements as functions of maturity and current temperature. It is further developed to handle the changes of boundary conditions and structure configurations such as multi-lift casting and removal of formwork. The prediction of concrete maturity and strength development in the structure is also added.

The basic input for the computer analysis includes: cement content and type, thermal properties of each conduction material and boundary, environmental conditions such as air temperatures and solar radiation intensity as a function of incidence angle, initial concrete tempera-ture and maturity, as well as the finite element mesh of the structure. The time step length for the analysis is normally chosen as one hour for the first 2 days or so and increased for the rest of the time.

6 Results

The main output of the thermal analysis of hardening concrete is the concrete temperature at each node of the finite element mesh at each time step. The temperature contour of the cross section at a particular moment and/or the temperature history of any point, for example, are readily available through some post processing.

Fig.1 is the temperature contour at age 72 hours in a fictitious concrete column of 2mx4m cross section cast in a steel formwork without insulation in an ambient temperature varying between 0 to 10°C. The fresh concrete temperature at casting is assumed to be the average air temperature of the day, i.e., 5°C.

Fig.2 shows the comparison of computed and measured temperature development in a 1mx1m test column cast in an indoor environment (Cook et al 1992). The cement content was 355kg/m^3, and the 28-day concrete strength was 35 MPa. The ambient air temperature during testing was about 28°C.

Fig.3 presents the comparison of computed and measured temperature development in a massive spine beam of the Cambie Bridge constructed in Vancouver, Canada in 1984 (Dilger 1985). The 1.6m thick spine beam was cast in one lift. The local weather during the first several days was mainly overcast, and the average minimum and maximum ambient air temperatures during the days were 2.5 and 9.5°C. After casting, the top concrete surface was covered with a thin thermal blanket to protect it from rain for about five days.

Fig.4 compares the predicted and measured (Tsukayama 1974) temperature distribution in a test beam cast in 3 lifts in 7-day intervals. It shows the temperature distribution along the vertical centre line of the cross section 24 hours after casting the third lift. The average ambient air temperature during the experiment was about 30°C, and 400kg/m³ of ordinary cement was used in the mix.

7 Conclusions and Remarks

The present study on thermal analysis and computer modelling of temperature development in hardening concrete clearly demonstrates that it is possible to predict with reasonable accuracy the temperature development in hardening concrete in the field. Consequently, one can readily estimate the concrete maturity and development of the compressive and tensile strength at any location of the structure. Since temperature distribution and history in hardening concrete are critical data in assessing both the possibility of thermal cracking and the development of concrete strength at early ages, the computer model is a very useful tool to concrete structure designers as well as contractors. It can also be used to help in construction planning, such as designing the curing measures in order to prevent the concrete from freezing in cold weather, or to achieve the desired strength at an early age, or to maintain the specified temperature differential limits in the structure.

Actually, the model and computer program are currently used in the design and construction planning of the 13.5km long Northumberland Strait Crossing, a concrete box-girder bridge connecting the Prince Edward Island to the mainland in Eastern Canada.

It should be pointed out that thermal analysis, especially a transient one as in hardening concrete, is a very complicated process, involving many uncertainties in both material properties and environmental conditions. According to Holman (1986) and many other experts, 20-30% of error is considered excellent accuracy in practical problems. The main source of error is from the input data rather than from the modelling or computation. Many key parameters, such as the thermal conductivities of concrete and formwork/insulation materials, convection heat transfer coefficient, ambient air temperature, etc., vary in wide ranges. Nevertheless, reference values from tests and studies by various researchers in the past can be found (Wang 1994), and the selection of them for a particular case should be based on the careful examination of the field conditions. Convection heat transfer coefficients at different surfaces for different wind speeds suggested by Kehlbeck (1975) are, for example, excellent guidelines.

The hydration heat properties such as the heat libration rate and total amount of heat are very much cement dependent. Even for the same type of cements of different manufacturers (origins), the total hydration heat and its development rate at early ages differ substantially. Therefore, it is most desirable to obtain the hydration heat characteristics by testing the cement to be used when a reliable prediction of concrete temperature development is required.

The current model does not take into account the effect of

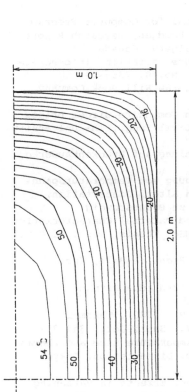

Fig.1. Temperature contour in a quadrant of a 2x4π column 72 hours after casting. (400kg/m³ of ordinary cement, steel formwork).

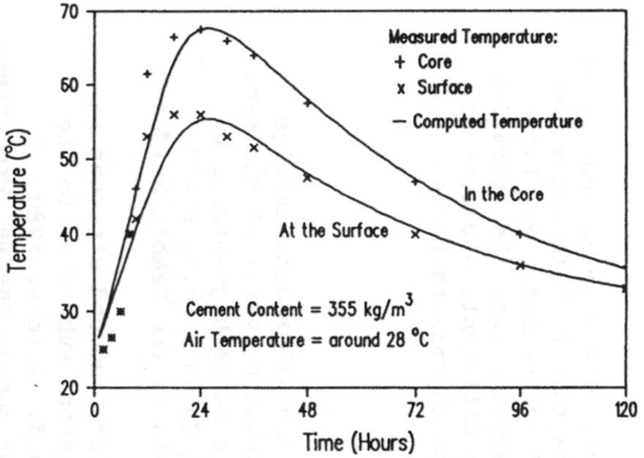

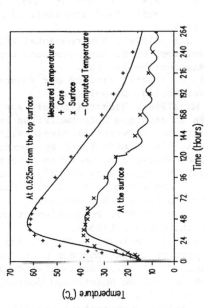

Fig.3. Comparison of computed and measured (Dilcer 1985) temperature in a spine beam.

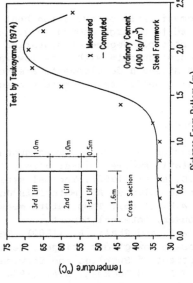

Fig.2. Comparison of computed and measured (Cook et al 1992) temperature in a column.

Fig.4. Temperature distribution along the vertical centre line of the cross section 24 hours after casting the third lift.

water/cement ratio on hydration heat release. The water/cement ratio is assumed to be high enough (e.g., not less than 0.4) to allow the maximum possible degree of hydration of all the cement. For high performance concrete with very low water/cement ratio, the hydration heat and strength developments are somewhat different, and further research is needed to accurately predict the thermal response.

8 References

Breitenbücher, W. (1989), **Zwangsspannungen und Rißbildung infolge Hydratationswärme**, Dissertation, Technischen Universität München.

Cook, W.D. et al (1992), Thermal stresses in large high-strength concrete columns, **ACI Materials Journal**, Vol.89, No.1, 61-68.

Dilger, W.H. (1985), **Temperature measurements in the Cambie Bridge, Vancouver**, internal report.

Dilger, W.H. and Ghali, A. (1980), **Temperature-induced stresses in composite box-girder bridges**, research report submitted to the Department of Supplies and Services, Canada, October 1980.

Duffie, J.A. and Beckman, W.A. (1974), **Solar Energy Thermal Processes**, John Wiley and Sons, Inc.

Elbadry, M.M. and Ghali, A. (1982), **User Manual for Computer Program FETAB: Finite Element Thermal Analysis of Bridges**, Research Report, Dept. of Civil Engg., The University of Calgary, Canada.

Elbadry, M.M. and Ghali, A. (1983), Temperature variations in concrete bridges, **ASCE J. of Struct. Eng.**, Vol.109, No.10, 2355-74.

Holman, J. (1986), **Heat Transfer**, 6th ed., McGraw-Hill Book Company, New York.

Hulsey, J.L. (1976), **Environmental Effects on Composite-Girder Bridge Structures**, Ph.D. Dissertation, Dept. of Civil Engg., The University of Missouri-Rolla, USA.

Kelbeck, F. (1975), **Einfluß der Sonnenbestrahlung auf Brückenbauwerke**, Werner-Verlag, Düsseldorf.

Laube, M. (1990), **Werkstoffmodell zur Berechnung von Temperaturspann-ungen in Massigen Betonbauteilen im Jungen Alter**, Dissertation, Technischen Universität Carolo-Wilhelmina zu Braunschweig.

Maes, M.A. (1980), **Effects of Environmental and Material Characteristics on the Behaviour of Concrete Structures**, MSc. Thesis, Dept. of Civil Engg., The University of Calgary, Canada.

Mindess, S and Young, J.F. (1982), **Concrete**, Prentice-Hall, Inc., Englewood Cliffs, New Jersey.

Neville, A.M. (1981), **Properties of Concrete**, 3rd ed., Pitman Publishing Ltd., London.

Thurston, S.J., Priestley, M.J.N. and Cooke, N. (1980), Thermal analysis of thick concrete sections, **ACI J.**, Sept.-Oct., 347-357.

Tsukayama, R. (1974), **Fundamental Study on Temperature Rise and Thermal Crack in Massive Reinforced Concrete**, Ph.D. Dissertation, Tokyo University (in Japanese).

Wang, C. (1994), **Temperature and Time-Dependent Effects in Hardening Concrete**, Ph.D. Dissertation, The University of Calgary, Canada.

Wang, C. and Dilger, W.H. (1994), Prediction of concrete strength development during construction, paper presented at the **1994 Annual Conference of the Canadian Society for Civil Engineering**, Winnipeg.

4 LOW-HEAT PORTLAND CEMENT USED FOR SILO FOUNDATION MAT - TEMPERATURES AND STRESSES MEASURED AND ANALYZED

M. YAMAZAKI
Kajima Corporation, Chofu, Tokyo, Japan
H. HARADA and T. TOCHIGI
Chichibu Cement Co. Ltd, Kumagaya, Saitama, Japan

Abstract
The increase in the size of concrete structures in recent years has brought with it a rise in the needs of low-heat type cement. This paper introduces the properties of concrete using low heat portland cement, deals with its application to the actual mass concrete structure and reports the results of the measurement of temperature and thermal stresses of the mass concrete.

The paper also deals with thermal stress analysis on the structure and a comparison between the measured and analyzed results.

Keywords: Low Heat Portland Cement, Mass Concrete.

1 Preface

For massive concrete structures, blast-furnace slag and/or fly ash are sometimes used with portland cement for their low heat generation in Japan. In recent years, however, concrete constructions have become larger and are demanding concrete with lower heat, higher strength and durability.

With these points as background, research and development on belite portland cement, in which more belite is contained than in moderate heat portland cement, is being carried out and its effectiveness on thermal and strength properties is being recognized. The basic properties of belite portland cement (low heat portland cement) and concrete are explained here, together with its application to an actual construction.

At the construction, 1500 m³ concrete with low heat portland cement was cast in a day for 20,000 ton Cement Silo Foundation Mat.

2 Introduction of Low Heat Portland Cement

The cement is composed only from clinker and gypsum. The cement contains no other pozzolanic materials such as fly ash, blast furnace slag, silica fume or natural pozzolans. The chemical composition and the physical properties of the cement are shown in Table 1 and Table 2, compared with other portland cements.

The chemical composition and the physical properties conform to the Moderate Heat Portland Cement of The Japanese Industrial Standard "Portland Cement"(JIS-R5210).

Concrete using low heat portland cement showed low strength at younger ages as 28 days, but high strength development in the long term.

Thermal Cracking in Concrete at Early Ages. Edited by R. Springenschmid. Published 1994 by E & FN Spon, 2-6 Boundary Row, London SE1 8HN, UK. ISBN: 0 419 18710 3.

Table 1　Mineral Composition of Cements

	C3S	C2S	C3A	C4AF	
low heat	27	58	2	8	(%)
moderate heat	44	33	4	12	(%)
ordinary	52	23	9	9	(%)

Table 2　Physical Properties of Cements

	Specf. gravity	Blaine fineness (cm^2/g)	Comp. Strength (MPa)				Heat of hydration (J/g)			
			3d	7d	28d	91d	7d	28d	91d
low heat	3.22	3350	7.4	11.3	31.6	59.7	202	266	313
moderate heat	3.21	3040	11.4	16.7	35.8	51.5	270	319	352
ordinary	3.16	3250	14.8	25.2	41.5	48.1	326	373	401

Compressive strength characteristics are shown in Fig. 1. The adiabatic temperature rise with low heat portland cement is low, particularly in the initial period compared with other concrete in Fig. 2.

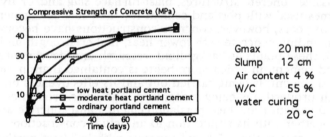

Fig 1　Strength of the Concrete using several kind of Cements

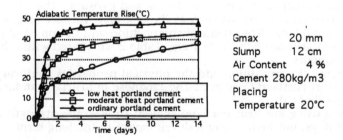

Fig. 2　Adiabatic Temperature Rise of the Concrete

3 Application to the Structure

3.1 Silo Foundation

The concrete structure in which the cement was applied was a foundation slab of 20,000 ton Cement Silo. The shape of the foundation mat is a circular disk, and the size is 29 meter in diameter, 2.4 meter thick, which supports seven hoppers and the silo barrel. The section of the silo is shown in Fig. 3. The hoppers and the silo barrel are constructed using ordinary portland cement, but for the foundation mat that has massive section low heat portland cement.

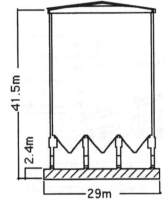

Table 3	Concrete Mix
Gmax	25 mm
Slump	15 cm
Air content	4 %
W/C ratio	57 %
s/a ratio	48 %
W	174 kg/m3
C	305 kg/m3
S	870 kg/m3
G	948 kg/m3
Add.	1068 ml/m3

Fig. 3   Section of the Silo

3.2 Concrete

The specified strength of concrete is 24 MPa at 56 days. The concrete mix and the strength properties are shown in Table 3 and Table 4.

Table 4   Strength Properties

	1d	3d	7d	28d	56d	91d
Compressive strength (MPa)	1.95	5.15	6.78	20.6	33.3	39.5
Tensile strength (MPa)	0.26	0.50	0.69	2.12	2.97	-
Young's modulus (*100 MPa)	64	103	134	228	270	290

3.3 Placement

Concrete was placed using two concrete pumps with boom. Temperature of the concrete cast was 26 degrees in centigrade. The placing was begun at 5:30 in the morning Sept. 22, continued until 19:00 in the evening.

3.4 Temperature and Stress Measurement

Temperature, concrete stress, and re-bar stress were measured at selected points. Measuring sensors are mainly placed at the center (bottom, mid-height, and the top of the slab) , and secondary at the peripheral position of the disk. Thermo-couples are placed a little more closely than other sensors. The measurement was started from the concrete placement and continued to the age of 56 days with four hours' intervals throughout the duration.

3.5 Results of Measurement

Temperature at the center of the thickness of the slab went up to 51 degrees centigrade from casting temperature which was about 26 degrees centigrade.

Concrete stress at mid height of the slab went up to 0.5 MPa compressive at young age, and to 1.3 MPa tensile in later age. Re-bar stress went up to 30 MPa tensile in maximum at the age of 6 days. Whole curves are shown in Fig. 4 to Fig. 5. No cracks were observed in the structure.

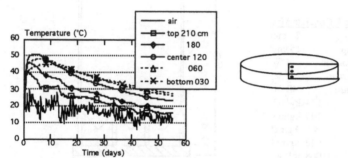

Fig. 4   Measured Results : Temperature History of the Concrete

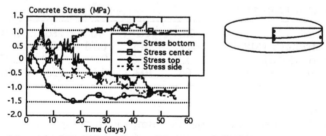

Fig. 5   Measured Results : Stress History of the Concrete

4 Analysis

4.1 Method of Analyses

Analyses were performed using solid element axis symmetrical FEM model program (CARC-MC). Analyzing model mesh is shown in Fig. 6. Time history of the analysis is summation of elastic solution of every individual interval steps.

The analyses were done in two cases, one is for low heat portland cement, and the other is for ordinary portland cement as a comparison. Input data used are shown in Table 5 and Table 6.

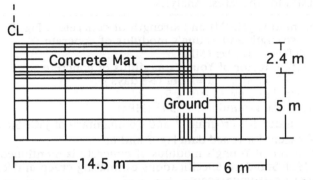

Fig. 6   Analyzing model of the Mat Slab and Ground

Table 5   Input Data for the Temperature Analysis

thermal conductivity :	2.9 W/(m.K)
coefficient of heat transfer:	13.9 W/(m².K)
atmospheric temperature :	age 0 day 26°C
	age 56 days 12°C
	(linearly interpolated)
adiabatic temperature rise :	Fig. 7

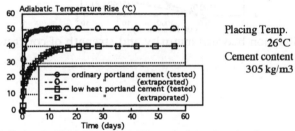

Fig. 7   Adiabatic Temperature Rise used in the Analyses

4.2 Results of Analysis

Results of the analyses are shown in Fig. 9 to 10.

Ordinary portland cement concrete gets high temperature and high tensile stress, but the low heat cement concrete gets comparative low temperature and low enough tensile stress to be free from cracks.

When ordinary portland cement concrete was used at the construction, some endeavor must be necessary to avoid concrete cracking.

Fig. 11 shows crack indexes of the two cement concrete structures at top and center point of the mat. Crack index is defined as tensile strength divided by tensile stress at that maturity age. Crack index lower than 1 means occurrence of crack. This shows the advantage of the low heat portland cement to the ordinary portland cement.

Table 6   Input Data for the stress Analysis

Relation between maturity(CUM) and strength of concrete : Fig. 8.
Relation between strength and Young's modulus of concrete :
    Ec=α*4695*SQRT(Fc)   Ec, Fc: (MPa)
        α : Depreciation factor of Young's modulus of concrete
        α =0.73 (for t <3 days)   α =1 (for t >5 days)
        linearly interpolated for (3 < t < 5)
Young's Modulus of the ground :   541 (MPa)

Material laws are under JCI Program[1] and Commentary of JSCE Standard Specifications[2] except for tested data Fig. 7 and 8.
    Depreciation factor of Young's modulus of concrete is mentioned in the Commentary of JSCE Standard Specifications estimating creep and/or relaxation of concrete at young age.

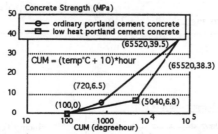

Fig. 8 Relation between maturity and strength of concrete

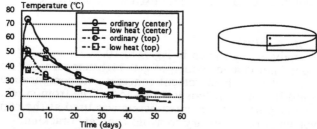

Fig. 9   Analyzed Results : Temperature History of the Concrete

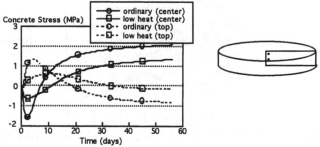

Fig. 10 Analyzed Result: Stress History of the Concrete

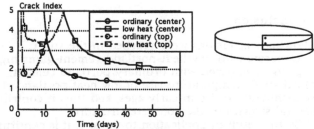

Fig. 11 Crack Index at top and center of the mat

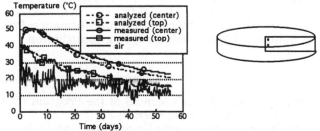

Fig. 12 Comparison of the Analyzed and Measured Results : Temperature

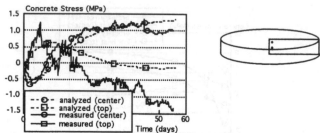

Fig. 13   Comparison of the Analyzed and Measured Results : Concrete Stress

4.3 Comparison between Measured and Analyzed Results
Comparison between measured and analyzed results are shown in Fig. 12 to 13
about the concrete temperature and stress at center and top of the slab.
Temperature histories fit good both at center and top of the slab. But for
concrete stress, agreement of the results are not so good at the top of the slab.

5 Conclusion

Following conclusions were reached in the study.
(a) Concrete with low heat portland cement gains higher strength in long
range age compared with ordinary portland cement.
(b) Adiabatic temperature rise of low heat portland cement concrete shows
less temperature rise in early ages, and thus it contributed the reduction of
thermal stresses in the concrete structure.
(c) By the result of application to the site, it is confirmed the low heat
portland cement concrete is suitable to the massive concrete structures with
these sizes.
(d) To get better fitting results, it seems to be better to input precise air
temperature history as a data for the analysis.

6 Acknowledgments

Appreciation is due to the persons who have contributed directly to
construction of the new cement silo: members at Chichibu Cement Kumagaya
Factory and members at Construction JV Office.
The authors also acknowledge to the members of Information Processing
Systems Dept. at Kajima Corp. for their analyzing contribution.

7 References

Japan Concrete Institute (1989) User's Manual of the Program for Calculation
    of Thermal Stress of Mass Concrete, Version 2
Japan Society of Civil Engineering (1991) Commentary on 15th Section: Mass
    Concrete, Standard Specification for Concrete Work

5 MODELLING OF HEAT AND MOISTURE TRANSPORT IN HARDENING CONCRETE

P.E. ROELFSTRA
Intron SME, Yverdon, Switzerland
T.A.M. SALET
Intron SME, Houten, The Netherlands

Abstract
A model to describe heat and moisture transport in hydrating materials is presented. It will be shown how results of this model, consisting of the development of temperature, humidity, and maturity, can be linked with mechanical processes like damage or fracturing.
Keywords: Modelling of Moisture Transport, Heat of Hydration, Prediction of Temperature, Water Migration.

1 Introduction

Temperature gradients in young concrete are often not the only cause of crack formation. Also the moisture distribution can play an important role by means of hygral shrinkage, especially at the surface of the structure and at the interfaces between young and older concrete. In addition, the moisture content has an influence on the rate of hydration and maturity, which means an influence on the evaluation of the mechanical properties. In order to study these phenomena and to be able to make realistic predictions for practice, a model based on heat and mass balance has been developed.

The heat balance is described by a diffusion equation with a non-linear source to take the rate of liberation of that of hydration into account. The material parameters involved in this equation have been made dependent on the moisture content. The mass balance is described by a non-linear diffusion equation with a sink for the consumption of water for the hydration process. This allows to simulate e.g. the case self desiccation.

The model of Kiessl (1984) for moisture transport has been used as a staring point, and has been extended with temperature and maturity dependent moisture storage functions.

The model has been implemented in a FE code called MLS(Multi-Layer-System), in order to solve boundary value problems for arbitrary geometries and variable environmental conditions. In this code a link has been made between the presented model and a mechanical model for ageing visco elastic and fracturing materials.

Thermal Cracking in Concrete at Early Ages. Edited by R. Springenschmid. Published 1994 by E & FN Spon, 2–6 Boundary Row, London SE1 8HN, UK. ISBN: 0 419 18710 3.

2 Mathematical Formulations of Coupled Heat and Moisture Transport

In this model it is assumed that there exists a unique relation between the evaporable moisture content w and the state parameters, temperature T, moisture potential h and degree of hydration a.
This relation is given by:

$$w \{h, T, \alpha\} \text{ in Kg/m}^3 \text{ (desorption isotherm)} \tag{1}$$

where
h: moisture potential,
T: temperature,
α: degree of hydration.

Kiessl (1984) has defined the moisture potential as follows:

$$h = \varphi \qquad\qquad \text{for } \varphi \leq 0.9 \tag{2}$$
$$h = 1.7 + 0.1 \log_{10} r \quad \text{for } \varphi > 0.9 \tag{3}$$

where:
φ: relative humidity,
r: equivalent pore radius in m.

This subdivision has been made to describe accurately the desorption isotherm at high moisture potentials. The equivalent pore radius indicates up to which size the capillary pores are filled with water. The transition of definition of the moisture potential is at 0.9, which corresponds with an equivalent pore radius of 10^{-8} m. The relative humidity is defined as:

$$\varphi = \frac{p}{p_a} \tag{4}$$

where
p: vapour pressure in Pa,
p_a: saturated vapour pressure in Pa.

The relation between the saturated vapour pressure and the temperature is given by:

$$p_a = 610.68 \, e^{\frac{17.08085 \; T}{234.175 + T}} \text{ in Pa} \tag{5}$$

where
T: temperature in °C.

The vapour concentration is related to the pressure with:

$$c = \frac{p}{R\,(\,273\,+\,T)} \quad \text{in Kg/ m}^3 \tag{6}$$

where
R: gas constant/molar weight of H_2O = 462 Nm/kg/K.
T: temperature in °C.

In the model the moisture flux is separated into two contributions:

$$m_{dw} = FKU\,\nabla\,w \qquad\qquad \text{(water)} \tag{7}$$

$$m_{dv,t} = \rho_w FDP\,\nabla\,h + \rho_w\,FDT\,\nabla\,T \qquad \text{(vapour, temperature) (8)}$$

where
∇ nabla operator
ρ_w: specific mass of water 1000 Kg/m^3,
FKU: transport coefficient with respect to the gradient of water concentration,
FDP: transport coefficient with respect to the gradient of vapour concentration,
FDT: transport coefficient with respect to the temperature gradient.

These coefficients are non-linear and are defined as follows:

$$FKU = \frac{FKU_0}{b}\,(b)^{\frac{w}{wl}}\left(\frac{T+\,20}{40}\right) \quad \text{in m}^2\text{/hour,} \tag{9}$$

where
FKU_0:   moisture diffusion for $w = w_l$ and T = 20° C, material property in m^2/hour,
b: parameter which determines the dependence on the water content,
w: water content in the material at saturation (with pressure) in Kg/m^3,
T: temperature in °C.

$$FDP = \frac{1}{\rho_w}\,(D\,c_a)\,\varepsilon_{s,\,FDT}e^{5000\left(\frac{1}{293}\,-\,\frac{1}{273\,+\,T}\right)} \quad \text{in m}^2\text{/hour} \tag{10}$$

where
ρ_w: specific mass of water, 1000 Kg/m^3,
c_a: vapour concentration at saturation in Kg/m^3,
T: temperature in °C,
$\varepsilon_{s,FDT}$ which accounts for micro structural effects.

Diffusion coefficient D is further specified as:

$$D = 0.083 \left(\frac{T + 273}{273} \right)^{1.81} \text{ in m}^2/\text{hour.} \tag{11}$$

Finally,

$$FDT = \frac{1}{\rho_w} \left(D \frac{dc_a}{dT} \right) \varphi \, \varepsilon_{s,FDT} \, e^{4750 \left(\frac{1}{238} - \frac{1}{273 + T} \right)} \tag{12}$$

Under transient conditions the mass balance is given by:

$$\dot{w} + Q\dot{\alpha} = \nabla \cdot \left[FKU \, \nabla w + \rho_w \, FDP \, \nabla h + \rho_w \, FDT \, \nabla T \right] \tag{13}$$

where
Q: amount of water needed for the hydration of cement in Kg/m^3,
$\dot{\alpha}$: rate of hydration.

By taking the partial differentials of the moisture rate and gradient the following differential equation is obtained:

$$\frac{\partial w}{\partial h} \dot{h} + \frac{\partial w}{\partial T} \dot{T} + \left\{ \frac{\partial w}{\partial \alpha} + Q \right\} \dot{\alpha} =$$

$$\nabla \cdot \left[\left(FKU \frac{\partial w}{\partial h} + \rho_w \, FDP \right) \nabla h + \left(FKU \frac{\partial w}{\partial T} + \rho_w \, FDT \right) \nabla T + FKU \frac{\partial w}{\partial \alpha} \nabla \alpha \right] \tag{14}$$

This equation is the starting point for the spatial and time discretisation with finite elements.

3 Links with Mechanical Properties

3.1 Visco elastic behaviour
The visco elastic behaviour is modelled with an ageing Maxwell chain. In the time integration of the boundary value problem the real time is weighted with a function for the moisture potential and a function (Arhenius) for the temperature.

3.2 Fracturing
The tensile strength has been made dependent on the maturity. Because this strength is linked with a softening diagram, the fracture energy has become maturity dependent.

4 Example

The example concerns a concrete wall with a thickness of 0.4 meter, poured with an initial temperature of 18°C. The temperature and the relative humidity of the surrounding air are supposed to be 15°C and 60°C respectively (constant). The wall is deshuttered 3 days after pouring and the surface is not cured afterwards. The data for the temperature evolution analysis is shown in figure 1:

$$\text{Heat capacity: } 2000 + \frac{Water}{158}(2222 - 2000) \quad kJ/m3/K$$

$$\text{Conductivity : } 1.85 + \frac{Water}{158}(2.25 - 1.85) \quad W/mK$$

$$\text{Heat of hydration: } H(t) = 81575.0 \; e^{-\left(\frac{22.00}{M(t)}\right)^{1.20}} \quad kJ/m3$$

$$\text{Maturity rate : } \dot{M}(t) = \frac{1}{1 + (5 - 5\,P)^4} \; e^{\frac{Q}{R}\left(\frac{1}{293} - \frac{1}{273+T(t)}\right)}$$

$$P = \text{Moisture Potential } (< 1)$$

$$\frac{Q}{R} = 4029 \qquad\qquad K : \; T(t) > 20 \; C$$

$$\frac{Q}{R} = 4029 + 177 *(20 - T(t)) \quad K : \; T(t) < 20 \; C$$

Figure 1: Data for the temperature development analysis

The adiabatic temperature development without influence of the moisture content is shown in figure 2.

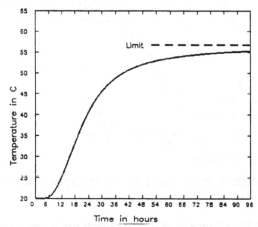

Figure 2: Adiabatic temperature development without influence of the moisture content

The moisture storage functions for a small and a high degree of hydration, and for two different temperatures, are shown in Fig. 3 and 4 respectively.

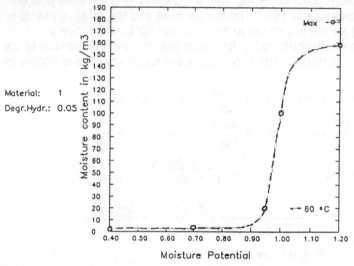

Figure 3: Moisture storage function for a small degree of hydration

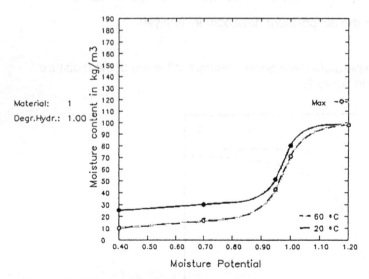

Figure 4: Moisture storage functions for a high degree of hydration at 20°C and 60°C.

Figure 5 shows the computed development of the temperature in the middle and at a point near the surface of the wall.

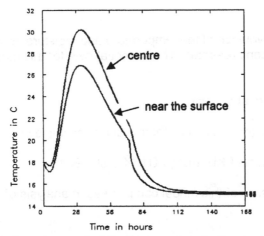

Figure 5: Development of the temperature in the centre and the near the surface of the wall.

The development of the moisture content in these points is shown in figure 6. The decrease of the moisture content on the centre of the wall is mainly caused by the water consumption for the hydration process. It can be seen in fig. 6 that near the surface, moisture is also transported to the surrounding air after deshuttering. These different developments entail different developments of maturity and related mechanical properties such as visco-elasticity and tensile strength.

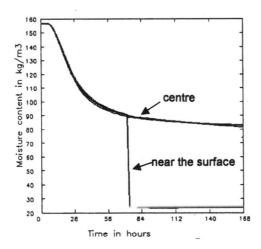

Figure 6: Development of the moisture content in the centre and near the surface of the wall.

5 Conclusions

A model to describe the evolution of the temperature and moisture content in hydrating concrete has been presented. The essential data for this model consists of:
Concrete composition,
Thermal conductivity and heat capacity,
Adiabatic heat development,
Moisture storage functions (desorption isotherms and cumulative pore size distribution),
Moisture transport coefficients FKU, FDP, FDT (Kiessl (1984)).

It is explained how the results of this model can be used in analysis of mechanical behaviour.

References

Kiessl, K. (1984), Kapillarer und dampfförmiger Feuchtetransport in mehrschichtigen Bauteilen, **Doctoral thesis**, Essen.

Acknowledgements
We gratefully acknowledge support of the Dutch Ministry of public Works for the development of MLS.

6 MODELLING OF TEMPERATURE AND MOISTURE FIELD IN CONCRETE TO STUDY EARLY AGE MOVEMENTS AS A BASIS FOR STRESS ANALYSIS

J.-E. JONASSON, P. GROTH and H. HEDLUND
Division of Structural Engineering, Luleå University of Technology, Luleå, Sweden

Abstract
This paper gives some essential models and features of young concrete with respect to moisture and thermal effects necessary when analysing stresses and making assessments of risks of early cracking. Data on heat of hydration for concretes made of different types of Portland cement are given. Hydration data are recorded during one year of production of cement, and the resulting variations can preferably be used to make sensitivity analyses in advance. The use of high performance concrete (HPC) means that shrinkage due to self-desiccation plays a more dominant role in early age behaviour. For three mixes of HPC the self-desiccation and the associated shrinkage is presented. Thereby, a linear relationship between shrinkage and relative humidity was found.
Keywords: Sealed Conditions, Heat of Hydration, Modelling of Temperature, Modelling of Moisture Field, Relative Humidity, Selfdesiccation, Shrinkage, Sorption Isotherms.

1 Introduction

The study of stresses in concrete induced by thermal volume changes in massive structures has been a very important area of research since the beginning of this century, and great progress have been made, see for instance Lövqvist (1946), Bernander (1973), Springenschmid (1984), and Emborg (1989). Today the interest is mainly focused on two areas: namely 1) the use of high performance concrete (HPC) and thereby the use of concrete mixes with lower water cement ratios, and 2) the use of more computerized systems to study the risk of cracking.

The main effect of low water cement ratios (≤ 0.40) is that there is a significant self-desiccation taking place at early ages causing an additional shrinkage effect in the concrete matrix. This means that humidity effects must be included in all theoretical and practical analyses of cracking in young concrete of HPC.

When using computers there is a need of relevant models to describe both the underlying thermal and moisture state as well as relevant constitutive behaviour to be able to make reliable assessments of the risks of cracking.

The aim of this paper is to present some recent developments in modelling and measurements of the basic behaviour of concrete with respect to moisture and

Thermal Cracking in Concrete at Early Ages. Edited by R. Springenschmid. Published 1994 by E & FN Spon, 2–6 Boundary Row, London SE1 8HN, UK. ISBN: 0 419 18710 3.

temperature effects necessary in the analysis of stresses and assessment of cracking in early age concrete.

At Luleå University of Technology there is also a development going on in the modelling of constitutive relationships for early age concrete based on discrete crack analysis and combined moisture and humidity effects, but this area is excluded in this paper.

2 Temperature and moisture field

2.1 General
For each concrete mix the hydration process is mainly depending on the moisture and thermal state, and this is in Jonasson (1984) modelled by the use of equivalent time of hardening as

$$t_{eq} = \int_0^t \beta_T \beta_w \beta_\Delta dt + \Delta t_{eq}^0 \qquad (1)$$

where t_{eq} is the equivalent time (sec), β_T is the temperature rate factor ($\neq 0$ for $T \neq$ a chosen reference temperature), β_w is the moisture rate factor (< 1 in the hygroscopic range; $= 1$ in completely wet state), β_Δ and Δt_{eq}^0 (sec) may be used to model different influences of admixtures or as one possible way to study variations in heat of hydration.

The modelling of time in terms of equivalent time in Eq. 1 means that there is a unique relationship with maturity, and all features of young concrete can therefore be related to t_{eq}.

The most world wide used description of β_w is from Powers (1948), and these findings were recently confirmed by tests of Norling-Mjörnell (1994).

The two-dimensional heat conduction inside hydrating concrete may for most applications be simplified as

$$\rho c \frac{\partial T}{\partial t} = \frac{\partial}{\partial x}(k_T \frac{\partial T}{\partial x}) + \frac{\partial}{\partial y}(k_T \frac{\partial T}{\partial y}) + \dot{Q}_c \qquad (2)$$

where ρc is the specific heat per unit volume ($J/(°C\ m^3)$), T is the temperature (°C), t is time (sec), x and y are spatial coordinates (m), k_T is the thermal conductivity ($W/(m°C)$), $\dot{Q}_c = dQ_c / dt =$ is the generated heat of hydration per unit volume (W/m^3), and Q_c is the totally generated heat energy per unit volume (J/m^3).

Even when moisture flow exists inside concrete, Eq. 2 can be used to calculate the temperature field separately. Namely, there is usually a difference in the order of magnitude between rates of moisture flow and thermal flux.

On the other hand, the moisture flow is strongly dependent on the long-term temperature gradients. In accordance with Hedenblad (1993) the two-dimensional moisture transportation can be formulated as

$$\frac{\partial w_e}{\partial t} = \frac{\partial}{\partial x}(\delta_v \frac{\partial v}{\partial x} + k_s \frac{\partial s}{\partial x}) + \frac{\partial}{\partial y}(\delta_v \frac{\partial v}{\partial y} + k_s \frac{\partial s}{\partial y}) - \dot{w}_n \tag{3}$$

where w_e is the evaporable water content (kg/m^3),δ_v is the moisture permeability with regard to humidity by volume (m^2/sec), v is the humidity by volume in the pores (kg/m^3), k_s is the moisture conductivity (kg/(m sec Pa)), s is the suction in the pore water (the pressure between the pore water and the ambient total pressure), and $\dot{w}_n$ is the rate of the formation of non-evaporable water due to hydration of the cement (kg/(m^3 sec)).

2.2 Sealed conditions
Especially in massive structures, the moisture state may be described as sealed or almost sealed conditions due to large dimensions of the structure. This is also the case when we actually cover the concrete for protection against environmental influences. The internal moisture state in HPC may be approximated as being sealed in the early age period for all types of hardening conditions due to very dense structure. So, in many applications the moisture flow as such can be neglected for the formation of stresses inside early age concrete. However, the self-desiccation must always be taken into consideration. This means that Eq. 3 for many practical situations can be regarded for the sealed conditions as

$$\frac{\partial w_e}{\partial t} = -\dot{w}_n \quad \text{or in the integrated form} \quad w_e = w_0 - w_n \tag{4}$$

where w_0 is the initial water content (kg/m^3).

According to the observations of Powers (1960) the non-evaporable water content made of pure Portland cement is related to the degree of hydration by

$$w_n = \alpha\ 0.25\ C \tag{5}$$

where α is the degree of hydration, and C is the cement content (kg/m^3).

By alternative definitions of the degree of hydration, the self-desiccation may be related to the total heat of hydration or to the chemically bonded amount of cement by

$$\alpha = \frac{q_{cem}}{q_u} = \frac{C_n}{C_\infty} \tag{6}$$

where q_{cem} is the generated heat per mass unit of cement (J/kg), q_u is the heat generated when the degree of hydration has reached unity (J/kg), C_n is the chemically bonded cement content (kg/m^3), and C_∞ is the chemically bonded cement content when the degree of hydration has reached unity (kg/m^3).

For the use of puzzolanic additives like silica fume and fly ash, Eqs. 4 - 6 no longer hold. According to recent findings in for instance Helsing-Atlassi (1993) it is possible to build models based on degree of reactions for different components of the binder, but these techniques are still under development. In the meantime analysis of free movements due to self-desiccation is always possible to perform by direct use of measured data on humidity and/or shrinkage under sealed conditions, see further section 2.5.

2.3 Heat of hydration

From numerous tests of strength growth at different curing temperatures on concrete made of pure Portland cement produced in Sweden, the following empirical expressions for the temperature effect on hydration rate has been established

$$\beta_T = \exp(\Theta(\frac{1}{293} - \frac{1}{T+273})) \quad \text{with} \quad \Theta = \Theta_{ref}\left(\frac{30}{T+10}\right)^{\kappa_3} \tag{7}$$

where $\Theta_{ref} = 5300$ K and $\kappa_3 = 0.45$ are fitting parameters obtained.

From one year of cement production in Sweden two concrete mixes ($w_0/C = 0.67$ and $w_0/C = 0.50$, respectively) have been tested under adiabatic and semi-adiabatic conditions. With the temperature influence according to Eq. 7, the resulting heat of hydration values are shown in fig 1. The scattering in all data is described by the coefficient of variation in the size of order of 10 %. This variation may be interpreted as a "natural" variation of features for cement as a commercial product.

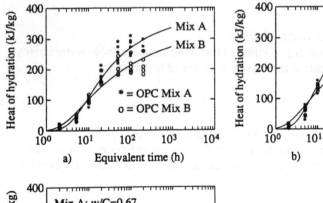

a)   Equivalent time (h)

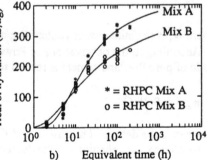

b)   Equivalent time (h)

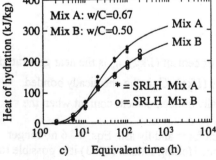

c)   Equivalent time (h)

Fig. 1
Measured heat of hydration for two concrete mixes and three types of cement. Fitting parameters for the solid lines, see table 1.

For use in numerical calculations, fitting of the data in fig 1 has been done according to the expressions

$$Q_C = q_{cem}C \quad \text{with} \quad q_{cem} = \alpha \, q_u \tag{8}$$

$$\alpha = \exp\left[-\lambda_1(\ln(1+\frac{t_{eq}}{t_1}))^{-\kappa_1}\right] \tag{9}$$

where λ_1, t_1 (h), and κ_1 plays the role of fitting parameters.

The final values for each individual cement, q_u, have been calculated as the sum of final heat per cement compounds in accordance with the Bougue formula, see for instance Neville and Brooks (1987). The results of the fitting procedure for the mean values are shown in table 1.

Table 1. Fitting parameters using Eq. 1, shown as solid lines in fig 1

Type of cement	$\dfrac{w_0}{C}$	q_u kJ/kg	λ_1	t_1 h	κ_1
a) OPC = ordinary Portland cement	0.67	460	1.18	6.58	0.85
	0.50	460	2.42	2.12	0.85
b) RHPC = rapid hardening Portland cement	0.67	468	1.30	5.09	1.10
	0.50	468	3.52	0.97	1.10
c) SRLH = sulphate resistant low heat Portland cement	0.67	406	1.40	8.43	1.07
	0.50	406	3.16	2.06	1.07

The variation of each mix within the standard deviation of the data in fig 1 may be studied by changes of the parameter t_1 within the range of ± 38 % (OPC), ± 18 % (RHPC), and ± 14 % (SRLH), respectively. These figures are a result of the scatter in the obtained data for each cement type separately, and thereby they are not inter-correlated. To take into account the observed scattering is important for the reliability when making assessment of thermal cracking based on general heat of hydration data.

For more specific mixtures (special cement types, lower water cement ratios, adding of silica fume and/or fly ash etc.), we do not have data enough today for more general analyses in advance. In such cases there is always a possibility to start with measurements on each concrete mix, see Ekerfors et al (1993) and Jonasson and Ronin (1993).

2.4 Sorption isotherms
The need of descriptions of sorption isotherms in analysing early age concrete is twofold:

1) to be able to connect different characterisations of moisture flow like the two necessary gradients in Eq. 3 (i. e. $\partial v / \partial x$ and $\partial s / \partial x$, respectively), and

2) to be able to connect self-desiccation in terms of water content to changes in humidity, and furthermore to the size of the shrinkage.

The first need is a general demand when studying moisture flow, but the study here of sealed conditions makes this part of less importance. On the contrary, the need of sorption isotherms for general analyses of self-desiccation is here quite obvious.

Based on the ideas of Powers (1960) to distinguish between gel pores and capillary pores in the cement matrix, a quantitative model on description of the sorption isotherms of cement paste, mortar and concrete has been developed. A brief background of the model is given in Jonasson (1985), where it is shown that both absorption and desorption isotherms can be reflected for a wide range of water cement ratios and ages of hardening, see examples in fig 2.

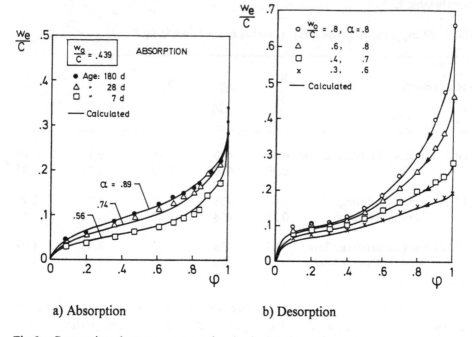

a) Absorption   b) Desorption

Fig.2   Comparison between measured and calculated sorption curves.
        From Jonasson (1985).

The only independent parameters in the model are the water cement ratio (w_0 / C) and the degree of hydration (α), both well known and necessary parameters in material based modelling of hardening concrete. The model in Jonasson (1985) is restricted to the use of pure Portland cement. However, some recent results in Helsing-Atlassi (1993) indicates that there is a possibility to use the same basic ideas when taking into account the influence of addition of silica fume, but this has not been done so far.

2.5 Tests on self-desiccation

As an alternative to theoretical analyses and as a base to build future models, tests on self-desiccation and shrinkage at sealed conditions have been carried out for some concrete mixes. The specimens had the size 100·100·250 mm, and they were sealed immediately after casting by covering the concrete with plastic foils. The curing temperature was 20 ±1 °C all the time. After the form stripping at about 10 to 12 hours since casting the specimens were sealed with a special epoxy glue to the thickness of about 2 to 3 mm. The measurements of relative humidity was made as soon as possible, which in most cases means before the form stripping time. The first shrinkage values was recorded shortly after the gluing of epoxy. Examples of results are shown in fig 3 for three water to binder ratios from 0.27 to 0.40. As expected, the shrinkage and humidity curves versus time differ between these three levels of water to binder ratios. However, the shrinkage at sealed conditions as a function of relative humidity appears to be almost equal for all mixes examined and approximately linear within the studied range. The linear curve in fig 3 is

$$\varepsilon_{RH} = -450 \cdot 10^{-6} (1 - \frac{RH}{100}) \tag{10}$$

where ε_{RH} is the free volume strain of the concrete, and RH is the relative humidity (%).

To get an indication on the effects of self-desiccation according to Eq. 10 one may regard that a drop of humidity with $\Delta RH = 10$ % approximately corresponds to a decrease in temperature with $\Delta T = 4.5$ °C. So, at high degree of restraint the self-desiccation may be the conclusive cause of cracking at early ages for high strength concrete, see fig 3.

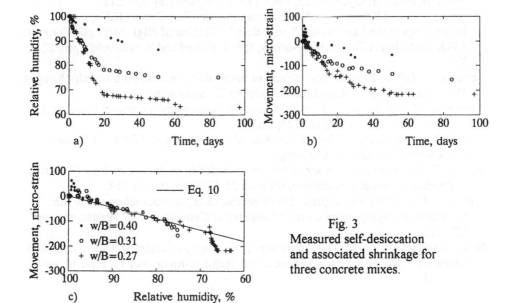

Fig. 3
Measured self-desiccation and associated shrinkage for three concrete mixes.

3. Acknowledgements

This work is supported by Cementa AB in Sweden. The work is supervised by Professor Lennart Elfgren, head of the Division of Structural Engineering at Luleå University of Technology.

4. References

Bernander, S. (1973) Cooling of hardening concrete by means of embedded cooling system (in Swedish), in **Nordisk Betong No 2**, Stockholm, 10 pp.

Ekerfors, K. Jonasson, J.-E. and Emborg, M. (1993) Behaviour of young high strength concrete, in Proceedings of the International Symposium **Utilization of High-Strength Concrete 1993,** 20-24 June Lillehammer, Norway, (eds I. Holand and E. Sellevold), pp 691-697.

Emborg, M. (1989) **Thermal stresses in concrete structures at early ages**. Doctoral thesis 1989:73D. Luleå University of Technology, Luleå.

Hedenblad, G. (1993) **Moisture permeability of mature concrete, cement mortar and cement paste**. Report TVBM-1014, Lund Institute of Technology, Lund.

Helsing-Atlassi, E. (1993) **A quantitative thermogravimetric study on the nonevaporable water in mature silica fume concrete**. Publication P-93:6, Chalmers University of Technology, Göteborg.

Jonasson, J.-E. (1984) **Slipform construction - calculations for assesssing protection against early freeezing**. Report Fo 4:84, Swedish Cement and Concrete Research Institute, Stockholm.

Jonasson, J.-E. (1985) Moisture fixation and moisture transfer in concrete, in **8th International Conference on Structural Mechanics in Reactor Technology,** Brussels, Belgium August 19-23, 1985. (eds J. Stalpaert), pp 235-242.

Jonasson, J.-E. and Ronin, V. (1993) Energetically Modified Cement (EMC), in Proceedings of the International Symposium **Utilization of High-Strength Concrete 1993,** 20-24 June Lillehammer, Norway, (eds I. Holand and E. Sellevold), pp 752-759.

Lövqvist, B. (1946) **Temperature effects in hardening concrete** (in Swedish). Kungliga Vattenfallsstyrelsen, Tekniskt meddelande No 22, Stockholm.

Neville, A. M. and Brooks, J. J. (1987) **Concrete Technology**. Longman Scientific & Technical, Longman Group UK Ltd, Essex.

Norling-Mjörnell, K. (1993) **Self-desiccation of concrete**. Report P 94:2, Chalmers University of Technology, Göteborg.

Powers, T. C. (1948) A discussion of cement hydration in relation to curing of concrete, in **Portland Cement Association, Bulletin 25**. Chicago, pp 179-188.

Powers, T C. (1960) Physical properties of cement paste, in Proceedings of the **4th International Symposium on the Chemistry of Cement, Vol II,** Washington, pp 577-612.

Springenschmid , R. (1984) Die Ermittlung der Spannungen infolge von Schwinden und Hydratationswärme in Beton, in **Beton und Stahlbetonbau, Vol 79, Heft 10,** pp 263-269.

7 INFLUENCE OF THE GEOMETRY OF HARDENING CONCRETE ELEMENTS ON THE EARLY AGE THERMAL CRACK FORMATION

G. De SCHUTTER and L. TAERWE
Magnel Laboratory for Concrete Research, University of Gent,
Belgium

Abstract
This paper focusses on the influence of the element geometry on the early age
thermal crack risk. The influence of the element geometry on early age therma
crack formation can be evaluated in terms of the maximum temperature rise and th
maximum temperature difference between element core and surface. By means o
numerical simulation it is shown that the massivity, defined as the ratio between th
volume of the concrete element and the surface, is not a general quantitativ
parameter. A new parameter is introduced which is called the equivalent thickness
defined for a separable heat flow area of the concrete element. Numerical simulati
on indicates that the equivalent thickness enables a good quantitative compariso
between different element shapes.
Keywords: Cracking of Different Element Geometries, Hardening Concrete,
Numerical Simulation.

1 Introduction

In hardening massive concrete elements the heat of hydration gives rise to conside
rable thermal gradients and thermal stresses, which might cause early age cracking
The problem of early age thermal crack formation is influenced by several pheno
mena. The heat of hydration constitutes the major factor, but its effect is influence
by thermal conditions, e.g. thermal characteristics, casting temperature an
environmental climate, and also by mechanical properties, e.g. Young's modulus
relaxation and plastic strain, all of which change during the hardening process (D
Schutter et al. 1992). This paper only focusses on the influence of the elemer
geometry on the early age thermal crack risk. Empirical rules state that early ag
thermal cracking occurs above a critical temperature rise (cracking due to externa
restraint) or above a critical temperature difference between element core an
surface (cracking due to internal gradient). The temperature control during th
construction of the Storebaelt-West Bridge e.g. is based on such empirical rulc
(Visser et al. 1992). Following these empirical rules the influence of the elemer
geometry on early age thermal crack formation can be estimated by investigating it
influence on the maximum temperature rise and the maximum temperature differen
ce between element core and surface.

Thermal Cracking in Concrete at Early Ages. Edited by R. Springenschmid. Published 1994
by E & FN Spon, 2–6 Boundary Row, London SE1 8HN, UK. ISBN: 0 419 18710 3.

For different geometries, like e.g. a wall and a cube, the same qualitative effects are noticed when size increases. Larger sizes yield a higher core temperature which is sustained for a longer period, and a higher maximum temperature difference between core and surface. However, the geometry of the cross section of the concrete element strongly influences the heat distribution caused by hydration after casting (Schickert 1990). In order to compare these effects for different shapes in a quantitative way it would be useful to dispose of a representative geometric parameter. The potential advantage of such a geometric parameter would be the possibility of obtaining a quick estimate of the maximum temperature rise and the maximum temperature difference between element core and surface by means of a simple one-dimensional calculation, even for three-dimensional heat flow problems. Thus, an approximate though effective evaluation of the early age thermal crack risk, even at an early state of the project becomes possible.

2 Numerical simulation

In order to examine the influence of the geometry of hardening concrete elements on the early age thermal crack formation numerical simulation is used. The temperature field in a hardening concrete member can be calculated as a function of time, by making use of the Fourier differential equation for non-stationary heat transfer combined with the appropriate boundary and initial conditions.

At the University of Gent a finite element computer program was developed that allows the calculation of the temperature field in a three-dimensional concrete member during hardening. The results are obtained using a new hydration model which is valid both for Portland cement and blast furnace slag cement. Furthermore, state-dependent material properties are considered (De Schutter et al. 1992). The accuracy of the developed calculation model was verified by means of experiments on hardening massive concrete cylinders subjected to variable ambient temperature conditions. A very good fit between experiment and simulation was obtained (Taerwe et al. 1993, Dechaene et al. 1994).

Besides the three-dimensional finite element code also a one-dimensional finite difference program was developed. This one-dimensional calculation tool enables, besides exact simulation results for one-dimensional heat flow problems, a quick qualitative evaluation of the influence of several parameters even for three-dimensional elements. As already stated in the introduction, a general geometric parameter would extend the validity of the one-dimensionally obtained results to a more quantitative level.

3 Massivity

3.1 Definition and examples
A frequently used parameter for the description of the element geometry is the massivity. The massivity M of a hardening concrete element is defined as the ratio between the volume V and the exposed surface S :

$$M = \frac{V}{S} \qquad\qquad (1)$$

Figure 1 summarizes the values of the massivity for different element shapes. The cube respectively hollow cube can be seen as a crude approximation of different concrete armour units for breakwater, namely grooved cube respectively HARO® (De Meyer and De Rouck 1990). The value given for an infinitely long massive respectively hollow cylinder can be used for the two-dimensional heat flow in e.g. a massive respectively hollow cylindrical column.

3.2 Usefulness

The usefulness of the massivity M as a general parameter for the estimation of the maximum temperature during hardening, θ_{max}, or the maximum temperature difference between core and surface, $\Delta\theta_{max}$, can be verified by means of numerical simulation. Consider for example a concrete composition with 300 kg blast furnace slag cement CEM III/B per m³. The casting temperature and ambient temperature are fixed at 15°C. For the formwork, a convection coefficient of 37500 J/m²h°C is applied. After removal of the formwork, four days after casting, the convection coefficient changes towards a value of 65000 J/m²h°C. The specific heat resp. heat conduction coefficient of the fresh concrete is 1150 J/kg°C resp. 12200 J/mh°C. During hydration the value of these thermal characteristics decreases (De Schutter et al. 1992), to reach a final value for hardened concrete of 1000 J/kg°C resp. 9600 J/mh°C. The heat production rate during hardening is calculated with the new hydration model mentioned earlier (De Schutter et al. 1992).

For this case, the maximum temperature occuring during hardening, θ_{max}, and the maximum temperature difference between element core and surface, $\Delta\theta_{max}$, can be calculated with the simulation method outlined in section 2. For the different element shapes mentioned in figure 1, the calculated values θ_{max} and $\Delta\theta_{max}$ are given in figure 2 as a function of the massivity M. For a given element shape, the values θ_{max} and $\Delta\theta_{max}$ are positively correlated with the massivity, as can be expected.

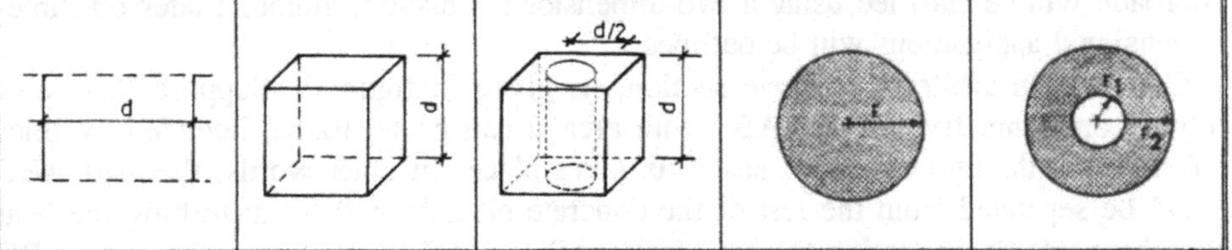

WALL	CUBE	HOLLOW CUBE	INFINITELY LONG MASSIVE CYLINDER	INFINITELY LONG HOLLOW CYLINDER
$M = d/2 = 0.5\,d$	$M = d/6 = 0.167\,d$	$M = 0.112\,d$	$M = r/2$	$M = (r_2 - r_1)/2$
$d_{eq} = d$	$d_{eq} = 2\,d/3$	$d_{eq} = 0.23\,d$	$d_{eq} = 3\,r/2$	$d_{eq} = r_2 - r_1$ (r₁ large enough)
$2M/d_{eq} = 1$	$2M/d_{eq} = 0.5$	$2M/d_{eq} = 0.97$	$2M/d_{eq} = 0.67$	$2M/d_{eq} = 1$

Fig. 1. Element shapes

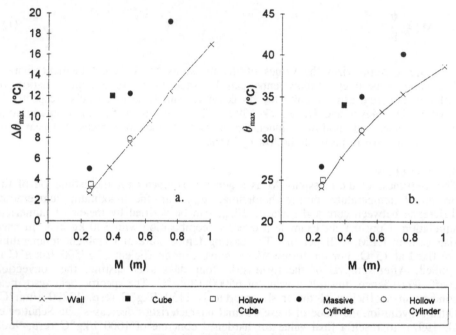

Fig. 2. Simulation results

Nevertheless, for different shapes there seems to be no unique relation between massivity and θ_{max} or $\Delta\theta_{max}$. Thus it can be concluded that the massivity of the hardening concrete element is not a sufficiently general quantitative parameter.

4 Equivalent thickness

4.1 Definition and examples

In this section a more refined and effective approach is developed. The new approach will be clarified using a two-dimensional situation, although later on three-dimensional applications will be outlined.

Consider an arbitrary concrete section, as given in figure 3. Suppose that, e.g. relying on symmetry, an area ABC with area Ω can be separated, bounded by lines AB and AC through which no heat flux takes place. In other words, the area ABC could be separated from the rest of the concrete element without disturbing the heat flow by perfectly insulating the boundaries AB and AC. Such a separable area ABC is called a heat flow area. The outer boundary BDC forms the exposed surface, with arc length s. For this heat flow area the massivity M_a can be calculated :

$$M_a = \frac{\Omega}{s} \tag{2}$$

In three dimensions this expression is extended into expression (1), with V the volume and S the exposed surface of the three-dimensional heat flow area. The point A in figure 3 is the most inner point of the heat flow area, considering the boundary conditions. In this most inner point the temperature will normally reach the maximum value θ_{max} due to heat of hydration effects. The point G is the centre of gravity of the heat flow area. Now consider the point D defined as the intersection point between the straight line AG and the exposed surface. Normally the maximum temperature difference $\Delta\theta_{max}$ between element core and surface can be calculated as the maximum temperature difference between A and D.

The equivalent thickness d_{eq} can now be defined as :

$$d_{eq} = \gamma_a \cdot M_a \tag{3}$$

in which γ_a is the so-called shape factor of the heat flow area, defined as :

$$\gamma_a = \frac{AD}{GD} \tag{4}$$

For a hardening wall (figure 4) a heat flow area BCEF can be separated, with exposed (unit) surface CE and perfectly insulated boundaries BF, BC and FE. The application of expression (3) in this case yields :

$$d_{eq} = \gamma_a \cdot M_a = \frac{AD}{GD} \cdot \frac{d}{2} = 2 \cdot \frac{d}{2} = d \tag{5}$$

This result clarifies why the product of the shape factor γ_a with the massivity M_a is called the equivalent thickness.

For the three-dimensional heat flow in a hardening concrete cube with side length d (figure 5) a pyramid shaped heat flow area ABCFE can be separated. It can easily be calculated that in this case :

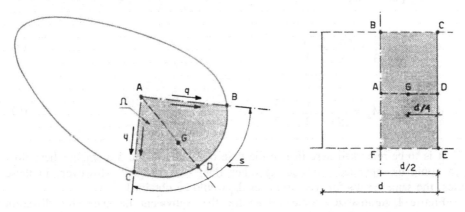

Fig. 3. Arbitrary section   Fig. 4. Wall

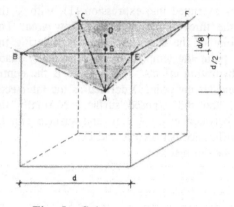

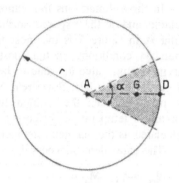

Fig. 5. Cube   Fig. 6. Massive cylinder

$$d_{eq} = \gamma_a \cdot M_a = 4 \cdot \frac{d}{6} = \frac{2d}{3} \qquad (6)$$

In the case of an infinitely long massive cylinder, a heat flow area as indicated in figure 6 could be separated. For each angle α the indicated section could serve as a heat flow area, completely in accordance with the definitions given above. However, the resulting equivalent thickness obviously is α-dependent. In order to accommodate for this inconvenience the definition of the equivalent thickness of an element should be extended with the clause that it has to be calculated for the smallest possible heat flow area. In the case of figure 6 this clause leads to the following results :

$$M_a = \lim_{\alpha \to 0} \frac{\dfrac{\alpha}{2\pi} \cdot \pi r^2}{\alpha r} = \frac{r}{2} \qquad (7)$$

$$AG = \lim_{\alpha \to 0} \frac{2}{3} r \frac{r\alpha}{r\alpha} = \frac{2r}{3} \qquad (8)$$

$$d_{eq} = \gamma_a \cdot M_a = \frac{r}{r/3} \cdot \frac{r}{2} = \frac{3}{2} r \qquad (9)$$

It is to be remarked here that in the case of figures 4 and 5 a smaller heat flow area could be separated, considering other lines of symmetry. However, in these cases, the same results for the parameter d_{eq} would be obtained.

Figure 1 summarizes some values for the equivalent thickness for different element shapes, calculated using the smallest possible heat flow area.

4.2 Usefulness

The usefulness of the equivalent thickness can be verified using the same simulation results as mentioned in section 3.2. In figure 7 the obtained simulation results are plotted as a function of the equivalent thickness d_{eq}. From figure 7a resp. 7b, it seems that, for a specified concrete and given boundary conditions, a unique relation exists between the maximum temperature difference $\Delta\theta_{max}$ respectively the maximum temperature θ_{max}, and the equivalent thickness d_{eq} defined in section 4.1. Considering the empirical temperature rules mentioned in section 1, the equivalent thickness enables an evaluation of the cracking tendency during hardening by means of a simple one-dimensional simulation of the temperature field, even in the case of three-dimensional concrete members.

From the values $2M/d_{eq}$, given in figure 1, it can be concluded that, in comparison with the massivity concept, the equivalent thickness method yields an alternative description of element geometry especially in the case of more-dimensional heat flow problems, e.g. cube and massive cylinder. For the hollow cube and hollow cylinder both methods hardly differ. This conclusion can be explained by remarking that the hollow elements are more or less wall shaped, yielding a heat flow very similar to the one-dimensional heat flow in a wall.

It is to be remarked here that the application of the equivalent thickness method for the evaluation of the tendency for early age cracking is bounded by the practical limits of the empirical temperature rules. The general applicability of these empirical rules still has to be verified. The advantage of the equivalent thickness method is the possibility of a quick evaluation, even at an early state of the project. However, in a more advanced state, a more fundamental evaluation of the cracking tendency should be based on stress calculations, considering evolutionary constitutive laws for hardening concrete.

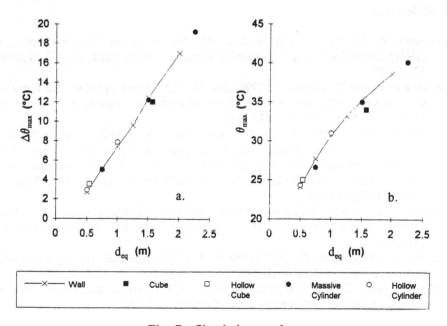

Fig. 7. Simulation results

5 Acknowledgment

This Research Project is financed by the Flemish Ministry of Public Works and by the Belgian National Fund for Scientific Research (N.F.W.O.). This financial support is greatly acknowledged.

6 Conclusion

- By means of numerical simulation it is shown that the massivity of a hardening concrete element, defined as the ratio of the volume over the exposed surface, is not a general quantitative parameter for the prediction of the temperature development.
- A new parameter is introduced which is called the equivalent thickness. It is defined for a so-called separable heat flow area of the concrete element. Numerical simulation indicates that the equivalent thickness enables a good quantitative comparison between different element shapes.
- The potential advantage of the equivalent thickness method is the possibility of obtaining a quick estimate of the maximum temperature rise and the maximum temperature difference between element core and surface by means of a one-dimensional calculation, even for three-dimensional heat flow problems. Considering empirical temperature rules, an approximate though effective evaluation of the early age thermal crack risk, even at an early state of the project, becomes possible.

7 References

Dechaene, R., De Rouck, J., De Schutter, G., Taerwe, L. and Van der Weeën, F. (1994) Thermal cracking in hardening concrete armour units. PIANC Bulletin, 82, 61-69.

De Meyer, P. and De Rouck, J. (1990) The HARO®, a new approach for breakwater armouring. Proceedings of the First International Conference on Ports and Marine Structures, Teheran, 540-568.

De Schutter, G., Taerwe, L. and Van der Weeën, F. (1992) Experiment based prediction of thermal characteristics of hardening concrete. Proceedings of the Third International Workshop on Behaviour of Concrete Elements under Thermal and Hygral Gradients, Wissenschaftliche Zeitschrift HAB Weimar, 3./4./5. Heft, 117-126.

Schickert, G. (1990) Testing temperature gradients of young concrete. Testing during Concrete Construction (ed. H.W. Reinhardt), RILEM Proceedings 11, 243-252.

Taerwe, L., De Schutter, G. and Dechaene, R. (1993) Thermal cracking in massive concrete members. Gent Works, 96, pp. 41-52.

Visser, J., Salet, T.A.M. and Roelfstra, P.E. (1992) Temperatuurbeheer van jong beton op basis van computermodellen. Cement, 9, 16-22.

PART THREE

DETERMINATION AND MODELLING OF MECHANICAL PROPERTIES

(Détermination et lois d'évolution
des propriétés mécaniques)

8 STIFFNESS FORMATION OF EARLY AGE CONCRETE

P. PAULINI and N. GRATL
Institut für Baustofflehre and Materialprüfung, Universität Innsbruck, Innsbruck, Austria

Abstract
This paper presents a multiphase model for the determination of stiffness moduli (E,K,G,ν) in early age concrete. The model is based on the kinetics of cement hydration and on the knowledge of the elastic properties of the mix design materials. The most important factors are the internal volume changes in early age concrete, especially that of free water. These changes and the degree of hydration are measured with the hydrostatic weighing method. The stiffness parameters of the original mix materials are found by determining their elastic and bulk moduli. Comparative calculations demonstrate not only the influence of stiffness, but also the applicability of the model to concrete.
Keywords: Bulk Modulus, Elastic Modulus, Multiphase Model.

1 Introduction

Stiffness formation of concrete is of interest in evaluating the risk of cracking in early age concrete as a result of internal stresses or building conditions. In standards the elastic modulus is usually defined as a variable of the compressive strength:

CEN EC2 $\qquad E_c = 9500 . (f_c + 8)^{1/3}$

France BAEL $\qquad E_c = 11000 . f_c^{1/3}$

USA ACI 318 $\qquad E_c = 5000 . f_c^{1/2}$ $\qquad\qquad E_c , f_c$ in MPa

These formulas consider neither the different stiffnesses of cement paste and aggregates nor fractional mix volumes. They also result in very differing values for early age concrete lower then 20 MPa compressive strength. They are therefore only helpful as an assessment for commonly used aggregates and mix designs.

Attempts to describe effective elastic material constants are often based on a two phase model and isotropic homogeneous materials. Reuss´ model of equal stresses provides the lower bound for the elastic modulus, Voigt´s Model of equal deformations the upper bound. Because both bounds cover a broad range they are not very precise. This was improved by combining both models for fractional volumes and by introducing empirical factors to take the compound effect into account. Several

Thermal Cracking in Concrete at Early Ages. Edited by R. Springenschmid. Published 1994 by E & FN Spon, 2–6 Boundary Row, London SE1 8HN, UK. ISBN: 0 419 18710 3.

authors applied these models to concrete rating material constants for the CSH-matrix and aggregates [1-4]. All these models consider only the elastic modulus and not the different poisson ratios. However, full identification of elastic materials is only possible with two of the four elastic constants E, K, G, v.

The introduction of multiphase models with a second elastic constant further defined the upper and lower bounds. This development occurred in solid state mechanics and was first applied to metals and compound materials [5-7]. The results were obtained using the variational principle of linear elastic theory for composite sphere geometry [6]. The Hashin-Shtrikman limits (H-S) were also deduced for multi phase models with random geometry [7]. While the description of the effective bulk modulus is possible with simple energy methods, it is more difficult for the effective shear modulus. Christensen und Lo solved the problem for different geometries with results within the H-S-limits [8,9].

Using these improved models on concrete, measured results may also lie beyond the H-S-limits [10]. Though the more accurate models have been used for calculations they were applied only to the two phase model matrix-aggregate. Therefore, Monteiro recently proposed a three phase model for concrete, proposing a transition zone [10, 11]. However, its volume is difficult to determine. This paper takes this approach further on the basis of a composite model which is applied successively to all concrete mix materials. To that end, two elastic constants have to be determined for each mix material.

2 Composite Model of Christensen and Lo

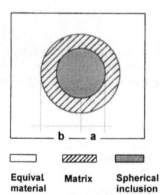

b — a

| Equival material | Matrix | Spherical inclusion |

Fig. 1, Three phase model

The composite model is based on a spherical inclusion surrounded by a matrix. Both media are enclosed by an equivalent homogeneous material whose unknown effective properties have to be determined. Therefore it is required that the equivalent material of Fig.1 should store the same strain energy as the combined sphere model. For this model elastic stress-strain states and elastic strain energy have been determined [8].

The effective bulk modulus K_c of the three phase model is the same as in the composite sphere model (Eq.1)

$$K_c = K_m + \frac{c\,(K_i - K_m)}{1 + \frac{(1-c)\,(K_i - K_m)}{K_m + \frac{4}{3}\,G_m}} \tag{1}$$

where the index m stands for matrix, i for inclusion and c=(a/b)³ means the concentration of the included volume. The effective shear modulus G_c is found using Eshelby's formula

$$U_c = U_0 + \frac{1}{2} \int_S (\sigma_i u_i^0 - \sigma_i^0 u_i)\, ds \qquad (2)$$

where U_c stands for strain energy of composite three-phase model, U_0 for strain energy outside the sphere b, $\sigma_i u_i$ for the stress-strain state inside the sphere b and $\sigma_i^0 u_i^0$ for the stress-strain state of the equivalent material inside the sphere b. The integration proceeds on the surface S of sphere b. The criterion for the effective shear modulus is the condition $U_c = U_0$ which leads to

$$\frac{1}{2} \int_S (\sigma_i u_i^0 - \sigma_i^0 u_i)\, ds = 0 \qquad (3)$$

G_c follows from the solution of the quadratic equation (Eq.4)

$$A\left(\frac{G_c}{G_m}\right)^2 + 2B\left(\frac{G_c}{G_m}\right) + C = 0 \qquad (4)$$

with the constants [9]

$$A = 8\left(\frac{G_i}{G_m}-1\right)(4-5v_m)e_1\, c^{\frac{10}{3}} - 2\left[63\left(\frac{G_i}{G_m}-1\right)e_2 + 2e_1e_3\right]c^{\frac{7}{3}} +$$
$$252\left(\frac{G_i}{G_m}-1\right)e_2 c^{\frac{5}{3}} - 50\left(\frac{G_i}{G_m}-1\right)(7-12v_m+8v_m^2)e_2 c + 4(7-10v_m)e_2e_3 \qquad (5)$$

$$B = -2\left(\frac{G_i}{G_m}-1\right)(1-5v_m)e_1\, c^{\frac{10}{3}} + 2\left[63\left(\frac{G_i}{G_m}-1\right)e_2 + 2e_1e_3\right]c^{\frac{7}{3}} -$$
$$252\left(\frac{G_i}{G_m}-1\right)e_2 c^{\frac{5}{3}} + 75\left(\frac{G_i}{G_m}-1\right)(3-v_m)e_2v_m c + \frac{3}{2}(15v_m-7)e_2e_3 \qquad (6)$$

$$C = 4\left(\frac{G_i}{G_m}-1\right)(5v_m-7)e_1\, c^{\frac{10}{3}} - 2\left[63\left(\frac{G_i}{G_m}-1\right)e_2 + 2e_1e_3\right]c^{\frac{7}{3}} +$$
$$252\left(\frac{G_i}{G_m}-1\right)e_2 c^{\frac{5}{3}} + 25\left(\frac{G_i}{G_m}-1\right)(v_m^2-7)e_2 c - (7+5v_m)e_2e_3 \qquad (7)$$

$$e_1 = (49-50v_iv_m)\ (\frac{G_i}{G_m}-1)\ +\ 35\frac{G_i}{G_m}(v_i-2v_m)\ +\ 35(2v_i-v_m) \tag{8}$$

$$e_2 = 5v_i\ (\frac{G_i}{G_m}-8)\ +\ 7\ (\frac{G_i}{G_m}+4) \tag{9}$$

$$e_3 = \frac{G_i}{G_m}\ (8-10v_m)\ +\ (7-5v_m) \tag{10}$$

The effective elastic modulus E_c and the poisson ratio v_c can be calculated with solutions K_c and G_c using Eq. 11 and 12.

$$E_c = \frac{9\ K_c\ G_c}{3K_c\ +\ G_c} \tag{11} \qquad\qquad v_c = \frac{3K_c\ -\ 2G_c}{2\ (3K_c\ +\ G_c)} \tag{12}$$

With the elastic constants from this composite three phase model we can also describe a multiphase material such as concrete. Starting with two materials and successively adding the remaining mix materials according to their fractional volumes we receive a multiphase model.

3 Multiphase Model for Concrete

Concrete is a multiphase material, as a minimum consisting of CSH-gel as matrix phase, and the inclusions: air pores, free capillar water, unreacted cement and aggregates. The volume fractions of these materials change on a linear basis during the reaction depending on the degree of hydration α (Fig.2). As a result of chemical bond forces the volume of CSH-gel is smaller than the starting volume. This chemical shrinkage

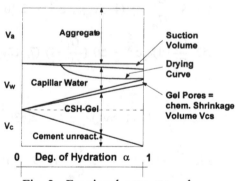

Fig. 2, Fractional concrete volume

volume V_{cs} is filled with gel pores and causes a suction effect with the same dimension on the concrete surface. Water or air is sucked in, according to the conditions of storage. This condition also determines the drying of free capillar water (drying curve). The water demand for full hydration and the chemical shrinkage volume follow the characteristic cement parameters specified in [12]. Proceeding from the original mixing values V_c , V_w and V_a the variable volumes of cement paste result in:

unreacted cement: $V_{cu} = V_c (1-\alpha)$ (13)

CSH-gel: $V_{csh} = V_c (1 + F_{wv} (1-F_s)) \, \alpha$ (14)

gel pore volume:
shrinkage volume: $V_{cs} = V_{gp} = V_c \, F_{wv} \, F_s \, \alpha$ (15)
suction volume:

capillar pore volume: $V_{cp} = V_w - V_c \, F_{wv} \, \alpha$ (16)

Where air is sucked in the capillar pore volume (Eq.16) is reduced by the sucked air volume (Eq.15). The volumetric water demand ratio F_{wv} amounts to 0.80-0.82, the shrinkage ratio F_s to 0.23-0.25, depending on the cement composition [12]. The variable CSH volume and therefore the material concentrations c_i follow from Eq. 13-16. The formulas for effective moduli can now successively be applied to the different mix volumes. In view's of the model assumption that an inclusion is being added to a connected matrix, the order of succession of the materials is of importance. The computing programme here followed the order specified above. The degree of hydration can be measured by one of the known methods. For measuring the shrinkage volume V_{cs} the hydrostatic weighing method described in [12,13] was used.

4 Elastic Moduli of Mix Materials

The water fraction and the moduli of the aggregates are of dominant influence on the concrete stiffness [1,3]. For model application, two elastic constants of each material have to be determined. For air and water these values can be taken from the literature. For cement, CSH-gel and aggregates, the material moduli have to be measured. Drilling cores can usually be used for aggregates to determine the elastic modulus and poisson ratio. Measurements of elastic moduli are particularly difficult for fine materials. In order to obtain a nondirectional value for sand and cement, a compression unit was developed which allows the determination of bulk modulus. Fig. 3 shows a diagram of this unit.

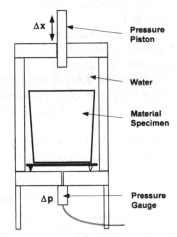

Fig. 3, Compression unit for bulk modulus

A material specimen of known volume is placed in the pressure vessel which is filled with water. Hydrostatic pressure is applied by a round smooth polished pressure piston. Its displacement Δx and the water pressure Δp are the measured values. The material compressibility is deduced from a pressure -volume change graph. The material bulk modulus can be calculated by comparing the compressibility of a control and a material measurement. First of all, the elastic compressibility of the pressure vessel itself -within the sample container- has to be measured and taken into account. For useful results a high material filling ratio is necessary. Materials with high bulk modulus give insufficiently accurate measurements. The unit is well suited for soft materials like early age concrete. The rise of bulk modulus can be measured at different times on the same material specimen. By measuring the displacement of the pressure piston without any load, the shrinkage volume as well as the degree of hydration can be determined.

5 Model Results

The multiphase model was programmed and tested in BASIC and EXCEL. Some results of these calculations are shown, making clear the influence of mixing parameters and material constants. The calculations are based on the mix design and material properties shown in table 1.

Tab. 1, Standard mix design and material constants

Material	Mass kg	Density kg/m3	Volume m3	E-Mod. GPa	K-Mod. GPa	Poisson Ratio
CSH-Gel	--	2340	--	40	23.0	0.21
Air	--	--	0.020	--	1. 10-4	0.50
Water	150	1000	0.150	--	2.04	0.50
Cement	300	3000	0.100	70	50.7	0.27
Aggreg.	1971	2700	0.730	50	33.3	0.25
Total	2421		1.000			

Changes in material fractions resulting from parameter variation are counted to the debit of the aggregates. Where the w/c ratio changed, the cement content was kept constant; similarly, where the cement content changed, the w/c ratio was kept constant. Unless it was varied as a parameter, the degree of hydration was set at 1.0.

As shown in Fig. 4, the degree of hydration is a dominant factor in the stiffness formation of early age concrete. Within 24 hours, cement can reach a degree of

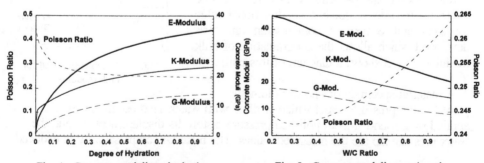

Fig. 4, Concrete moduli vs. hydration   Fig. 5, Concrete moduli vs. w/c-ratio

hydration between 10 and 50% . Hydration kinetics thus constitute the controlling parameter for the stiffness formation of early age concrete. Contrary to what we find in the literature, the Poisson ratio of the model becomes monotonically decreasing. Fig. 5 shows the relationship between the w/c-ratio and the final stiffness of the concrete. The influence of the aggregate stiffness and fractional aggregate volume against the degree of hydration on elastic modulus development in concrete is shown in Fig. 6 and 7. As expected, increasing aggregate stiffness and fractional aggregate volume raise the stiffness of concrete.

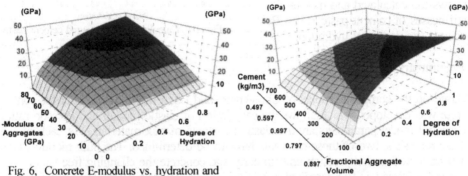

Fig. 6, Concrete E-modulus vs. hydration and E-modulus of aggregates

Fig. 7,   Concrete E-modulus vs. hydration and cement content, Eagg=50 GPa

6 Comparison with measurements

Comparative measurements were made on 40/40/160 mm specimen with cement paste and concrete. The bulk modulus was determined as described earlier. The Poisson ratio was calculated from measured values of elastic and bulk modulus. Hydration of cement was recorded with the hydrostatic weighing method and the degree of hydration was calculated from chemical shrinkage volumes [12,13]. Fig. 8 shows the reaction process of a PZ 35F cement (w/c=0.35) and the recorded stiffness values. The starting reaction is characterized by a strong swelling period during the first day. The bulk modulus after one day is therefore lower than at the beginning. Fig. 9 shows the reaction behaviour of the same cement type, but with a swelling phase of

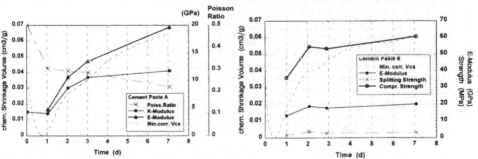

Fig. 8, Measured cement paste A reaction and moduli (w/c = 0.35)

Fig. 9, Measured cement paste B reaction, E-modulus and strength, (w/c = 0.35)

10 hours. It is striking that, after three days, the measured strength and also the elastic modulus are lower than after two days. This effect can be explained by the unusual reaction process, which does not increase monotonically but shows counter reactions after 2, 4 and 5 days. In Fig. 8 and 9, the chemical shrinkage volume is based on the maximum swelling volume (initial period). Fig. 10 shows the calculated and measured stiffnesses of cement paste A and B, against the degree of hydration.

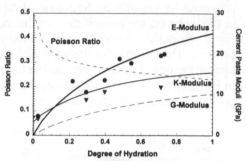

Fig. 10, Calculated and measured cement paste moduli vs. hydration

7 Summary

Based on the composite model of Christensen and Lo, a multiphase model for concrete was developed which allows the calculation of elastic moduli and Poisson ratio. Essential model parameters are the elastic constants of mix materials. For each of these materials, two of those constants have to be determined. The bulk modulus was measured with a compression unit. In early age concrete the changing free water content is decisive. The degree of hydration, calculated from hydrostatic weighing recordings, is used to take account of the variable volume fractions. Reaction kinetics of cement, mix design and material constants are essential for the stiffness formation of early age concrete. These influences are demonstrated by model calculations . The applicability of the multiphase model was demonstrated by comparing stiffness measurements of cement pastes with model results.

Acknowledgement
Financial support for this work by the Österreichische Industriellenvereinigung (VÖI) and University Innsbruck research fund is gratefully acknowledged.

References
1 Hirsch T.J.: **ACI**, Vol.59, No.3, 1962, pp.427-451
2 Hansen T.C.: **ACI**, Vol.62, No.2, 1965, pp.193-216
3 Manns W.: **Beton**, H9+10, 1970, pp.401-405,455-460
4 Alfes C.: **ACI SP-132**, Vol.2, 1992, pp.1651-1671
5 Paul B.: **Trans.Metall.Soc. AIME**, Vol.218, No.1, 1960, pp.36-41
6 Hashin Z.: **J.Appl.Mech.**, Vol.29, No.1, 1962, pp.143-150
7 Hashin Z., Shtrikman S.: **J.Mech.Phys.Solids**, Vol.11, 1963, pp.127-140
8 Christensen R.M., Lo K.H.: **J.Mech.Phys.Solids**, Vol.27, No.4, 1979, pp.315-329
9 Christensen R.M.: **Mechanics of composite materials**. J. Wiley, New York, 1979
10 Monteiro P.J.M.: **CCR**, Vol.21, 1991, pp.947-950
11 Nielsen A.U., Monteiro P.J.M.: **CCR**, Vol.23, 1993, pp.147-151
12 Paulini P.: **Zement-Kalk-Gips**, Nr.10, 1988, pp.525-531
13 Paulini P.: **Proc. 9th ICCC**, New Delhi, 1992, Vol.IV, pp.248-254

9 FROM MICROSTRUCTURAL DEVELOPMENT TOWARDS PREDICTION OF MACRO STRESSES IN HARDENING CONCRETE

S.J. LOKHORST and K. van BREUGEL
Delft University of Technology, Delft, The Netherlands

Abstract
A rheological model is presented that combines simple material laws and a description of the microstructure of hardening concrete. Although the material properties are fixed, the division of concrete into different structural components and the stiffening of the microstructure by addition of new components enable to simulate a deformational behaviour that is highly influenced by the course of the hydration process. Creep and stress-relaxation observed on the macro level have become a mixture of elastic-, and time-dependent behaviour and of the effects of redistribution of stresses over old and new hydration products on the micro level. With the model, trends from creep and relaxation experiments can be predicted.
Keywords: Creep Prediction, Modelling of Creep, Modelling of Relaxation, Relaxation Prediction.

1   Introduction

The probability of cracking of hardening concrete due to thermal stresses can be calculated if the development of strength and stress and their statistical distribution are known. With regard to the prediction of strength during hardening much progress has been made in recent decades. Various models have been proposed, that relate the degree of hydration, or the state of the microstructure, with strength. However problems still arise in the field of stress prediction. These problems are mainly caused by the complicated deformational behaviour of hardening concrete. A possible method to increase the understanding of stress development at early-ages, exists of the application of rheological models with explanatory capacities.

Hardening concrete is a continuously changing material. Therefore, it has been suggested that explanations for the deformational behaviour of hardening concrete can be found by taking into account the effects of the hydration process and the development of the microstructure. From this point of view, van Breugel [1] and Laube [6] proposed models for creep and relaxation in which the degree of hydration plays an important role. A rheological model is presented that combines simple material laws with a detailed description of the microstructural development in hardening concrete.

Thermal Cracking in Concrete at Early Ages. Edited by R. Springenschmid. Published 1994 by E & FN Spon, 2–6 Boundary Row, London SE1 8HN, UK. ISBN: 0 419 18710 3.

2   Development of the microstructure during hydration

2.1 General

The load bearing capacity of the cement paste microstructure depends on
the amount of contact zones between hydrating cement grains. Suggestions
of this tenor were made by Granju & Maso [4], Jambor [5] and van Breugel
[2]. The latter elaborated these ideas quantitatively and developed a
computer-based numerical program (HYMOSTRUC). In the following para-
graphs the microstructural development and its effects on the macro
behaviour and properties of concrete will be briefly outlined.

In fresh cement paste, cement grains are surrounded by water (fig.
1). The particle-to-particle distance depends on the water/cement ratio
and the particle size distribution. During hydration a cement grain
gradually dissolves and a porous shell of hydration products (cementgel)
is formed around the grain. As hydration proceeds the expanding
particles come into contact, and a structure with the ability to carry
load, is formed.

From this view on microstructural development the deformational
behaviour of hardening concrete can be described qualitatively. In this
description the structural components of the concrete are accounted for
explicitly. Besides the discerning of aggregate and cement paste, the
cement paste itself is also divided into different components, by
regarding it as a structure composed of basic structural elements
(hydration products). During hydration the configuration of this
structure changes because the number of elements increases. The elements
are presumed to be made of one type of material with fixed properties,
and thus independent of the time of formation. Bearing this in mind, the
deformation of the structure does not only depend on the properties of
the elements, but also on the configuration of the structure itself i.e.
on the number and spatial orientation of the structural elements.

A proposal for a quantitative characterization of the
microstructure will be given in the next section. The underlying
calculations have been performed with HYMOSTRUC.

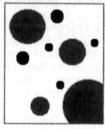 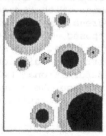 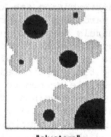 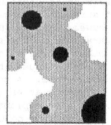

| initial stage | early hydration | "clusters" (first embedding) | "bridging" between clusters |

Fig. 1: Hydrating cement grains

2.2 Characterization of the microstructure

As shown in fig. 1 small cement particles will become embedded in the
shell of a central particle. The smaller particles can be embedded
partly or completely, depending on the degree of hydration and the
distances between the particles. As long as the central particle is not
embedded in the outer shell of a larger particle it is called a free

(central) particle. For a system consisting of a free central particle
and embedded particles the term cluster is used.

The contacts between clusters can be formed by cement particles
that are not completely embedded yet. These particles, the so-called
"bridging" particles exceed the shell thickness of the central particle
and can also be partly embedded in another neighbouring cluster (fig.
1). The minimum size of a particle that is able to bridge the distance
between two (or more) neighbouring particles depends on the thickness of
the shell of the clusters *and* the distance between the clusters.

The aforementioned clusters and bridge particles will now be used
to characterize the development of the microstructure and its
properties. These two basic structural elements determine the stiffness
of the microstructure. The total volume of the bridging particles -
being the weakest links - can be regarded a measure for the strength.

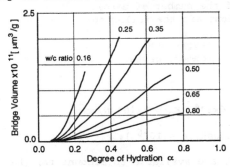 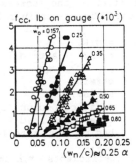

Fig. 2: The volume of the "bridging" particles as a measure for
strength (HYMOSTRUC predicts a shortage of water for wcr 0.16 and 0.25)

2.3 HYMOSTRUC calculations

HYMOSTRUC is a computer-based numerical model for simulation of
hydration and microstructural development of Portland cement-based
materials. For a more detailed description, see [2].

With HYMOSTRUC the volumes of both the clusters and the bridging
particles can be calculated as a function of time and degree of
hydration. This was done for cement pastes with various water/cement
ratios. The criterion used for the determination of the minimum
interparticle distance that has to be bridged is a function of the
thickness of the (outer) shell of a central particle and the water/-
cement ratio. The results are shown in fig. 2.

Taplin [10] determined the development of the strength as a function of the degree of hydration of cement pastes with similar water/-
cement ratios. From fig. 2 it can be seen that the bridge volume can be
used as a measure for the strength.

3   Rheological model

3.1 Description of the model

The above mentioned view on the development of the microstructure
is used as a basis for a rheological model for hardening concrete. In
this model only two types of elements have been used. The aggregate is
modeled as an elastic spring element. It is embedded in the cement

paste by placing it both in parallel and in series with cement paste elements (fig. 3). These cement paste elements consist of a variable number of parallel vertical bars with equal properties and dimensions, between two horizontal platens, with infinite stiffness. In the model, which is an extension of an earlier presented model [7], 4 cement paste elements have been used. Two cement paste elements represent the bridging particles in the hydrating cement paste, the other two represent the growth of the basic skeleton i.e. the clusters. The number of bars increases with hydration and is proportional to the volumes of the clusters and the bridges that can be calculated with HYMOSTRUC. The stress-strain relations of the bars are linear elastic and time-dependent. For the time-dependency a simple creep formula is used. It is emphasized that no ageing coefficient is used. The effects of ageing are simulated explicitly with the development of the micro-structure (i.e. by the increase of the number of bars).

The used stress-strain relations are the following:

$$\varepsilon_a \quad = \sigma/E_a \qquad\qquad\qquad \text{(aggregate)}$$
$$\varepsilon_p \quad = \sigma/E_p + a\, t^n\, \sigma \qquad \text{(cement paste)}$$

ε strain;
σ stress;
E Elastic modulus ($E_a = 6\times10^4$, $E_p = 3\times10^4$ N/mm^2)
t time under load [hours];
a basic creep rate (constant, $a \approx 10^{-6}$);
n $= 0.3$.

Three coefficients (κ, λ and μ) are available to account for the relative volumes of the components of the model (aggregate, bridges and cluster).

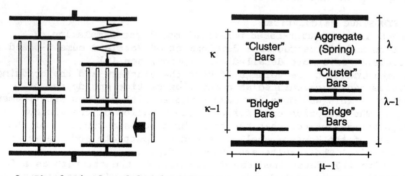

Fig. 3: Rheological model (the progress of hydration is simulated by inserting new "cluster" and "bridge" bars).

3.2 Macro behaviour

The macro stress-strain relations of the model depend on the constitutive relations of its components and the way the components are put together.

In a creep test, redistribution of stresses in the cement paste elements will take place. An external load will be distributed over the components of the model according to their stiffness. As the stress-strain relations are assumed to be linear, the stress increments in all bars of a paste element are equal. The total stress in the bars, however will generally not be the same because of their different age and, consequently, their different stress-history. Immediately after loading the bars start to creep with a rate that depends on the stress in the bar and the time under load. Compatibility in the cement paste elements requires that the deformation of all bars is the same, and therefore a redistribution of stress will take place. In this way, new bars, that were initially inserted stress free, gradually become load bearing without application of an external load. This redistribution of stresses will cause a decrease of the rate of the time-dependent deformation on the macro level. This restraining influence increases with the rate at which the number of bars increases.

In a relaxation test the redistribution of stresses is different from a creep test. An external deformation will induce elastic stresses in the bars. Due to the time-dependent deformation of a stressed bar, the elastic stress will decrease. If the length of the bars of one paste element is kept constant no redistribution can be expected due to relaxation. However, if two cement paste elements - one representing bridges and the other representing clusters - are placed in series, the connected horizontal platens are able to move and therefore deformation of the elements, and the bars is possible, and consequently redistribution of stresses can take place.

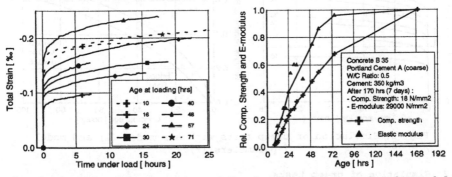

Fig. 4: Results of creep tests and development of strength and E-modulus

4   Creep

4.1 Creep tests
In order to check the reliability of the model 8 creep tests were performed on hardening concrete in compression at ages of 10 to 100 hours after casting. The load was constant for a period of 7 to 24 hours. The stress/strength ratio at the time of application of load was 30%.

For the experiments concrete specimens of 100x 100x 400 mm^3 were used. After casting the specimens were stored in a climate room at 20°C until about 30 minutes before testing. In order to avoid drying shrinkage during the creep test the specimens were wrapped in PE-foil immediately after demoulding. Furthermore a dummy specimen was installed together with each creep specimen to monitor the load-independent deformations due to temperature variation and chemical

shrinkage. During storage and testing all specimens hardened
approximately isothermal. For further details on the experimental set-
up and results we refer to [8].

From fig. 4 two trends can be observed. Firstly, the elastic and
total strains increase with the age at loading. Secondly, the rate of
deformation increases with age. Exceptions to these trends are the
deformations of the youngest and the eldest specimens. We have reason
to believe that the youngest specimen was overloaded. The lower
deformation of the eldest specimen in respect to the previous specimen,
however points to a change of the deformational behaviour.

The above mentioned trends seem to contradict with the results of
creep tests in tension by Laube and by Rostasy, Gutsch & Laube. In [6]
and [9] they found that the creep coefficient (φ) decreased with
increasing age at loading. These tests, however, represent the
behaviour of hardening concrete in a longer period (7 days) and with
larger intervals (several days) between the ages at loading. These
intervals were so large that the trends and the breaking of the trends
observed in our tests, could not be observed. Therefore, the results of
both tests do not have to contradict.

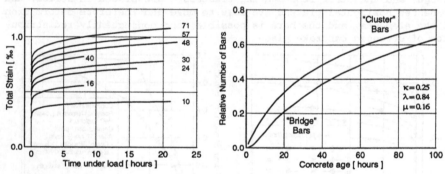

Fig. 5: Simulation of creep tests with Bar model (l) and model
parameters (r)

4.2 Simulation of creep tests

With the presented model the 8 creep tests were simulated. Fig. 5 shows
that the trends can be predicted fairly well. As the computer program
is still in development it is not yet possible to predict absolute
strains.

The required input for the model consists of material parameters,
the number of bars in the cement paste elements and the volumecoeffi-
cients (κ, λ and μ). The material parameters were mentioned in section
3. Only the coefficient "a" -the basic creep rate- was not determined
yet. This coefficient was chosen 4×10^{-6} to obtain a ratio of the
elastic and time-dependent deformations similar to the ratio found in
the experiments. Fig. 5 shows the relative number of bars (100% =
complete hydration) and the volume coefficients. λ and μ divide the
concrete in aggregate and cement paste content. The number of bars and κ
divide the cement paste in bridging particles and clusters. For the
determination of the bridge- and cluster volumes HYMOSTRUC was used.

5   Relaxation

With the Bar model macro stress relaxation as a function of the age at
loading has been predicted (fig. 6) and results have been compared with
trends found in literature. For these calculations the same model
parameters and configuration as described in section 4.2 have been used.
As the model predicts the macro creep deformations properly up to about
3 days after casting, no relaxation calculations were made at later
ages. A deformation was applied at 4 ages and kept constant during 24
hours. The stress/strength ratio at the time of application of the
deformation was the same in all cases.

Data on relaxation experiments in which the deformation is kept
constant are rather scarce. However, such experiments are very
important for tuning relaxation models. In general, it was found that
relaxation decreases with increase of the age at loading. This was
found by Wierig [11] for concrete in compression and by Laube and by
Rostasy, Gutsch & Laube for concrete in tension. Figure shows that the
bar model predicts this trend as well. However, the effect of age is
not as pronounced as found in literature. An explanation for this might
be found in the absence of the effects of chemical shrinkage and temperature. According to Laube, these load-independent deformations were a
multiple of the applied deformation and had to be compensated continuously to avoid additional stresses. This means that the deformation was
not constant during his experiments. For calculations with the model,
the modeling of the internal load-independent deformations and an
external compensation would give rise to additional stress redistribution and consequently to a different macro stress-relaxation.

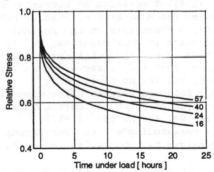

Fig. 6: Prediction of relaxation as a function of the age at loading.

7   Conclusions

Creep tests on hardening concrete were performed under constant load and
a load level of 30% of the actual strength at loading. From the results
it can be concluded that, up to a certain age, the elastic deformation
and the time-dependent deformation increase with the age at loading.
Furthermore it can be seen that the hardening process has a restraining
influence on the rate of the time-dependent deformation. However, at
later ages (> 60 hrs for the concrete under view) this behaviour
changes.

In the proposed rheological model for hardening concrete, the concrete is described as a material that changes continuously with the course of the hydration process. The particular aim was to develop a model that enables to simulate the effect of microstructural changes on the rate of time-dependent processes like creep an relaxation. The presented results show that the model is able to predict trends that were found in creep and relaxation experiments. For a more comprehensive description of the early-age behaviour the effects of temperature and chemical shrinkage must be taken into account as well. Furthermore, in order to extend the applicability of the model to later ages, more attention must be paid to the description of the microstructural changes at later ages. These parameters are being investigated at the moment in a research program on stress development and the risk of cracking of hardening concrete.

8   Acknowledgements

The assistance of Mr F. Schilperoort in carrying out the experiments is highly appreciated. The financial support of the Dutch Technology Foundation (STW), the Ministry of Traffic, Public Works and Water Management and the COAB Foundation is gratefully acknowledged.

9   Literature

1   van Breugel, K. (1980) **Relaxation of Young Concrete** Report 5-80-D8, Delft University of Technology.

2   van Breugel, K. (1991) **Simulation of Hydration and Formation of Structure in Cement Based Materials**, Thesis, Delft.

3   Ghosh, R.S. (1973) A Hypothesis on the Mechanism of Maturing Creep of Concrete, **Matériaux et constructions**, Vol.31, 23-27.

4   Granju, J.L. & Maso, J.C. (1980) Loi des résistance en compression des pâtes de ciment Portland conservées dans l'eau. **Cement and Concrete Research**, Vol.10, No.5, 611-621.

5   Jambor, J. (1963) Relation Between Phase Composition, over-all Porosity and Strength of Hardened Pozzolana Pastes. **Magazine of Concrete Research**, Vol.15, No.45, 131-142.

6   Laube, M. (1991) **Werkstoffmodell zur Berechnung von Temperaturspannungen in massigen Betonbauteilen im jungen Alter**, Thesis, Braunschweig.

7   Lokhorst, S.J. & van Breugel, K. (1993) The Effect of Microstructural Development on Creep and Relaxation of Hardening Concrete, in **Creep and Shrinkage of Concrete**, 5-th Rilem Symposium, Barcelona.

8   Lokhorst, S.J. & van Breugel, K. (1994) Rheologisches Modell für jungen Beton auf Grund der Entwicklung der Mikrostruktur. **Deutscher Ausschuss für Stahlbeton**, 29. Forschungskolloquium, Delft.

9   Rostasy, F.S., Gutsch, A. & Laube, M. (1993) Creep and Relaxation of Concrete at Early Ages- Experiments and Mathematical Modeling, in **Creep and Shrinkage of Concrete**, 5-th Rilem Symposium, Barcelona

10   Taplin, J.H. (1959) A Method for Following the Hydration Reaction in Portland Cement Paste, **Australien Journal of Applied Science**, Vol.10, No.3.

11   Wierig, H-J.(1971) Einige Beziehungen zwischen den Eigenschaften von grünen und jungen Betonen und denen des Festbetons, **Beton**,21.

10 EFFECT OF CREEP IN CONCRETE AT EARLY AGES ON THERMAL STRESS

H. UMEHARA and T. UEHARA
Nagoya Institute of Technology, Nagoya, Japan
T. IISAKA and A. SUGIYAMA
Meijo University, Nagoya, Japan

Abstract
Creep in concrete at early ages is one of the most important physical
properties which affect thermal stress. In this study, creep tests of
concrete at early ages are conducted using two kinds of equipments;
one for compressive creep and the other for tensile creep. The stress
level, period of loading and temperature are selected as parameters
of creep tests. The equation to calculate creep strain is derived
from the testing results using a viscoelastic rheological model,
and adapted to the analytical method for calculating thermal stress.
Furthermore, the thermal stress analysis considering creep effect is
applied to the reinforced concrete box culvert in which thermal
stresses are measured, and creep model is examined by comparing
analytical results of thermal stress to measured results.
Keywords: Creep, Effective Elastic Modulus, Modelling of
Creep, Structural Members (Box Culvert).

1 Introduction

It is difficult to calculate thermal stress in massive concrete
structures accurately using finite element method or other simple
methods, if the physical properties such as thermal conductivity,
coefficient of thermal expansion, and Young's modulus are not
estimated correctly. Especially, creep in concrete at early ages
seems to be one of the most important physical properties which
affect thermal stress. However, it is hard to predict the reduction
of thermal stress caused by creep, because in concrete at early ages
temperature varies, Young's modulus increases with time, and stress
varies from compression to tension. Furthermore, there are few data
of creep strain measured in concrete at early ages.

Under these circumstances, the authors conducted the compressive
and tensile creep tests on concrete at early ages and constructed a
structural formula for creep using a viscoelastic rheological model.
Then, they selected a reinforced concrete box culvert, in which
measurement of thermal stresses has been done and conducted the
thermal stress analysis in which creep was taken into consideration
and the investigation was made by comparing the results with those
collected in the measurement on site.

Thermal Cracking in Concrete at Early Ages. Edited by R. Springenschmid. Published 1994
by E & FN Spon, 2–6 Boundary Row, London SE1 8HN, UK. ISBN: 0 419 18710 3.

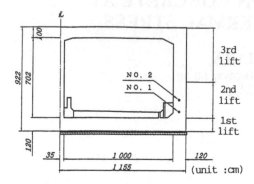

Fig.1. Concrete box culvert

Table 1. Mix proportion of concrete

W/C	Air	Composition (kg/m³)			
(%)	(%)	W	C	S	G
56	4.0	157	281	835	1010

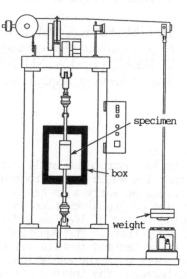

Fig.2. Tensile creep testing machine

2 Measurement of thermal stresses

Measurement of thermal stresses was made on the reinforced concrete box culvert as shown in Fig.1. This structure was constructed by dividing it in 3 lifts. The time interval between two lifts is one month. The mix proportion of concrete is shown in table 1. Cement used in concrete is normal portland cement and concrete is placed at 19°C. Measurement of tempratures and stresses was made at two locations in the bottom wall.

3 Creep tests of concrete at early ages

3.1 Outline of the tests
In concrete at early ages, compressive stress generally works like the thermal stress, and it turns to the tensile stress as time passes. Therefore, two testing machines are used in this study; The one is the compressive creep testing machine and the other the tensile creep testing machine as shown in Fig.2. The creep testing machines are of the lever type; the compressive one has a lever ratio of 1:40 and a maximum load of 50 KN and the tensile one has a lever ratio of 1:50 and a maximum load of 80 KN. Each of the testing machines has a box which allows to set the temperature and humidity at any level and the test can be made by placing the specimen within it.
The specimen used in the test has a circular cylindrical shape $10\phi \times 20$cm. The specimens are made of the same materials and mix proportion as those of the box culvert and the strain gauge and thermostat of the built-in type are buried in the specimen.

Table 2. Variables for creep test

	Compression			Tension			Temperature (°C)
	Stress (N/mm²)	A	B	Stress (N/mm²)	A	B	
I	1.0 1.0	1 3	5 5	—— ——	—— ——	—— ——	20,40,80 40
II	0,1.5,2.5	1	1	0.2	3	5	20
III	1.5	1	1,2,3	0.2	3,4,5	5	20
IV	1.5	1	1	0.2	3	5	20,30,40

A: Age of loading (day)　B: Period of loading (day)

　　The creep tests are conducted by dividing them into 4 series as shown in Table 2.　Series I is intended to investigate the compressive creep.　The tests are conducted at 3 different temperatures of the specimen, namely 20, 40 and 80°C when loaded with a compressive stress of 1.0 N/mm² at the age of 1 day and at a temperature of the specimen of 40°C only when loaded with a compressive stress of 1.0 N/mm² at the age of 3 days.　In the same time, other series are all intended to deal with the tensile creep.　In Series II, the tests are conducted at 3 different levels of compressive stress of 0, 1.5 and 2.5 N/mm².　The load is applied at a temperature of 20°C and the age of 1 day, and released at the age of 2 days, then a tensile stress of 0.2 N/mm² is applied at the age of 3 days and kept for 5 days.　In Series III, the tests are conducted for the periods of loading of 1, 2 and 3 days and the compressive stress of 1.5 N/mm² is applied at a temperature of 20°C.　Then, after one day unloaded period, the tensile stress of 0.2 N/mm² is applied for 5 days.　In Series IV, the tests are conducted at the temperature of 20, 30 and 40°C and the compressive stress of 1.5 N/mm² is applied at the age of 1 day.　The load is released at the age of 2 days and the tensile stress of 0.2 N/mm² is applied at the age of 3 days and kept for 5 days.　Throughout all the series of the tests, the compressive stress of 1.5 N/mm² and the tensile stress of 0.2 N/mm² are about 25% of the strength at that time.

3.2 Results of the tests
The results of the tests regarding the influence of the temperature in Series I are shown in Fig.3.　In the diagram, the time passed after loading is taken on the abscissa and on the ordinate, the creep compliance J which represents the creep strain for unit stress is taken.　It is made clear from Fig.3 that J when tested at the temperature of 40°C at the age of 5 days is about 1.3 times bigger than in the case of 20 °C and J when tested at 80°C becomes 1.8 times as compared with that in the case of 20°C, thus the compressive creep strain increases when the temperature of the specimen rises.

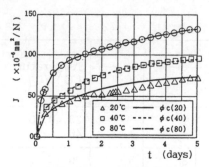

Fig.3. Effect of temperature at
compressive loading

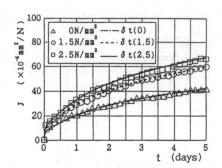

Fig.4. Effect of compressive
stress

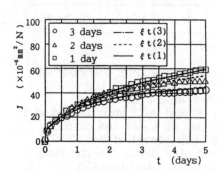

Fig.5. Effect of period of
compressive loading

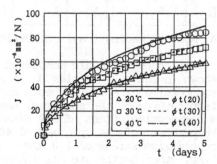

Fig.6. Effect of temperature
at tensile loading

From the result of the test regarding the age of loading, J tends to
decrease in the case when the load is applied at the age of 3 days as
compared with the case of the age of 1 day.

The results of tests of Series II, III and IV are presented in
Fig.4, 5 and 6, respectively. It is made clear from Fig.4 that the
higher the compressive stress is, the higher the tensile creep
strain is. In Fig.5, it is shown that when the period of
compressive loading becomes longer, the tensile creep strain tends to
decrease. Also, from Fig.6, J at the temperature of 30℃ is about
1.1 times and that at 40℃ is about 1.3 times higher than J at
20℃. Therefore, similarly to the cases of compressive creep, it is
made clear that the higher the temperature rises, the higher the
tensile creep strain is.

3.3 Structural formula of creep
By applying a viscoelastic rheological model to the results of creep
tests, an investigation is made to formulate a structural formula
for creep of concrete at early ages. The viscoelastic rheological
model is represented by a combination of Maxwell model and Voigt
model. The number of elements which can generally be used into
practice would be up to around 5. Thereupon, in the cases of
compressive creep, for the results of tests of Series I with the

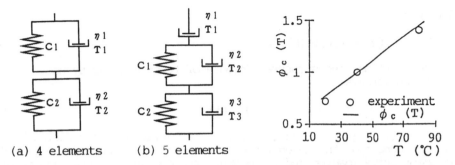

|(a) 4 elements | (b) 5 elements |

Fig.7. Creep model

Fig.8. Temperature coefficient

compressive stress of 1.0 N/mm² at 40 °C and in the cases of tensile creep for the results of tests of Series II to IV with the common compressive stress of 1.5 N/mm² loaded for 1 day at 20°C, various models composed of elements whose number does not exceeds 5 are investigated. As result of these investigations, in the case of compressive creep, the 4-element model as illustrated in Fig.7 (a) gives with the test results at 40 °C the best results as shown in Fig.3. The creep compliance for the period of loading of t days in this case ($J'_c(t)$) (mm²/N) is represented the following formula:

$$J'_c(t) = \{ 26.96(1 - e^{-24.7t}) + 71.99(1 - e^{-0.575t}) \} \times 10^{-6} \tag{1}$$

On the other hand, in the case of tensile creep, the 5-element model as illustrated in Fig.7 (b) conforms with the test results the best as shown in Figs.4 to 6. The creep compliance for the period of loading of t days ($J'_t(t)$) is represented by the following formula:

$$J'_t(t) = \{ 28.74(1 - e^{-0.801t}) + 8.13(1 - e^{-45.38t}) \\ + 4.468 t \} \times 10^{-6} \tag{2}$$

The formulas (1) and (2) represent the phenomena only under certain conditions of compressive and tensile creep respectively. Therefore, the effect of the temperature is to be represented by multiplying the formula (1) with the temperature coefficient $\phi_c(T)$ which is a linear function of the temperature T(°C). $\phi_c(T)$ is calculated applying the method of least squares as shown in Fig.8 to get the following formula.

$$\phi_c(T) = 0.0112T + 0.552 \tag{3}$$

It is made clear that the creep compliance which is obtained by multiplying the formula (1) with the formula (3) almost coincides with the results of tests as shown in Fig.3. Likewise, to the age τ (day) at which the load is applied, the loading age coefficient $\xi_c(\tau)$ as represented by the following formula is obtained:

$$\xi_c(\tau) = -0.307\log \tau + 1.0 \tag{4}$$

Thus, the structural formula of compressive creep is written as the following:

$$J_c = \xi_c(\tau) \cdot \phi_c(T) \cdot J'_c(t) \tag{5}$$

In the meantime, the strucural formula of tensile creep by the following formula:

$$J_t = \delta_t(\sigma) \cdot \xi_t(\tau) \cdot \phi_t(T) \cdot J'_t(t) \tag{6}$$

Where $\delta_t(\sigma)$, $\xi_t(\tau)$ and $\phi_t(T)$ are the variables representing the influences of the intensity of the compressive stress, the period of compressive loading and temperature respectively on the results of tests conducted with the compressive stress of 1.5 N/mm², at the temperature of 20 ℃ for the period of compressive loading of 1 day. Representing each of them as the linear function of the compressive stress σ (N/mm²), the period of compressive loading τ (day), and the temperature T(℃), the following formulas are gained:

$$\delta_t(\sigma) = 0.017\sigma + 0.701 \tag{7}$$

$$\xi_t(\tau) = -1.107\log \tau + 1.538 \tag{8}$$

$$\phi_t(T) = 0.0257T + 0.487 \tag{9}$$

Comparison of the results of calculation obtained by the formula (6) with the results of the tests is shown in Figs.4 to 6. It shows that each of them matches with the results of the test.

4 Temperature and stress analysis

4.1 Temperature analysis
The temperature of the reinforced concrete box culvert is obtained by applying the 2-dimensional finite element method to the divided elements as shown in Fig.9. The values of thermal characteristics

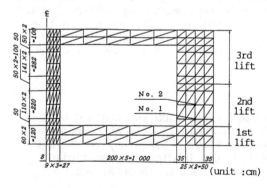

Fig.9. Divided elements

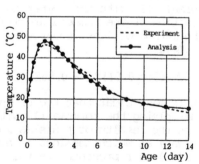

Fig.10. Comparison of temperature

used in this analysis are described in the Standard Specifications for Concrete of Japan Society of Civil Engineers.

As an example, the temperature at the point of measurement No.1 obtained as the result of analysis and the same actually measured are shown in Fig.10. As seen clearly in the diagram, both values conform very well, though at the highest point, the temperature obtained by the analysis is higher than the measured value by 2 to 3℃.

4.2 Stress analysis

By analyzing the thermal stress of the wall-shaped structure, the compensation plane method which is described in the Specifications is normally used. In order to take the creep characteristics into account when this method is applied, the effective modulus of elasticity is used by multiplying a reduction factor with the Young's modulus generally assuming that it is reduced by creep. Thus, in this study the effective modulus of elasticity is calculated from the structural formula of creep by the following method.

The Young's modulus at an age of t days $E_0(t)$ is represented in terms of the stress $\sigma(t)$ and the strain $\varepsilon(t)$ by the formula $E_0(t)=\sigma(t)/\varepsilon(t)$. Accordingly, the effective modulus of elasticity $E_e(t)$ at that time in which the creep strain $\varepsilon_c(t)$ is taken into account is represented as the following:

$$E_e(t)=\sigma(t)/\{\varepsilon(t)+\varepsilon_c(t)\}=E_0(t)/\{1+J\cdot E_0(t)\} \quad (10)$$

Where $J=\varepsilon_c(t)/\sigma(t)$

In the stress analysis where the value of $\sigma(t)$ differs with the elements, each element has a different modulus of elasticity. In case $\sigma(t)$ is the compressive stress, J_c is substituted into J in the formula (10) to find $E_e(t)$ and in case it is the tensile stress, J_t is substituted instead. By the way, the modulus of elasticity $E_0(t)$ is the measured compressive modulus of elasticity of the concrete specimen made of the same materials and with the same mix proportion. The tensile modulus of elasticity is assumed to be equal to the compressive one, since no tensile test has been conducted.

Investigations by the stress analysis using three kinds of modulus of elasticity are made; the effective modulus of elasticity described in the Specification E_e, the effective modulus of elasticity E_c shown

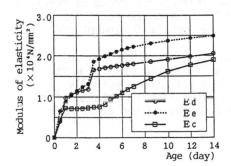

Fig.11. Comparison of modulus
of elasticity

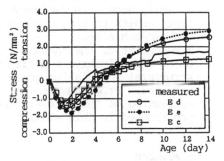

Fig.12. Comparison of
stress

in the formula (10) and the effective modulus of elasticity E_d
obtained by multiplying with the reduction factor $\beta(t)$ which is
employed in the Specification to the modulus of elasticity E_0.
In Fig.11, comparison among these moduli of elasticity is presented.
By the way, E_c is the value at the point of measurement No.1. The
effective modulus of elasticity described in the Specifications E_e
is represented by the following formula as a function of the
compressive strength $f'_c(t)$;

$$E_e(t) = \beta(t) \times 4.79 \times 10^3 \times \{f'_c(t)\}^{1/2} \tag{11}$$

where, $\beta(t)=0.73$(up to 3days), 0.87(4days), 1.0(after 5 days)
    As an example, the analyzed value and the measured one at the point
of measurement No. 1 are shown in Fig.12. When comparison is made
between E_e and E_d, it is known that the analyzed value calculated
with E_d gives smaller results by about 10% both on compressive and
tension sides. When comparison between E_d and E_c is made, the
analyzed value calculated with E_c is smaller than those calculated
with E_d by about 10 to 20% on the compression side and by about 50%
on the tension side. And in case E_e is used, it is made clear that
the analyzed value gives the closest value to the measured one
compared with those by other analyses, though it indicates a slightly
smaller value on the tension side. Therefore, from Fig.11, the
modulus of elasticity of concrete at early ages is reduced by more
than 50% due to creep. From the facts as mentioned above, it is made
clear that consideration of creep of concrete at early ages is
indispensable in the stress analysis of concrete and that the thermal
stress can be estimated with high accuracy applying the creep model
developed in this study.

5 Conclusion

The results gained in this study are summarized as the followings;
(a) It is clear that the higher the temperature rises, the higher the
    compressive creep as well as the tensile creep are and the shorter
    the period of compressive loading and the higher the compressive
    stress at that time are, the higher the tensile creep strain is.
(b) It is clear that as the structural formula for creep of concrete
    at early ages, the 4-element model conforms the best with the
    test results on the compression side and the 5 element model on
    the tension side.
(c) The results of the thermal stress analysis on the box culvert
    using the effective modulus of elasticity obtained from the
    structural formula for creep conform with those actually
    measured. From these facts, good prediction is obtained for
    estimation of the thermal stress by applying the creep model
    constructed in this study.

6 References

Standard specification of concrete structures (1991), JSCE

11 BASIC CREEP AND RELAXATION OF YOUNG CONCRETE

G. WESTMAN
Division of Structural Engineering, Luleå University of Technology,
Luleå, Sweden

Abstract
Viscoelastic behaviour of early-age normal strength and high strength
concrete has been studied. A creep model for young concrete has been
developed for use in e g thermal stress calculations.
    This paper deals with the elastic, creep and relaxation behaviour at
early ages. Four concrete mixtures were examined: two mixtures of high
performance concrete and two concrete mixtures designed for use in
future marine structures.
Keywords: Creep, Relaxation, Viscoelastic Model, Thermal Stress
Analysis.

1 Introduction

Earlier research has shown the importance of accurate modelling of the
mechanical behaviour at early ages. Elastic and creep properties,
strength development, maturity development, heat of hydration and free
thermal volume change have been examined at early ages for different
mixes. These properties have great importance in analysis of risks of
thermal cracking.
    The material model used in this study consists of three rheologic
elements in series. One element describing the fracture behaviour at
high tensile stresses, a visco-elastic element describing elasticity
and creep (Maxwell Chain model), and an element representing thermal
displacement, see Fig. 1.

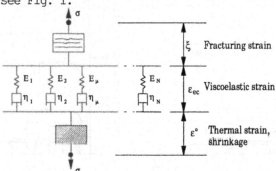

Fig.1. Rheologic model for the analysis of thermal stresses.

Thermal Cracking in Concrete at Early Ages. Edited by R. Springenschmid. Published 1994
by E & FN Spon, 2–6 Boundary Row, London SE1 8HN, UK. ISBN: 0 419 18710 3.

The properties of the elements have been calibrated by means of the tests for normal and high strength concrete and have been implemented in computer programs for structural analyses.

In the paper thermal stresses are calculated disregarding fracturing strain and shrinkage. Therefore, the comparison between the four concrete mixtures with regard to stress development made in this context merely is one way to compare the basic creep.

2 Creep

2.1 Creep tests

Creep compression tests have been carried out on young concrete subjected to loading at ages in the range of 0.55 to 7 days. The specimens were cylinders 80 mm in diameter and 300 mm in height.

Immediately after pouring the specimens were placed in water of 20°C. At varying age the concrete was subjected to a constant load corresponding to 20% of the cube strength at the current age. Before load application, each specimen must attain a compressive strength of about 3 MPa. This implies , depending on concrete quality an earliest loading age within the range of 8-12 hours after pouring. However, at this time the compressive strength develops rapidly and applicated load will quite soon be too small to cause further significant creep in the concrete. Therefore a minimum loading age of 13-15 hours was chosen in this study.

Mixture No 1 (w_0/B=0.30, SF/C=0.05) and No 2 (w_0/B=0.27, SF/C=0.10) are the high strength concrete and mixtures Nos 3 and 4 (w_0/B=0.40, SF/C=0.05) are designed for use in future marine structures. A high creep disposition at very early ages may be observed for the high strength concrete which, however, rapidly turns into a stiffer response. At loading ages of 48 hours and more the response is almost the same for the four different mixtures, see Fig. 2.

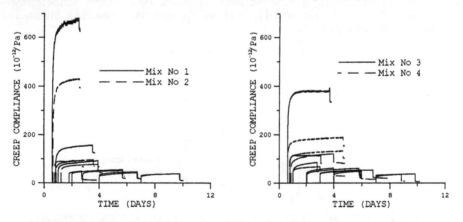

Fig.2. Examples of results from compression creep tests of young concrete. The applied loads constituted 20% of the strength at the respective ages.

2.2 Theoretical model of creep at early ages

The Triple Power Law by Bazant and Chern (1985) has been shown to have several advantages in creep modelling. It has therefore been used in this study. The Triple Power Law has been extended to

$$J(t,t') = \frac{1}{E_0} + \frac{\varphi_1}{E_0}(t'^{-m} + \alpha)\left[(t-t')^n - B(t,t';n)\right] + \frac{\Psi_1(t')}{E_0} + \frac{\Psi_2(t,t')}{E_0} \quad (1)$$

were E_0, φ_1, m, α and n are material coefficients and $B(t,t';n)$ is the binomial polynomial of the Triple Power Law (for the so called Double Power Law $B(t,t';n) = 0$) and $\Psi_1(t')$ and $\Psi_2(t,t')$ are additional functions. If the material coefficients are calculated as proposed by Bazant and Chern, the Triple Power Law gives higher values of the basic creep for loading ages of 48 hours and more than the test results. Therefore, the material coefficients m and n are used for curve fitting and the creep coefficient φ_1 is defined as proposed by Bazant, see Neville et al (1983)

$$\varphi_1 = 0.3 + 152.2\left(\frac{1}{f_{cyl}^{28}}\right)^{1.2} \quad (2)$$

Calculations with the original Triple Power Law renders lower instantaneous deformations for a load duration of $t-t' = 0.001$ days (1.4 minutes) than the test results. Also the very early age viscous behaviour when the load has been applied is underestimated, see Fig. 3.

To overcome these shortcomings the Triple Power Law is here supplemented with a function $\Psi_1(t')$, Eq (3), taking care of the strong early-age dependence of the instantaneous deformation (load duration 1.4 minutes) and $\Psi_2(t,t')$, Eq (4), which models the early-age creep when the load has been applied, see Fig. 4.

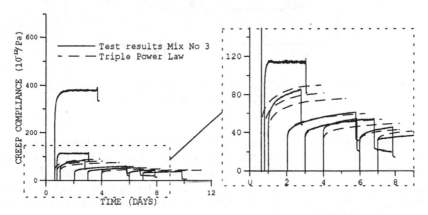

Fig.3. Example of calculation with Triple Power Law compared with results from creep tests.

$$\psi_1(t') = \gamma_1 \left(\frac{t_1 - t'}{t_1 - t_0} \right)^{a_1} \tag{3}$$

$$\psi_2(t,t') = \gamma_2 \left[1 - \exp\left(-\frac{t-t'}{t_2} \right)^{a_2} \right] \left(\frac{t_3 - t'}{t_3 - t_0} \right)^{a_3} \tag{4}$$

where t′ : equivalent age when the load is applied, [days]
t_0 : apparent setting time of the concrete, [days]
t_1, t_3: time limits for adjustments at early ages, [days]
t_2, a_2: parameters for the development of the time
    function, [days, -]
t-t′ : actual time period after loading, [days]
γ_1 : initial value of function $\Psi_1(t')$ at t′= t_0 , [-]
a_1 : parameter modifying the shape of $\Psi_1(t')$, [-]
γ_2 : end value of function $\Psi_2(t,t')$ for t′= t_0 , [-]
a_3 : parameter modifying the end value of $\Psi_2(t,t')$, t′>t_0, [-]

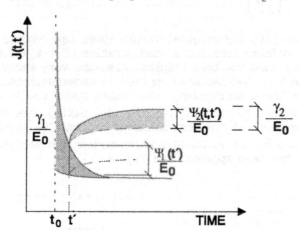

Fig.4. Additional functions $\Psi_1(t')$ and $\Psi_2(t,t')$ for the
early-age creep response shown schematically.

2.3 Results
As can bee seen in Fig. 5. the creep response can be closely predicted
with the extended version of the Triple Power Law. Added in the figure
is a curve showing the instantaneous deformation 0.001 days after
loading computed by the function $\Psi_1(t')$. In the figure results of the
theoretical calculations of the creep responses are compared with the
test results for mixtures No 1 and No 3.

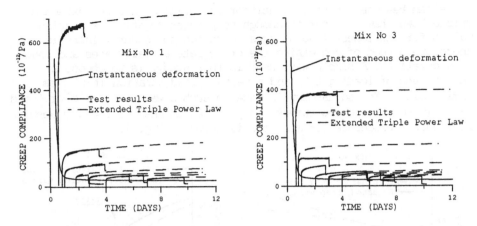

Fig.5. Calculations with the extended Triple Power Law compared with test results.

3 Relaxation

3.1 Conversion of creep data into relaxation data

By subdivision of the time t in discrete times $t_1 \ldots t_N$ yielding time steps $\Delta t_r = t_r - t_{r-1}$ the relaxation function R(t,t´) is solved from the compliance function J(t,t´), see e.g. Bazant and Wu (1974). In this study the extended version of the Triple Power Law is used in a program developed at Luleå University of Technology for the calculation of relaxation values, see Emborg (1990). In Fig. 6. creep values and corresponding relaxation values calculated with and without $\Psi_1(t´)$ and $\Psi_2(t,t´)$ in Eq (1) are shown.

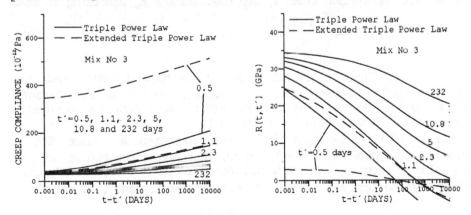

Fig.6. Creep and corresponding relaxation values calculated with and without functions $\Psi_1(t´)$ and $\Psi_2(t,t´)$.

As can bee seen in Fig. 6. neither the original nor the extended Triple Power Law is accurate enough to prevent negative relaxation values for the lowest loading ages (wich can not occur). There is therefore a need of decreasing the creep rate at early ages and long load durations. A simple way, used in this study, is to choose a certain age of loading t_1', and a certain load duration $(t-t')_1$ as limit values. The adjustment is made in such a way that for $t'<t_1'$ and $t-t'>(t-t')_1$ all creep curves are parallel to that of $t'=t_1'$. Results from such a procedure for mix No 3 is shown in Fig. 7.

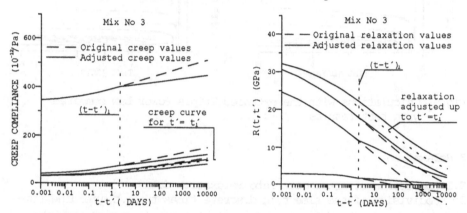

Fig.7. Theoretically obtained creep values with adjusted response to avoid negative values of $R(t,t')$ and corresponding relaxation values.

3.2 Procedure of obtaining values of relaxation modulus

The relaxation function $R(t,t')$ may be approximated by a Dirichlet series with relaxation times τ_μ and coefficients E_μ according to Eq(5).

$$R(t,t') = \sum_{\mu=1}^{N} E_\mu(t')\exp\left(-\frac{t-t'}{\tau_\mu}\right) \tag{5}$$

$E_\mu(t)$, the stiffness of the μth spring in the Maxwell Chain model, may be determined by the method of least square, see for instance Bazant and Wu (1974). However, the function $R(t,t')$ must first be obtained from $J(t,t')$ as described in section 3.1.

Values of the relaxation times τ_μ must be chosen in advance. The individual exponential terms in Eq (5) look like steps spread out over a period of about one decade when plotted in $\log(t-t')$ scale. In this study the smallest $t-t'=0.001$ (from test results) and τ_μ is set to

$$\tau_\mu = 10\tau_{\mu-1} \qquad \text{with } \tau_1 = 0.005 \text{ days} \tag{6}$$

3.3 Results

By simulating a step load history creep data are computed by means of the chosen Maxwell Chain model. Hence, it is possible to check the accuracy of the conversion procedure and the least-square fitting approach. As can be seen in Fig. 8. the agreement between input data and data obtained from the chosen Maxwell Chain model (8 elements in this case) are very good. Similar agreement is obtained for all four mixtures.

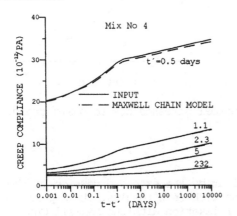 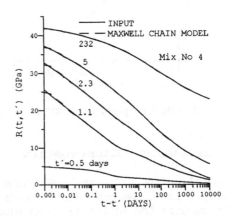

Fig.8. Creep values and conversion into relaxation values compared with results obtained from the Maxwell Chain model with eight elements.

4 Stress analyses

Input temperatures for the stress analyses are obtained from theoretical computations simulating a concrete structure poured between inflexible supports.

Data with respect to the mechanical behaviour of the four concretes is obtained from the creep tests described in chapter 2 and transformed into relaxation according to section 3. Thermal stresses are calculated using the same temperature curve for the different concrete mixtures, where fracturing strain and shrinkage are disregarded. Therefore, the comparison between the four concrete mixtures with regard to stress development made here only take basic creep into consideration.

If the time t is subdivided by discrete times t_r (r = 1,2...n) into time steps $dt_r = t_r - t_{r-1}$ the stress increment during the time step t_{r-1} to t_r are given by (see e.g. Emborg (1990))

$$\Delta\sigma_r = E_r''\left(\Delta\varepsilon_r - \Delta\varepsilon_r''\right) \tag{7}$$

The used temperature curve and corresponding thermal stresses are shown in Fig. 9. for the four concrete mixtures. The difference between mixture No 1 and No 2 is more significant than expected.

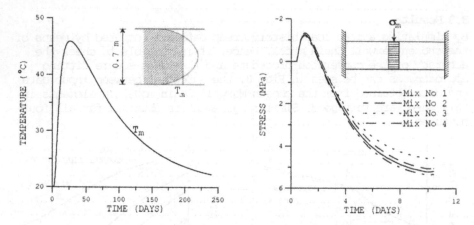

Fig.9. Used temperature curve and thermal stresses due to basic creep.

5 Conclusions - future research

Modelling the early age creep behaviour by the extended version of the Triple Power Law gives good agreement with test results. It is possible to convert the early-age creep into relaxation with the Maxwell Chain model, and then compute thermal stresses. It is however important to calibrate the total behaviour in so called temperature relaxation tests (see eg Emborg (1990))- which will be done in the future.

6 Acknowledgements

This work has been supported by the Swedish Council for Building Research, the Swedish National Board for Technical Development, Cementa AB, Elkem A/S, Euroc Beton, NCC, Skanska and Strängbetong. The work was supervised by Professor Lennart Elfgren, head of the Division of Structural Engineering at Luleå University of Technology.

7 References

Bazant, Z.P and Chern J.C. (1985) Triple Power Law for concrete creep. **J Eng Mat.** Vol III, No 4, pp 63-83.
Bazant, Z.P. and Wu S.T. (1974) Rate-type creep law for aging concrete based on Maxwell Chain. **Materials and structures** (RILEM, Paris), Vol 7, No 37, pp 45-60.
Emborg, M. (1990) **Thermal stresses in Concrete Structures at Early age.** Doctoral thesis 1989:73 D. Division of Structural Engineering, Luleå University of Technology.
Neville A.M. et. al. (1983) Creep of plain and structural concrete. **Constructionn Press,** Longman Group Limited, New York, 361 pp

12 ESTIMATION OF STRESS RELAXATION IN CONCRETE AT EARLY AGES

H. MORIMOTO and W. KOYANAGI
Department of Civil Engineering, Gifu University, Gifu, Japan

Abstract
This paper reports the results of compressive and tensile relaxation tests on concretes at early ages. Tests were performed using a universal testing machine capable of controlling strain for the compression tests and a high-rigidity loading frame fabricated by combining shape steel for the tension tests. Concrete specimens were loaded at 1, 3, 7, 14, and 21 days at 20, 40, and 60°C.
Based on the test results, the effects of age of concrete, initial stress level, and testing temperature on stress relaxation are fully discussed and equations for estimating stress relaxation at early ages are proposed.
Keywords: Relaxation, Mass Concrete, Stress Level Effect of Relaxation, Temperature Effects on Relaxation.

1 Introduction

Past studies on thermal stress analysis have generally utilized creep properties, due to the lack of data accumulated on stress relaxation properties. McHenry (1943), Raphael (1952), Carlson (1959), and Arutyunyan (1966), presented a method for the determination of stress relaxation based on the creep function. Thermal stress, however, is generated by the restraint of thermal strain, and decreases over time, due to the visco-elastic properties of concrete. It is, therefore, considered more appropriate to employ the relaxation function directly expressing the amount of stress relaxation, rather than the creep function when computing the amount of thermal stress relaxation.

In this paper the results of compressive and tensile relaxation tests as well as the effects of the initial stress level, loading age, and testing temperature are described. Empirical equations for estimating compressive and tensile relaxation at early ages are also presented.

Thermal Cracking in Concrete at Early Ages. Edited by R. Springenschmid. Published 1994 by E & FN Spon, 2–6 Boundary Row, London SE1 8HN, UK. ISBN: 0 419 18710 3.

2   Testing apparatus

2.1 Compressive stress relaxation
The layout of the testing apparatus is shown in Fig. 1. The compressive relaxation tests were performed using a universal testing machine capable of controlling the strain of specimens. The amount of the stress relaxation of specimens was detected from the output of the potentiometer used to indicate the loads.

2.2 Tensile stress relaxation
The tensile relaxation tests were performed using a high-rigidity loading frame shown in Fig. 2. The amount of the stress relaxation of specimens was detected through wire strain gages attached to the bolts.

3   Procedure

Two test series were carried out, one for studying the effects of the loading age and of the initial stress (Series A), and the other for the effects of the testing temperature (Series B).

Series A
The loading ages were 1, 3, 7, 14, and 21 days. The strain applied to the specimens was indirectly determined by the initial stress level. The initial stress levels in terms of initial stress-to-strength ratios were 30, 60, and 80% in both of the compressive and the tensile relaxation tests. The sizes of the specimens for compressive and tensile relaxation tests were 10 × 10 × 40 cm and 10 × 10 × 86 cm, respectively. After being demolded, the specimens were cured at a temperature of 20°C and a relative humidity higher than 90% until testing, and were coated with paraffin wax before testing, to avoid any loss of moisture. The testing temperature was controlled at 20°C.

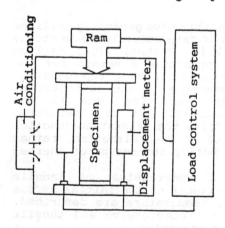

Fig. 1. Testing apparatus
        of compressive
        relaxation test

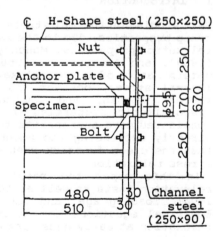

Fig. 2. Testing apparatus
        of tensile
        relaxation test

Drying shrinkage of a separate specimen was also measured under the same conditions.

Series B
The testing temperatures for Series B were controlled at 20, 40, and 60 °C. The tests were performed when the compressive strength of the specimens had reached 10 MPa and 20 MPa. In this series only compressive relaxation was tested with an initial stress level of 50%. The specimens were cured until testing at their respective testing temperatures.

4   Materials and proportioning of concrete

The materials used were ordinary portland cement, river sand (ρ= 2.58, F.M.: 2.53), and crushed gravel (ρ= 2.60, Maximum size 25 mm). The mix proportions of concretes used for Series A and B are presented in Table 1. Table 2 gives the strengths and Young's moduli of the concrete for Series A at the age of loading.

5   Relaxation functions

The authors adopted the same form of an empirical formula as the relaxation function for both compressive and tensile relaxation:

$$\frac{\sigma_t}{\sigma_i} = \frac{A+Ct}{A+t}$$

where σ_t = residual stress at t hr after loading
    σ_i = initial stress
    A, C = experimental constants

Table 1. Mix proportion

Series	W/C (%)	s/a (%)	Unit weight(kg/m³)			
			W	C	S	G
A	50	44	173	346	793	996
B	59	45	166	280	824	1040

Table 2. Properties of concrete(Series A)

Age (days)	Compressive strength(MPa)	Tensile strength(MPa)	Young's modulus(GPa)
1	4.3	0.42	9.8
3	17.4	1.54	19.6
7	27.5	2.29	23.5
28	38.1	2.45	28.4

Here, C denotes the ultimate residual stress to the initial stress ratio ,and A denotes the elapsed time when the amount of relaxation reaches 1/2 of the ultimate value, and it is named as 1/2 relaxation time.

6   Results and discussion

6.1 Compressive stress relaxation

Series A
Fig. 3 shows the test results with loading ages of 3 day. The amount of drying shrinkage measured were so small that its effect on the test results was not taken into account in all the tests. The relaxation functions determined by means of the least square method to be discussed later in this paper are also shown.   These figures reveal that stress relaxation rapidly proceeds during the first hour after loading, up to 25-40% of the ultimate value, and reaches the ultimate value at approximately 100 hr after loading.   In other words, stress relaxation terminates in a significantly shorter period than creep. No effects of the initial stress levels were observed in any of the loading ages.   This indicates that the amount of stress relaxation is proportional to the initial stress, at least in the initial stress level range under 80%.   Table 3 presents the ultimate amount of relaxation and the 1/2 relaxation time with respective loading ages.   The ultimate amount of relaxation and 1/2 relaxation time tend to decrease as the loading age increases.   Though slight differences are observed in the range of 6-12 hr after loading, generally the relaxation function indicated in the figure adequately express the stress relaxation of concrete.   Figs. 4 and 5 show the values of A and C obtained from the experiments.   The following are the equations determined by means of the least square method, for calculating A and C.

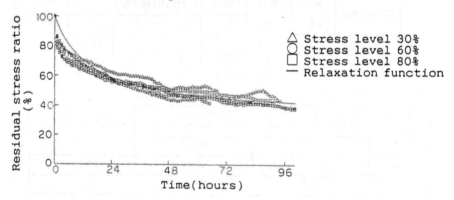

Fig. 3. Compressive relaxation curve

A=-8.25 log t +49.74   (t< 168hr)
 =7.43   (t≧ 168hr)
C=0.25 log t -0.75   (t< 168hr)
 =0.07 log t +0.18   (t≧ 168hr)

Table 3. 1/2 Relaxation time(r) and ultimate amount(Q)

Age (days)	r (hours)	Q (%)
1	25~30	95~100
3	10~14	60~65
7	5~15	40~50
14	7~8	30~40
21	7~15	35~45

Table 4. 1/2 Relaxation time(r) and ultimate amount(Q)

Age (days)	r (hours)	Q (%)
1	0.37	26
3	0.22	14
7	0.48	15
14	0.30	13
21	0.32	13

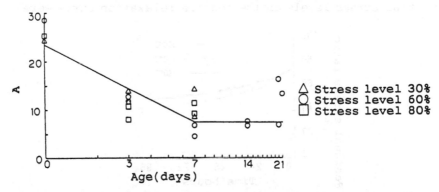

Fig. 4. Value of the constant A(compressive relaxation)

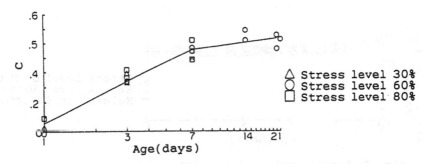

Fig. 5. Value of the constant C(compressive relaxation)

Series B
Fig. 6 shows the result of the tests on specimens with strengths
of 20 MPa at their respective testing temperatures. Whereas
creep is empirically proved to increase as the testing
temperature rises, these test results demonstrate that stress
relaxation is scarcely affected by the temperature, at least in
the range under 60°C.

6.2 Tensile stress relaxation (Series A)

Fig. 7 shows the test results with loading ages of 3 day. The
ultimate amount of relaxation and 1/2 relaxation time with
respective loading age are tabulated in Table 4. The tensile
relaxation terminates in even shorter periods than compressive
relaxation. It took only 2-3 hrs to reach the ultimate values.
The effects of the loading age on the relaxation were small when
compared with compressive one. The ultimate amount of
relaxation was approximately 25% in specimens loaded at 1 day
and approximately 15% in specimens loaded at 3 days and later.
The effects of loading age on the 1/2 relaxation time were not
clear. Similarly to compressive relaxation, the effects of
initial stress levels on the tensile relaxation curve were

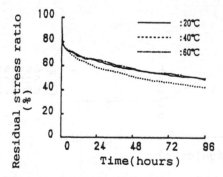

Fig. 6. Effect of temperature on compressive relaxation

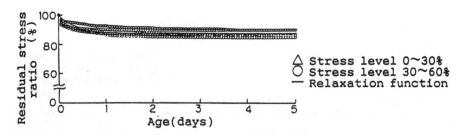

Fig. 7. Tensile relaxation curve

scarcely observed. Thus the tensile relaxation is also considered to be proportional to the initial stress. The relaxation function indicated in the figure agree well with the measured values. Figs. 8 and 9 show the values of A and C. The following are the equations to calculate A and C determined by means of the least square method.

A=0.32
C=0.10 log t +0.39   (t< 72hr)
 =0.85   (t≧ 72hr)

6.3 Comparison of compressive and tensile stress relaxations (Series A)

Fig. 10 show the compressive and tensile relaxation curves with loading ages of 3 day. The ultimate amount of relaxation and 1/2 relaxation time under tension were substantially smaller than those under compression. In other words, tensile relaxation is smaller and terminates in a shorter period than compressive relaxation. This suggests a certain difference exists in the mechanisms of compressive and tensile relaxation.

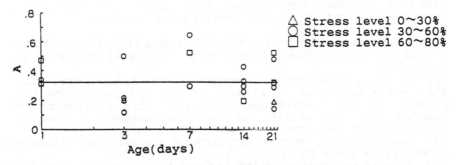

Fig. 8. Value of the constant A(tensile relaxation)

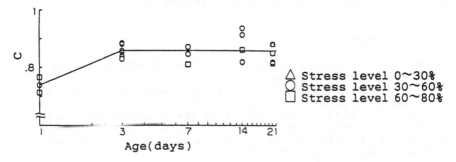

Fig. 9. Value of the constant C(tensile relaxation)

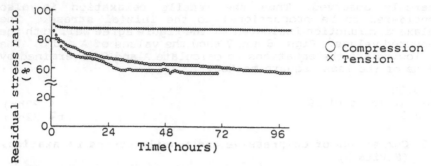

Fig. 10. Comparison of compressive and tensile relaxation

7   Conclusions

The compressive and tensile relaxation tests were performed to study the effects of loading age, initial stress levels, and testing temperature on compressive and tensile relaxation . The differences in compressive and tensile stress relaxation were also elucidated.   The equations to estimate stress relaxation based on the test results were presented.   The conclusions obtained from this study are summarized as follows:

(1) Compressive and tensile relaxations are proportional to the initial stress , at least in the range of initial stress levels under 80%.
(2) Stress relaxation is dependent on the loading age. Both the   ultimate amount of relaxation and 1/2 relaxation time decrease as the loading age increases.   The effects of the loading age   are   smaller   on   tensile   relaxation   than   on compressive relaxation.
(3) The effects of testing temperature are marginal, at least in the range under 60°C.
(4) Tensile relaxation is smaller and terminates in a shorter period than compressive relaxation.
(5) Compressive and tensile relaxation can be estimated by Eq. 1, 2, and 3, and Eq. 1, 4, and 5, respectively.

8   References

Arutyunyan, L. Kh. (1966) **Some problems in the theory of creep.**, Pergamon press.
Carlson, R. W. and Thayer, D. P. (1959) Surface cooling of mass concrete to prevent cracking. J. ACI., 29, 107-120.
McHenry, D. (1943) A new aspect of creep in concrete and its application to dezign. Proc. ASTM., 43, 1069-1043.
Raphael, J. M. (1952) The development of stresses in Shasta Dam. Trans. ASCE., 118, 289-309.

13 STRESSES IN CONCRETE AT EARLY AGES: COMPARISON OF DIFFERENT CREEP MODELS

I. GUENOT
Laboratoire Central des Ponts et Chaussées, Paris, France
J.M. TORRENTI
Commissariat à l'Énergie Atomique, Saclay, France
P. LAPLANTE
Technodes S.A., Ciments Français, Guerville, France

Abstract
Creep plays a significant part in the evolution of thermal stresses and in the risk of cracking in concrete at early ages. Creep must be taken into account in the calculation of mechanical effects brought about by temperature fields resulting from the exothermal hydration reaction of cement. This article proposes a comparative analysis of different creep models taken from the literature and inserted in a uniaxial calculation of total strains (creep strain included) of a concrete member. The calculations are carried out in parallel using respectively the principle of superposition (linear viscoelasticity) and the incremental model, and tested on data and results from an existing experimental case.
Keywords: Basic Creep, Maturity/Equivalent Time, Modelling of Creep, Superposition/ Incremental Model, Temperature Effects on Creep.

1 Introduction

Thermal cracking at early ages in massive concrete structures (dams, nuclear power plants, large bridges) is an important problem that structural engineers endeavour to control. Significant modelling efforts have been made in recent years at the LCPC within the CESAR-LCPC finite-element code, the TEXO and MEXO modules of which allow, respectively, the simulation of early-age temperature distribution and that of the resulting thermal stresses [Torrenti 1992]. One of the present shortcomings of the MEXO module is that it does not take the delayed behaviour of concrete at early ages into account, especially in case of unloading, for which the creep strain is partially reversible. Yet, this can have major consequences on the behaviour of structures subjected to thermal stresses at early ages because, if impeded shrinkage, the concrete goes from a compressed state to a tensioned one (see Springenschmid "cracking bench" tests after [Emborg 1989]).

With regard to creep calculations, linear viscoelasticity and the incremental model are competing [Acker 1989]. The first one is based upon the superposition principle: the global response of a concrete subjected to any loading history is obtained by the algebraic addition of the responses to each elementary loading or unloading. The second one defines the creep rate by:

$$\dot{\varepsilon}_{creep} = f(\sigma, \varepsilon_{creep}, T, m, \omega) \tag{1}$$

Thermal Cracking in Concrete at Early Ages. Edited by R. Springenschmid. Published 1994 by E & FN Spon, 2–6 Boundary Row, London SE1 8HN, UK. ISBN: 0 419 18710 3.

where σ denotes the current applied stress, T the temperature, m the maturity, ω the water content at the considered point and ε_{creep} the creep strain having already occurred. So that, with each stress increment, we are brought back to the calculation of an equivalent time, i.e. of the time that would have been needed by the considered volume element under the present conditions and under a stress equal to the present stress to reach the present creep strain ε_{creep}. These two methods coincide on the domain of constant loadings; the second one offers however advantages because of its easier numerical implementation and enables a better prediction in case of discontinuous loading variations (accounting for recovery phenomena), even if the assumption (applied in the rest of this article, even though criticised [Bazant 1987]) is made that the creep rate is zero in the case of large unloading; the incremental model has, moreover, since been improved concerning this point [Acker 1992].

2 The experimental framework [Laplante 1993]

2.1 Test specimens
The test specimens were cylindrical columns (diameter = 300 mm, height = 1200 mm) of concrete cast in metallic moulds. Just after demoulding (the concrete being only a few hours old), they were covered with adhesive aluminium sheets in order to prevent any hygrometric exchange with the surrounding medium [Attolou 1989]. Thus, only the basic creep and the endogenous shrinkage will be taken into consideration in our calculations.

2.2 Types of concrete investigated
The tests were carried out respectively on two types of concrete, a plain concrete (PC) and a very high performance concrete (VHPC) having the compositions given in Table 1.

Table 1. Composition of concretes investigated

	PC	VHPC
Gravel 5/20 mm (kg/m³)	1200	1216
(semi-crushed silico-calcareous Seine aggregate)		
French OPC HP cement (kg/m³)	342	398
Silica fumes (kg/m³)	---	40
Total water (litre/m³)	171	133
(including the superplasticizer water)		
Superplasticizer (litre/m³)	---	19.3
(melamine with 31% dry extract)		

2.3 Temperature history
By means of thermocouples embedded in the concrete, the temperature measured on the axis in the core of the column is monitored continuously from the casting time (see Fig.1). The history of temperatures T(t) is assimilated with a series of constant temperature plateaus T_i, $t_i \leq t < t_{i+1}$, i=0,1...

2.4 Stress history
The columns underwent, as of the early age of the concrete, a uniaxial loading history by means of a completely slaved hydraulic press. The respective stress histories σ_k,

$t_k \leq t < t_{k+1}$, k=0,1... are shown in Figure 1.

2.5 Measurement of total strains

The instrumentation of the columns included TML KM.100B type strain sensors embedded in the concrete on the column axis.

2.6 Advance of hydration reaction

This is characterised by the degree of hydration d_i, $t_i \leq t < t_{i+1}$, i=0,1..., obtained from the release of heat (measured during a quasiadiabatic test) which has occurred at the age t_i.

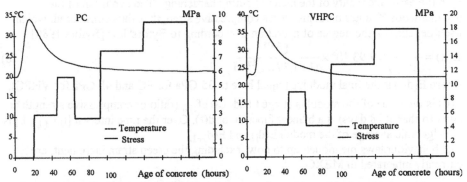

Fig. 1. Respective histories of loading (compressions are positive) and temperature

3 Principles of numerical simulation

3.1 Assumptions

Temperature has two effects on concrete creep (acceleration of creep, but also of the chemical reaction of hydration of the cement which reduces the creep); the relative importance of these effects is still not well known at early ages.

We assume that at early ages temperature does not modify the intrinsic properties of the concrete; to take the effect of temperature on creep into account, the periods of time at variable temperature are thus converted into equivalent periods of hydration at the reference temperature of 20°C, by means of a maturity function based on the Arrhenius law:

$$t_e = \int_{t_o}^{t} \exp\left[\frac{E}{R}\left(\frac{1}{20+273} - \frac{1}{T(\tau)+273}\right)\right]d\tau \tag{2}$$

(For instance, loading started at $t_e \approx 29$ hours after the casting time).
This equivalent time thus controls here both the evolution of hydration and creep strains.

3.2 Numerical process
3.2.1 MEXO type calculation

Assuming the independence of the various types of strain (elastic, thermal, shrinkage and creep), and idealising the problem as a uniaxial one, the total strain increment at the age t is written as:

$$\dot{\varepsilon}_{total}(t) = \dot{\varepsilon}_{elastic}(t) + \dot{\varepsilon}_{thermal}(t) + \dot{\varepsilon}_{shrinkage}(t) + \dot{\varepsilon}_{creep}(t) \tag{3}$$

where we have, by a step-by-step numerical process, over the time interval $[t_i;t_{i+1}]$:

$$\varepsilon_{thermal(i+1)} - \varepsilon_{thermal(i)} = \dot{\varepsilon}_{thermal}\Delta t = \alpha(T_{i+1} - T_i)\Delta t \tag{4}$$

$$\varepsilon_{shrinkage(i+1)} - \varepsilon_{shrinkage(i)} = \dot{\varepsilon}_{shrinkage}\Delta t = -\beta(d_{i+1} - d_i)\Delta t \tag{5}$$

$$\varepsilon_{elastic(i+1)} - \varepsilon_{elastic(i)} = \dot{\varepsilon}_{elastic}\Delta t = E_i^{-1}(\sigma_{i+1} - \sigma_i)\Delta t \tag{6}$$

$$\varepsilon_{total(i+1)} - \varepsilon_{total(i)} = \dot{\varepsilon}_{total}\Delta t \tag{7}$$

where α denotes the coefficient of thermal expansion (taken here as equal to $12.10^{-6}/K$ for the PC and the VHPC), β is the endogenous shrinkage coefficient (taken here as equal to 30.10^{-6}), E_i is the Young's modulus and $\Delta t = t_{i+1}-t_i$.

Taking a "flash" Young's modulus into account in the calculations is not compatible with the early-age reality of the material during hardening. The evolution of the instantaneous Young's modulus versus compressive strength is thus considered, which itself depends on the degree of hydration d, according to Byfors' law [Byfors 1980]

$$E(d) = \frac{1}{E(d_\infty)} \times 9.93.10^3 \times \frac{f_c^{2.675}}{(1+1.37 \times f_c^{2.204})} \tag{8}$$

where $E(d_\infty)$ is the final modulus (equal here to 35 GPa for PC and 45 GPa for VHPC), $E(d)$ is the value of the modulus at age t and $f_c(t)/f_{c28}$ (ratio of compressive strength at age t to that at 28 days) is a bilinear function of d(t). Over the time interval $[t_i;t_{i+1}]$, E_i is the algebraic average of the moduli $E(d_i)$ and $E(d_{i+1})$.

The behaviour laws mentioned up to now, excluding the creep strain increment, are currently integrated in MEXO.

3.2.2 Creep strain increment
The creep strain increment is such that:

$$\varepsilon_{creep(i+1)} - \varepsilon_{creep(i)} = \dot{\varepsilon}_{creep}\Delta t \quad \text{and} \quad \varepsilon_{creep}(t) = \Delta\sigma(\tau) \times N_c(t,\tau) \tag{9}$$

Several creep coefficients N_c, found in the literature and inventoried below, are tested. For a given creep coefficient and a given concrete, an ageing linear viscoelastic calculation (applying the superposition principle), as well as an incremental calculation (based upon the equivalent time method) are carried out.

Notations: τ (τ_e) is the age (the hydration equivalent age) at loading; t (t_e) is the current age (the hydration equivalent age); E_{28} is the 28-day Young's modulus; $f_c(\tau)$ is the compressive strength at age at loading τ and f_{c28} the 28-day one.

CEB code [CEB 1991]: This model is adopted in European rules (Eurocode). In the case of basic creep, the creep coefficient is written as

$$N_c(t,\tau) = \frac{1}{E_{28}} \times \left(\frac{16.8}{f_{c28}^{0.5}}\right) \times \left(\frac{1}{0.1+\tau^{0.2}}\right) \times \left(\frac{t-\tau}{1500+t-\tau}\right)^{0.3} \tag{10}$$

This model takes constant creep kinetics into account (term 1500), which is not evident at very early ages. The following two models try to improve this aspect by making the kinetics depend on the age at loading.

Model of P. Laplante [Laplante 1993]: this creep law is taken from a global uniaxial mechanical behaviour model of hardening concrete.

$$N_c(t_e,\tau_e) = C_c(\tau_e)\frac{(t_e - \tau_e)^{A_c}}{B_c(\tau_e)+(t_e - \tau_e)^{A_c}} \tag{11}$$

The creep coefficient is expressed directly in equivalent time. Maturity is defined by

$$\mu(t_e) = \exp\left[-(\frac{t_i}{t_e})^\gamma\right].$$ (12)

where t_i and γ are parameters depending on the mix design of the concrete.
The final basic creep strain coefficient C_C is equal to:

$$\begin{cases} C_c(\tau_e) = C_{c\infty}[4.6 - 12.1 \times \mu(t_e)] & \text{if } \mu(t_e) \leq 0.3 \\ C_c(\tau_e) = C_{c\infty}[1.26 - 0.26 \times \mu(t_e)] & \text{if } \mu(t_e) > 0.3 \end{cases}$$ (13), (14)

The coefficient B_C depends on the age at loading:

$$B_c(t_e, \tau_e) = B_{c\infty}[0.91 \times \mu(t_e) + 0.08]$$ (15)

A_C characterises the kinetics. The values of the parameters are given in Table 2.

Table 2. Creep law coefficients

	PC	VHPC
A_C	0.35	0.35
$B_{C\infty}$	17.6	13
$C_{C\infty}$	44.6.10-6	17.10-6
t_i	49.17	42.41
γ	0.7	1.07

Model of R. Le Roy [Le Roy 1993]:

$$N_c(t,\tau) = \frac{131e^{-6}}{f_c(\tau)^{0.5}} \times \frac{(t-\tau)^{0.5}}{1.47e^{3.4\frac{f_c(\tau)}{f_{c\infty}}} + (t-\tau)^{0.5}}$$ (16)

This model, undergoing development for application to high and very high performance concrete, takes kinetics depending on $f_c(\tau)$ into account.

4 Discussion

The compared evolutions of total strains, calculated from the different creep models and according to the different methods discussed above, in the case of plain concrete and very high performance concrete, are shown in Figures 2 to 7.

We also compared the experimental results with a calculation "without creep" which gives an idea of the imprecision of MEXO simulations at early age: the difference between the two ranges from about 50 microstrains at the end of the first loading step to over 150 before the return step at zero stress (similar differences for PC and VHPC).

In general, the differences noted between experimental data and simulations change sign with time: the models overestimate, then underestimate (or vice versa) the concrete creep. If the hierarchy between the models varies with time, it also varies with the type of concrete [from the third loading step, the Laplante incremental model gives total strain values higher than those given by the CEB incremental model in the case of PC (see figure 3); the opposite is true in the case of VHPC (see Figure 6)]. While the Le Roy model is very pertinent at very early ages (up to 170 hours), the CEB model (which considers constant kinetics of the creep phenomenon) gives the best results for each concrete and each calculation method used.

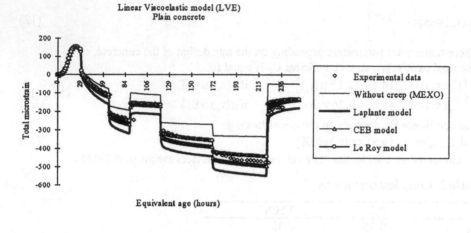

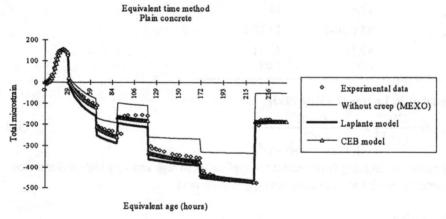

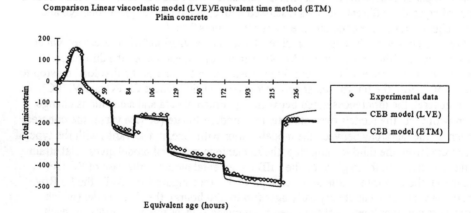

Fig. 2 to 4 Comparison experimental data/simulations for plain concrete

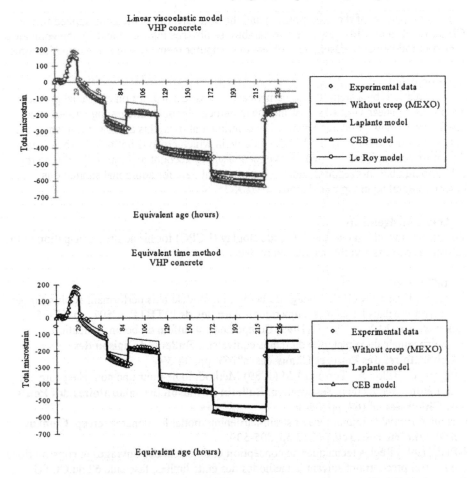

Linear viscoelastic model
VHP concrete

Total microstrain

Equivalent age (hours)

Equivalent time method
VHP concrete

Total microstrain

Equivalent age (hours)

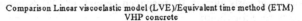

Comparison Linear viscoelastic model (LVE)/Equivalent time method (ETM)
VHP concrete

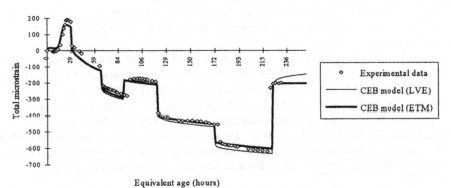

Total microstrain

Equivalent age (hours)

Fig. 5 to 7 Comparison experimental data/simulations for VHP concrete

The comparison of the superposition and the equivalent time methods applied to the CEB model shows that they give comparable results (see Figures 4 and 7), the equivalent time method however posing no problem of computer memory size and calculation time.

5 Conclusions

Our simulations by a step-by-step numerical process, carried out on a uniaxial problem according to linear viscoelasticity and the incremental model, integrating various creep coefficients, show good agreement with experimental data; this approach is in any case better than that offered by the MEXO module in which creep is not taken into account. An improvement is still needed to take recovery into account and transcribe this to the 3-D domain; but this requires suitable experimental data for additional simulations, before integrating creep calculations in MEXO.

6 Acknowledgements

The authors would like to thank Claude Boulay (LCPC) for his active participation in the set-up of the tests and the acquisition of data.

7 References

Acker P., Eymard R.(1992) Fluage du béton : un modèle plus performant et plus simple à introduire dans les calculs numériques. **Annales de l'ITBTP**, n°507, pp. 57-75.

Acker P., Lau M.Y., Collet F.(1989) Comportement différé du béton: validation expérimentale de la méthode du temps équivalent. **Bulletin de liaison des Laboratoires des Ponts et Chaussées**, n°163, pp. 34-39.

Attolou A., Belloc A., Torrenti J-M.(1989) Méthodologie pour une nouvelle protection du béton vis-à-vis de la dessiccation. **Bulletin de liaison des laboratoires des Ponts et Chaussées**, n°164, pp. 85-86.

Bazant Z.P.(1987) Limitations of strain-hardening model for concrete creep. **Cement and concrete research**, vol.17,pp. 505-509.

BPEL(1991) Règles techniques de conception et de calcul des ouvrages et constructions en béton précontraint suivant la méthodes des états limites, fascicule 62 du CCTG.

Byfors J. (1980) Plain concrete at early ages. **CBI research report**, n°3:80, Swedish Cement and Concrete Institute, Stockholm.

CEB (1991) Evaluation of the time-dependent behaviour of concrete.

Emborg M.(1989) Thermal stresses in concrete structures at early ages. **PhD**, Lulea University of technology, Sweden.

Laplante P.(1993) Propriétés mécaniques des bétons durcissants : analyse comparée des bétons classiques et à très hautes performances. **Thèse de doctorat de l'ENPC**, Paris.

Le Roy R.(1993) Travail du groupe AFREM (sous-groupe fluage) en cours.

Torrenti J-M., Paties C., Piau J-M., Acker P., de Larrard F.(1992) La simulation numérique des effets de l'hydratation du béton. **Colloque StruCoMe**, Paris.

Torrenti J-M, De Larrard F., Guerrier F., Acker P., Grenier (1994) Numerical simulation of temperatures and stresses in concrete bridges at early ages : the French experience. **RILEM Symp. on Avoidance of thermal cracking in concrete at early ages**, Munich.

14 YOUNG CONCRETE UNDER HIGH TENSILE STRESSES - CREEP, RELAXATION AND CRACKING

A. GUTSCH and F.S. ROSTÁSY
Technical University of Braunschweig, Germany

Abstract
For the estimation of thermal cracking detailed knowledge about the behaviour of early age concrete under high tensile stresses is necessary. For the calculation of restraint stresses during the hydration process the crack development and the pronounced viscoelasticity must be taken into account. Test results of axial tensile tests, creep and relaxation tests under high tensile stresses wil be reported. Finally a model based on linear creep theory is presented which is applicable for the prediction of creep and relaxation at early age concrete.
Keywords: Degree of Hydration, High Tensile Stresses, Cracking, Creep, Relaxation, Principle of Superposition

1 Introduction

In massive concrete structures high temperatures are generated in course of the hydration of concrete. In case of restraint high tensile stresses may develop with the consequence of cracking. For the estimation of the occurrence of cracking, models for the material properties of concrete at early age are essential. Such models comprise:

- temperature development $T(t)$ in the structure
- degree of hydration $\alpha(T(t))$
- mechanical properties, such as the tensile strength $f_{ct}(\alpha)$, compressive strength $f_c(\alpha)$ and modulus of elasticity $E_{ct}(\alpha)$
- crack development
- creep and relaxation under high tensile stresses

The material models must be developed on basis of experiments. For this axial tensile tests, creep and relaxation tests under high initial tensile stresses in the ascending branch of the stress-strain relationship were performed.

Thermal Cracking in Concrete at Early Ages. Edited by R. Springenschmid. Published 1994 by E & FN Spon, 2–6 Boundary Row, London SE1 8HN, UK. ISBN: 0 419 18710 3.

2 Experiments

2.1 Concrete composition

Previous reseach has shown that the development of the material properties depends on the degree of hydration, Breitenbücher (1989), Emborg (1989), Laube (1990). Due to the nonlinearity of the hydration process, mechanical properties should be described in terms of the degree of hydration.

Recent experiments were performed with three different concrete mixes. Tab. 1 shows the compositions of the concrete mixes. The PZ 35 F is a rapid-hardening, ordinary PC. The HOZ 35 L is a slow-hardening GBFS-PC with 65 % ground blast furnace slag content. The specimens were stored at 20°C and were sealed to prevent drying-out.

Tab. 1: Composition of concrete mixes

Concrete	1	2	3
Cement type	PZ 35 F	HOŻ 35 L	PC 45
Cement content C (kg/m^3)	270	390	320
W/C	0.65	0.47	0.42
C/A/FA/SF	1/6.85/0.22/0	1/4.58/0/0	1/5.71/0.13/0.07

FA: fly ash, SF: silica fume, quartz. aggreg., d_{max} = 16 mm

2.2 Axial short-term tests

In order to model the short term stress-strain behaviour, concrete specimens were tested to failure in axial tension. The specimens were symmetrically pre-notched to locate the cracking zone. The tests were performed with a constant total strain rate of 2.0 ^{o}/oo h^{-1}. The effective age at the on-set of test was varied: t_{el} = 1, 2, 3, 7 and 28 days.

2.3 Creep and relaxation tests

Creep and relaxation tests under high tensile stresses were performed to enhance the knowledge about the viscoelastic behaviour at early concrete age.

A test frame was built which allows tests under real temperature conditions of massive structures, Laube (1989). The specimens (120 x 16 x 16 cm^3) were loaded at the age of t_{el} = 1, 2, 3 and 7 days. The stress-independent deformations were measured on identical compensation specimens hardening under the same temperature history.

Up to now, isothermal creep and relaxation tests (T = 20°C) with initial stresses on the ascending branch of the stress-strain relationship were performed.

3 Results and models

3.1 Degree of hydration

The degree of hydration $\alpha(T(t))$ can be expressed as the ratio of the heat release $Q(T(t))$ to the maximum heat release max Q, Eq. (1).

$$\alpha(T(t)) = \frac{Q(T(t))}{\max Q} = \frac{\Delta T_{ad}(t)}{\max \Delta T_{ad}} \tag{1}$$

With the maturity function based on Arrhenius' law of kinetics the temperature history $T(t)$ of a real structure can be taken into consideration. A relation between the equivalent age t_e and the degree of hydration $\alpha(T(t))$ can be established, Laube (1990).

3.2 Development of the material properties

The relationship between the degree of hydration and the mechanical properties, Rostásy (1993), were verified for the three different concrete mixes. Only the results of the axial tensile strength tests on cylinders are reported here. Fig. 1 shows the dependence of the tensile strength on the degree of hydration.

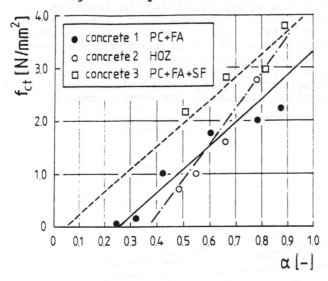

Fig. 1:   Tensile strength for different concrete mixes dependent on the degree of hydration

The tensile strength follows a straight-line relationship with the degree of hydration. The slope of the straight lines and their intersection point α_0 depend on the constituents and composition of concrete. The value α_0 denotes the end of the dormant phase and the on-set of solid-matter properties of young concrete. The coefficient of variation of the tensile stength is about 0.1, it is independent on α.

On basis of such tests the dependence of the important mechanical properties on the degree of hydration can be modelled. In Fig 2 the relevant relationships are depicted.

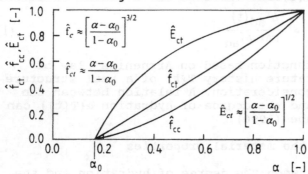

Fig. 2: Normalized tensile and compressive strength, modulus of elasticity vs. degree of hydration

3.3 Stress-strain relationship

In order to determine the spontaneous strain response to stress or stress response to imposed strain, the short-term stress-strain line in axial tension and compression must be known. Fig. 3 shows the ones for axial tension. The distribution of the axial strains across the unnotched width of specimen was measured with strain gauges, $l_0 = 30$ mm, positioned in the crack zone. Their mean value ε is plotted in Fig. 3 vs. the nominal stress in the crack zone.

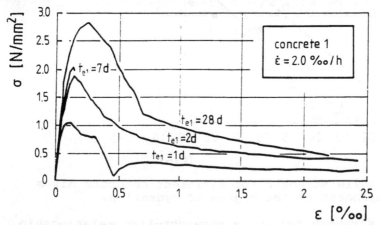

Fig. 3: Stress-strain relationship for different equivalent ages at the on-set of test

The total mechanical strain consists of two components:

$$\varepsilon(t_{el}) = \varepsilon_{el}(t_{el}) + \varepsilon_r(t_{el}) \tag{2}$$

with:

ε_{el} elastic strain dependent on the stress, Young's modulus and effective age t_{el}

ε_r inelastic cracking strain dependent on the stress and effective age t_{el}, arising at a stress of roughly $0.3 \cdot f_{ct}$

The test results agree with those of e.g. Elfgren (1984) and Duda (1991). Fracture energy increases with the degree of hydration. Suitable models for the stress-strain line were proposed e.g. by Duda (1991) and Rostásy et al. (1993).

3.4 Creep and relaxation

The axial tension creep tests were performed for different equivalent ages t_{el} and stress-strength ratios $\sigma_1/f_{ct}(t_{el})$ with the above mentioned concrete mixes (first column of Fig. 4). In Fig. 4 the test results are depicted. The creep strain is expressed by the creep function $\varphi(t-t_1,\alpha_1)$ which is the ratio of the creep strain $\varepsilon_c(t-t_1,t_{el})$ to the elastic strain $\varepsilon_{el}(t_{el})$.

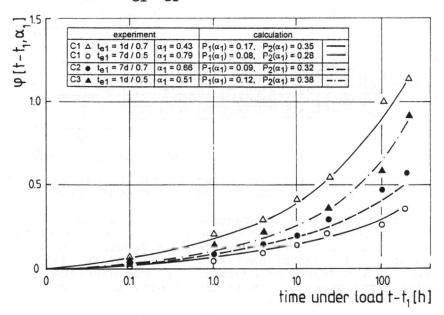

Fig. 4: Creep function dependent on the age at loading, test results (dots) and model (lines)

For all different concrete mixes creep increases with decreasing equivalent age t_{el}, respectively the degree of hydration α_1, at first loading. Additional results have shown that even for high stress-strength ratios of $\sigma/f_{ct}(t_{el}) = 0.7$ and 0.9 no creep failure occured within the time under load of 168 h. In these tests, the total strain had attained a value corresponding to twice the strain at the peak stress f_{ct} in the short-term test.

The results of the axial tension relaxation tests are shown in Fig. 5. Relaxation increases with the decrease of the age t_{el}, respectively the degree of hydration α_1 for the different concrete mixes. The initial stress-strength ratio seems to have less influence on relaxation.

After termination of the long-term tests the residual tensile strength was measured. It was found that the latter does not significantly differ from that of unloaded specimens of the same age. The load history does not seem to have a significant influence on strength.

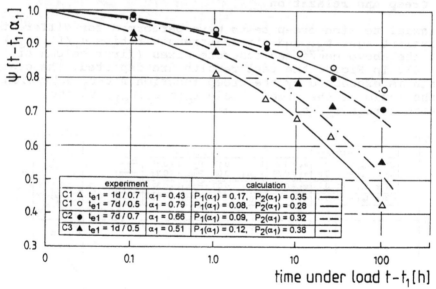

Fig. 5:   Relaxation function dependent on the age at first loading, test results (dots) and model (lines)

Taking into account the spontaneous cracking strain $\varepsilon_r(\sigma, t_{el})$ at loading, linear viscoelasticity is assumed for modelling creep and relaxation. Creep is described by Eq. (3), Laube (1990). The parameters P_1 and P_2 of Eq. (3) are functions of the degree of hydration α_1. The parameters can be derived from creep and relaxation tests. The creep function is (with $t_k = 1$ h):

$$\varphi(t-t_1,\alpha_1) = \frac{\varepsilon_c(t,t_1)}{\varepsilon_{el}(t_1)} = P_1(\alpha_1)\left[\frac{(t-t_1)}{t_k}\right]^{P_2(\alpha_1)} \tag{3}$$

The creep function can be transformed into the relaxation function, Wittmann (1974), Eq. 4.

$$\psi(t-t_1,\alpha_1) = \frac{\sigma(t,t_1)}{\sigma(t_1)} = \exp\left[-P_1(\alpha_1)\left(\frac{(t-t_1)}{t_k}\right)^{P_2(\alpha_1)}\right] \quad (4)$$

Fig. 4 and 5 show that the tests can be satisfactorily described by the model, Eq. (3) and (4).

In the structure the stresses and strains vary with time and location. The total strain $\varepsilon(t,t_1)$ under sustained and time-variant stresses can be calculated if the validity of the classic principle of superposition is assumed. Eq. (5) presents the total strain.

$$\varepsilon(t_n,t_1) = \sum_{i=1}^{n-1}\left(\frac{\Delta\sigma_i}{E_i}[1+\varphi(t-t_i,\alpha_i)] + \Delta\varepsilon_r(\Delta\sigma_i,t_i)\right) \quad (5)$$

In this equation $\Delta\varepsilon_r(\Delta\sigma_i,t_i)$ is the spontaneous and plastic cracking strain response to the stress increment $\Delta\sigma_i$ at t_i.

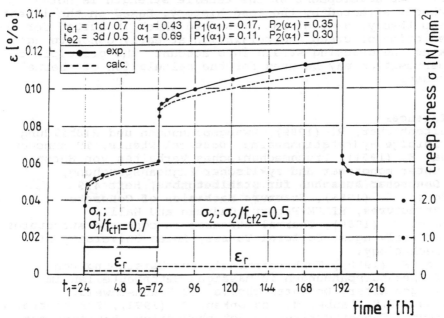

Fig. 6: Results of a creep test with two load steps, experiment and model

In Fig. 6 the results of a creep test comprising two stress steps are shown. Furthermore the theoretical lines determined with Eq. (5) are shown. The test results are well described by Eq (3) and (5). This statement is also true for the stress-relaxation of specimens subjected to a

strain history consisting of strain steps. Further
investigations are needed, especially with respect to the
formulation of a justified superposition principle.

4   Conclusions

The existence of a linear relationship between the development of the tensile strength and the degree of hydration
could be shown for different concrete mixes. In the
ascending branch of the stress-strain relationship under
tension a loss of stiffening caused by microcracking was
noticed above a stress $\sigma = 0.30 \cdot f_{ct}$. Creep and relaxation
tests showed the pronounced viscoelasticity when the load
is applied at early age. Test results showed that the
initial stress/strength-ratio does not exert a very
significant influence on creep and relaxation. The total
strain in a creep test after a certain time under load
exceeds the ultimate strain ε_u in a short time tensile
test. The development of the tensile strength is not
markedly affected by the load history. Based on the linear
creep theory, a model was presented which is suited for the
prediction of creep and relaxation at early concrete ages.
Taking the cracking strain into account, the principle of
superposition may be used for the calculation of thermal
stresses.

References

Breitenbücher, R. (1989), Zwangspannungen und Rißbildung
    infolge Hydratationswärme, **Doctoral Thesis**, TU München.
Duda, H. (1991), Bruchmechanisches Verhalten von Beton
    unter monotoner und zyklischer Zugbeanspruchung,
    Deutscher Ausschuß für Stahlbetonbau, Heft 419
Elfgren, L. (1989), **Fracture Mechanics of Concrete
    Structures**, RILEM-Report, Chapman and Hall.
Emborg, M. (1989), Thermal stresses in concrete structures
    at early ages, **Doctoral Thesis**, Lulea University of
    Technology.
Laube, M. (1990), Werkstoffmodell zur Berechnung von
    Temperaturspannungen in massigen Betonbauteilen im
    jungen Alter, **Doctoral Thesis**, TU Braunschweig.
Rostásy, F.S. Laube, M. and Onken, P. (1993), Zur Kontrolle
    früher Temperturrisse in Betonbauteilen, **Bauingenieur
    No. 68**, pp 5-14.
Rostásy, F.S., Gutsch, A. Laube, M. (1993), Creep and
    relaxation of concrete at early ages - experiments and
    modelling, in **Creep and Shrinkage of Concrete, RILEM
    Proceedings 22**, Chapman and Hall
Wittmann, F. (1974), Bestimmung physikalischer Eigen-
    schaften des Zementsteins, in **Heft 232 Deutscher
    Ausschuß für Stahlbeton**.

15 THE THERMAL STRESS BEHAVIOUR OF CONCRETE BASED ON THE MICROMECHANICAL APPROACH

H. OHSHITA
Department of Civil Engineering, Hiroshima University,
Higashi-hiroshima, Japan
T. TANABE
Department of Civil Engineering, Nagoya University, Chikusa-Ku,
Nagoya, Japan

Abstract
Thermal stress analysis in early age concrete has several problems to be clarified. One of the problems is the characteristic of water migration since for early age concrete, much water is present when compared with hardened concrete. In the beginning, a mathematical model which satisfies the force equilibrium and flow continuity condition is introduced. The flow continuity condition is presented in this paper with the volumetric contraction by hydration which generates a tensile stress in the voids. Further experimental work measuring pore water pressure which result in water migration was carried out and it has been shown that the proposed model is capable of predicting the characteristic of pore water pressure. Finally, pore water migration effect on the creep problem is investigated.
Keywords: Water Migration, Pore Water Pressure, Effective Stress

1 Introduction

In recent years, civil engineering construction of deeper underground structures and marine structures are in creasing. In order to construct a concrete structure which is required to satisfy the quality and durability, the deformational behavior of concrete which is under water pressure and ground pressure, has to be understood in detail. Concrete is regarded as a material that changes its mechanical characteristics drastically with elapse of time especially in early ages. Moreover, as concrete in early ages has much water in the voids compared with the hardened concrete, the variation of characteristics and water migration are expected to significantly affect the behavior of concrete. In this study, a mathematical model for early age concrete which is assumed to be a two phase porous material of solid and liquid is developed and used for analysis. This mathematical model satisfies the force equilibrium and flow continuity condition in which volumetric contraction caused by hydration is considered in particular. To calibrate the proposed model, a triaxial compressive test was carried out at the age of 3 days and the temperature gradient, stress, strain as well as pore water pressure were observed, and the prediction by the analytical model was compared with the test data. Finally application to the creep problem is conducted and discussed.

2 Mathematical Modeling of Early Age Concrete by Saturated Porous Two Phase Material

In the analysis, concrete is regarded as a porous material which is composed of aggregate, cement paste and water. Aggregate is considered as a perfectly elastic material while cement

Thermal Cracking in Concrete at Early Ages. Edited by R. Springenschmid. Published 1994 by E & FN Spon, 2–6 Boundary Row, London SE1 8HN, UK. ISBN: 0 419 18710 3.

paste is assumed to behave as an elasto-plastic permeable material. Both the force equilibrium and flow continuity equations of pore water proposed by Contri, Majorana and Schrefler[1] and Lewis and Schrefler[2] are modified to incorporate the time effect and the material nonlinearity. Material nonlinearity is considered using the modified yield criterion of Drucker Prager type[3], and calculated by finite element method.

2.1 Formulation of Force Equilibrium Equation

With the presence of pore water pressure p, the relation of effective stress $\{\sigma'\}$ and total stress $\{\sigma\}$ will become

$$\{\sigma\} = \{\sigma'\} - \{m\}p \tag{1}$$

where the sign of tensile stress is taken as positive and $\{m\} = \{1 \ \ 1 \ \ 1 \ \ 0 \ \ 0 \ \ 0\}^T$. The incremental effective stress-strain relationship for concrete can be written as

$$d\{\sigma'\} = \left[D_T^{ep}\right]\left\{d\{\varepsilon^T\} - d\{\varepsilon^{pr}\} - d\{\varepsilon^t\}\right\} \tag{2}$$

where $\left[D_T^{ep}\right]$ is the elasto-plastic stiffness matrix of concrete where the voids are not saturated with water, $d\{\varepsilon^T\}$ is the incremental total strain of concrete, $d\{\varepsilon^{pr}\}$ is the incremental strain of solid phase resulting from the incremental pore water pressure $\{dp\}$ and $d\{\varepsilon^t\}$ is the incremental thermal strain. Elasto-plastic stiffness matrix of concrete can be written with the use of average elasto-plastic stiffness matrix of solid phase $\left[D_S^{ep}\right]$ and porosity ξ as follows.

$$\left[D_T^{ep}\right] = (1 - \xi)\left[D_S^{ep}\right] \tag{3}$$

$$d\{\varepsilon^{pr}\} = -\left[D_S^{ep}\right]^{-1}\{m\}dp \tag{4}$$

Then by using the principle of virtual work and appropriate shape function, the equilibrium equation can be written in differential form as

$$K_T\frac{d\{\bar{u}\}}{dt} - L\frac{d\{\bar{p}\}}{dt} - A\frac{d\{\bar{T}\}}{dt} - \frac{d\{f\}}{dt} = 0 \tag{5}$$

where matrices K_T, L, A are the tangential stiffness, effect of pore water pressure and volumetric change of solid phase and effect of temperature, respectively. The vector $\{f\}$ denotes the effect of external force. These matrices can be written as follows.

$$K_T = (1 - \xi)\int_\Omega B^T D_S B \, d\Omega \ , \ \ L = \int_\Omega B^T \xi\{m\}\bar{N} \, d\Omega$$

$$A = (1 - \xi)\int_\Omega B^T D_S\{m\}\alpha\bar{N} \, d\Omega \ , \ \ f = \int_\Omega N^T\{b\} \, d\Omega + \int_\Gamma N^T\{t\} \, d\Gamma \tag{6}$$

Here, N and $\bar{N}$ are the shape functions for displacement, pore water pressure and temperature, while B is the stain-displacement metrix.

2.2 Formulation of Flow Continuity Equation

Water head h can be written as

$$h = z + \frac{p}{\gamma} \tag{7}$$

where z is the vertical coordinate of the point which is positive for upward direction, γ is the specific gravity of fluid. From the mass conservation law, the volumetric change in a unit volume(ΔV) is equal to the difference of the amount of in-flow q and out-flow from this volume, which can be written as

$$\Delta V = q - \nabla^T \{v\} \tag{8}$$

where $\{v\}$ is the flow velocity, which is considered to follow Darcy's law. The factors which contribute to the volume change(Eq.(8)) can be summarized as follows.
(a) due to total strain change

$$\frac{\partial \varepsilon_v}{\partial t} = \{m\}^T \frac{\partial \{\varepsilon\}}{\partial t} \tag{9}$$

(b) due to the change of volume of particles caused by changes of hydrostatic pressure,

$$(1-\xi)\{m\}^T D_S^{-1}\{m\}\frac{\partial p}{\partial t} \tag{10}$$

where ξ is the porosity which depends on the rate of hydration of cement,
(c) due to the change of fluid volume,

$$\frac{\xi}{k_f}\frac{\partial p}{\partial t} \tag{11}$$

where k_f is the bulk modulus of fluid including water vapor,
(d) due to the change of particle size by effective stresses,

$$-\{m\}D_S^{-1}\frac{\partial \{\sigma'\}}{\partial t} \tag{12}$$

(e) due to the change of fluid volume caused by temperature,

$$-3\xi\mu\frac{\partial T}{\partial t} \tag{13}$$

where μ is the thermal expansion coefficient of water,
(f) due to the volumetric contraction by hydration,

$$\frac{\eta\gamma_p}{\rho_w}\frac{dC_H}{dt} \tag{14}$$

where $C_H \cdot \gamma_p, \rho_w$ and η denote the hydrated cement weight in a unit volume of cement, the water cement ratio of complete hydration, water density and the coefficient to convert the volumetric contraction by hydration, respectively. Eq.(14) is transformed into the volumetric contraction in a unit volume of concrete as follows.

$$\frac{\eta\gamma_p}{\rho_w}\frac{dC_H}{dt}\frac{C}{\rho_c} \tag{15}$$

where C, ρ_c denote the initial cement weight in a cubic meter of concrete and cement density, respectively

Substituting the six factors of Eqs.(9) to (13) and Eq.(15) into Eq.(8) and applying the Galerkin's method, Eq.(8) becomes in differential form as

$$-H\{\bar{p}\} - L^T \frac{d\{\bar{u}\}}{dt} - S\frac{d\{\bar{p}\}}{dt} - W\frac{d\{\bar{T}\}}{dt} - \frac{d\{g_p\}}{dt} + \{f_p\} = 0 \tag{16}$$

where

$$H = \int_\Omega \left(\nabla \bar{N}\right)^T k' \nabla \bar{N} \, d\Omega \quad , \quad S = \int_\Omega \bar{N}^T \frac{\xi}{k_f} \bar{N} \, d\Omega \quad , \quad W = \int_\Omega \bar{N}^T \left\{3(1-\xi)\alpha - 3\xi\mu\right\} \bar{N} \, d\Omega$$

$$f_p = \int_\Omega \bar{N}^T q \, d\Omega - \int_\Omega \left(\nabla \bar{N}\right)^T k' \nabla \gamma \, z \, d\Omega + \int_\Gamma \bar{N}^T \left(\{v\}^T \bullet n\right) d\Gamma \tag{17}$$

$$g_P = \int_\Omega \bar{N}^T \frac{\eta \gamma_p}{\rho_W} \frac{C}{\rho_C} C_H \, d\Omega \quad , \quad s = \frac{\xi}{k_f}, k' = \frac{k}{\gamma}$$

where k is the permeability of concrete.

2.3 Hydration Process of Cement Paste

Here, we refer to the study by Kawasumi et al[4]. Assuming velocity of hydration is controlled by the concentration of cement in unhydrated water, the hydration process can be written as

$$\frac{dC_H}{dt} = k_0(1-n_0)t^{-n_0}(W - \gamma_p C_H)(C - C_H) \tag{18}$$

where W is the initial water weight in a cubic meter of concrete. The term $\gamma_p C_H$ denotes the hydrated water weight. Concrete age t is expressed by day and k_0, n_0 are material constant in terms of concrete age and temperature. The solution of this differential equation is obtained in the explicit form as

$$C_H = \frac{1 - exp\left[(\gamma_p C - W)k_0 t^{1-n_0}\right]}{1 - \gamma_p {C}/{W} exp\left[(\gamma_p C - W)k_0 t^{1-n_0}\right]} \times C \quad , \text{ for } W/C \neq \gamma_p \tag{19}$$

$$C_H = \frac{\gamma_p k_0 t^{1-n_0}}{1 + \gamma_p k_0 t^{1-n_0}} \quad , \text{ for } W/C = \gamma_p \tag{20}$$

The porosity ξ is calculated by reducing the volume of hydrated water from the initial volume V_{w0} of water considering the hydration contraction.

$$\xi = V_{w0}(1 - \gamma_p C_H / W) \tag{21}$$

2.3 Formulation of Coupled Equation

The coupled equation which satisfies the force equilibrium and flow continuity condition can be written in the following matrix form.

$$\begin{bmatrix} 0 & 0 \\ 0 & -H \end{bmatrix} \begin{Bmatrix} \{\bar{u}\} \\ \{\bar{p}\} \end{Bmatrix} + \begin{bmatrix} K_T & -L \\ -L^T & -S \end{bmatrix} \begin{Bmatrix} \dfrac{d\{\bar{u}\}}{dt} \\ \dfrac{d\{\bar{p}\}}{dt} \end{Bmatrix} = \begin{Bmatrix} \dfrac{d\{\bar{f}\}}{dt} + A\dfrac{d\{\bar{T}\}}{dt} \\ W\dfrac{d\{\bar{T}\}}{dt} - \{f_p\} + \dfrac{d\{g_P\}}{dt} \end{Bmatrix} \tag{22}$$

3 Experimental Work

3.1 Experimental Method

A triaxial compressive experiment was carried out by placing the concrete specimen into the mold measuring pore water pressure, as shown in Fig.1. Cylindrical type concrete specimen with dimensions 10cm diameter and 20cm height were used. The mold is composed of two forms whose interface has the leakage preventable rubber plate and which are anchored rigidly by bolted joint. In this way, specimen is restrained for lateral direction and the condition is perfectly undrained.

Pore water pressure was measured by a water pressure gauges installed in a drainage hole at the bottom of the mold. The deformation of vertical direction was measured by the displacement meter which is installed at the top of the load cell, as shown in Fig.2. The mix proportion of the concrete is given in Table 1 and water cement ratio was decided to take the maximum continuous diameter of voids to be larger.

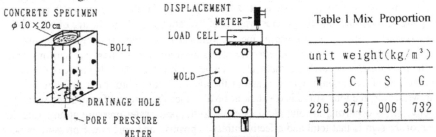

Fig.1 Mold Apparatus   Fig.2 Setting of Displacement Meter

Table 1 Mix Proportion

unit weight(kg/m³)			
W	C	S	G
226	377	906	732

3.2 Loading Method

The loading apparatus is a universal testing machine whose capacity is 25ton and a mold is placed in it, as shown in Fig.3. Load is applied until about 70 percent of the capacity of pore pressure meter. Further, loading speed is 0.02mm/sec.

Table 2 Experimental Parameters

Loading Age(day)	Pre-loading Age(day)	Curing Period(day)	Pre-loading Stress Ratio(%)
3	1	2	0 60 90

3.3 Experimental Parameters

Experimental parameters were loading age, pre-loading age and the ratio of pre-loading stress and uniaxial compressive strength, as shown in Table 2. The applied loading age is 3 days, pre-loading age is 1day for 3days of applied loading, the ratio of pre-loading stress and uniaxial compressive strength is 0, 60 and 90%. Curing of concrete specimen was done such that all specimens are kept in water where temperature is constant at 20°C. Pre-loading was given to make initial inside crack distribution different among each specimens.

Fig.3 Universal Testing Machine

4 Volumetric Contraction by Hydration

The product by hydration of cement and water is mainly composed of cement gel. It is said that the reduction of the volume of hydration from the initial volume is about 25% when hydration is completed, and it is generally called " **Contraction by Hydration** ". The effects of contraction by hydration on the stress state of concrete while curing is still under the investigation. However, the very probable stress state of concrete can be considered to be tensile in pore water and compressive (called "initial stress state") in the solid phase. Detail of this phenomenon and further discussion are given in succeeding section.

To perform the analysis, we need to know the various material parameters such as Young's modulus, temperature gradient and permeability. Young's modulus were decided from the experimental results obtained at the age of 0.5, 1, 1.5, 2 days by interpolation and extrapolation based on the assumption of linearity. The values are 1800 at 0.5 day, 5500 MPa at 1 day, 6500 at 1.5 days and 8000 MPa at 2 days. The temperature gradient used are obtained by experiment. The permeability is assumed to be constant at $2.0 \times 10^{-8} cm/sec$. Further the material parameters k_0 and n_0 are decided such that the progressive rate of hydration is 50 % at the age of 6 days[4].

The numerical analysis is performed giving the volumetric contraction of hydration and temperature. The pore water pressure and effective stress as well as total stress are shown in Fig.4 and 5. The pore pressure occurring in the voids has a negative value. The value increases exponentially as the rate of hydration progresses, and at the age of 3 days pore water pressure becomes 2.2MPa. Moreover, the total stress is constant at 0MPa since the external force is not applied, so the effective stress occurring in solid phase is equal to the pore water pressure at any ages. The definition of the sign is that total and effective stress is positive and pore water pressure is negative for tensile stress. It may be seen that the contraction by hydration of the voids will be restrained by the surrounding solid phase. Therefore as shown in the next chapter, it may be seen that the sudden increase of water migration occurs at the strain in which the negative stress occurring in the voids is cancelled out by a part of the applied load, and then the positive pore water pressure is considered to be generated.

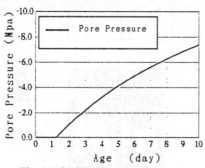

Fig.4 Initial Stress
(age-pore pressure relationship)

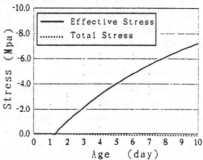

Fig.5 Initial Stress
(age-total,effective stress relationship)

5   Analysis of Pore Water Pressure

In the experimental work, the parameters are loading age , pre-loading age and the ratio of pre-loading stress and uniaxial compressive strength, as shown in section 3. The experiment revealed that the more water content and the less the combined force between the particles of concrete, the greater the pore water pressure which occurs in the voids.

To calibrate the proposed model, we need to know the various material parameters such as uniaxial compressive strength, young's modulus of elasticity, permeability, the cohesion and angle of internal friction for failure surface of plastic analysis. Uniaxial compressive strength are decided in the same manner with young's modulus. The values are 12.4MPa at 3 days. Young's modulus are decided as described in section 4. The values are 9300MPa at 3 days. The permeabilities are $1.05 \times 10^{-7} cm/sec$ at 3 days. In the proposed model in plastic region, the varying failure surface of the Drucker-Prager type is used. Therefore, the material parameters such as initial cohesion and internal friction are decided. The initial cohesion values are taken as fc'/2 and the final value of angle of friction are taken as 27 degrees according to the experimental results.

Fig.6 and 7 show the comparison between the experimental and analytical results for total stress-strain and pore water pressure-strain relationship, respectively. In these figures, the experimental results are marked with circles at 0%, squares at 60% and triangles at 90% where the percentage is the ratio of pre-loading stress and uniaxial compressive strength, the analytical results are shown with solid line, dotted line and broken line, respectively.

As shown in Fig.7, for the experimental results it may be seen that the sudden increase of the positive pore water pressure occurs at the strain in which the negative stress occurring in the voids while curing is equal to be a part of the applied load and when the specimen is under perfectly saturated condition. The gradients between pore water pressure and strain due to the difference of the ratio of pre-loading stress are the almost same. For the analytical results, a good agreement with the experimental results can be seen on the whole, but particularly in case the of 60 and 90% of the ratio of pre-loading stress, the strains in which the pore water pressure occurs, are different between the experimental and analytical results. As shown in Fig.6, for the analytical results, a good agreement with the experimental results can be seen on the whole. Therefore, the proposed model is acceptable in evaluating the pore water pressure, but the evaluation of the initial stress occurring by the contraction by hydration must be studied in detail in the future as the initial stress greatly influences the characteristics of hardened concrete.

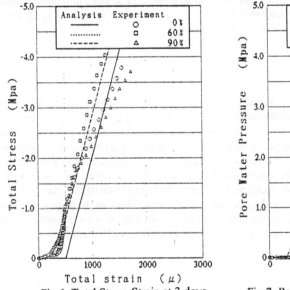

Fig.6 Total Stress-Strain at 3 days

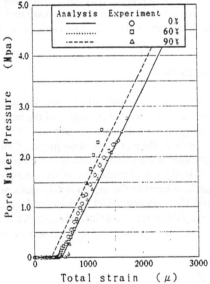

Fig.7 Pore Water Pressure-Strain at 3 days

6 Analysis of Creep Strain

The creep experiment of concrete at the age of 0.69 days performed at the Central Research Institute was analyzed and the comparison between both results were performed. The strength and Young's modulus of concrete used in the analysis are 2.8MPa and 6400MPa, respectively. Permeability is assumed to be constant at $5.0 \times 10^{-6} cm/sec$. The measurement of the creep strain was started after the load intensity as same as the one (1MPa) for experiment was applied. The boundary conditions of the flow analysis are that the pore water pressure at the surface of the specimen is equal to the atmospheric pressure.

Fig.8 shows the comparison between experimental and analytical result for the creep strain. In this figure, the experimental results are marked with circles and the analytical results are shown in solid lines, and 15μ at the time of 1 day shows the elastic strain. As shown in Fig.8, for the experimental result, it can be seen that the sudden increase of the creep strain occurs in a few hours, and then the gradual increase occurs. On the other hand, for the analytical result, a good agreement can be seen in a few hours after loading , but after that stage the creep strain becomes constant. This sudden increase of the creep strain may be caused

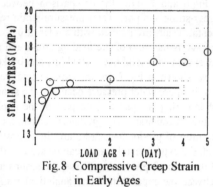

Fig.8 Compressive Creep Strain in Early Ages

by the condition that after load intensity was applied, pore pressure is converted to the effective stress instantaneously and then after the water migration is completed, the incremental strain vanishes. Moreover, the gradual increase of creep strain may be caused by the effect of the viscous flow of the solid phase. It should be noted that the proposed model evaluates the sudden increase of the creep strain, in other words initial creep strain, but to evaluate the succeeding creep strain we have to analyze the solid phase as a viscous flow material.

7 Conclusion

In this study, a mathematical model for early age concrete is developed and used for analysis. In a proposed model, the evaluation of the initial stress state occurring by the volumetric contraction by hydration is considered. And a triaxial compressive experiment has been conducted in order to evaluate the capability of the proposed model, and the characteristics of pore water pressure in early ages is discussed according to the comparison between the experimental and analytical results. Finally, the characteristic of the creep in early age concrete is discussed according to the comparison between the experimental results at the Central Research Institute and the analytical results. The proposed model shows a good agreement with the experiment on the pore water pressure and initial creep behavior. However, the evaluation of the initial stress must be studied in detail in the future as the initial stress greatly influences the characteristics of hardened concrete.

8 References

1 Contri, L., Majorana,C. E. and Schrefler, B. A.,:Proceeding of International Conference on Concrete of Early Ages, Vol.1, 1982, pp.193-198. (Ecole Nationale des Ponts et Chaussees, Paris 6-7-8 April, 1982)

2 Lewis, R. W., Schrefler, B. A.,:**A Fully Coupled Consolidation Model of the Subsidence of Venice**, Water Resources Research: Vol.14, pp.223-230, 1978

3 Wu, Z. S. and Tanabe, T.,:**A Hardening-softening Model of Concrete Subjected to Compressive Loading** :,Journal of Structural Engineering, Architectual Institute of Japan. Vol.36B, pp.153-162, 1990

4 A. Haraguchi, M. Kawasumi, T. Tanabe and T. Okazawa : **Study on Concreting Schedule of Kuroda Dam**, Part1 1 (Mechanical and Thermal Properties of Concrete), CRIEP Report No.375561, July, 1976

16 NUMERICAL SIMULATION OF THE EFFECT OF CURING TEMPERATURE ON THE MAXIMUM STRENGTH OF CEMENT-BASED MATERIALS

K. van BREUGEL
Delft University of Technology, Delft, The Netherlands

Abstract
It is well-known from both the engineering practice and experimental research that low curing temperatures in the early stage of hydration will result in a higher ultimate strength. It is widely accepted today that this strength reduction must be attributed to the temperature dependency of the microstructure. In this paper it will be explained how this temperature effect can be simulated numerically with the computer-based simulation program HYMOSTRUC. The background of this programm is briefly outlined and examples of strength predictions of concretes cured at different temperatures are presented and discussed. The relevance for the practice, in particular in view of the prediction of the risk of early-age thermal cracking, will be dealt with.
Keywords: Curing Temperature, Microstructure, Numerical Simulation, Risk of Cracking, Strength Influenced by Curing Temperatures.

1 Introduction

Strength development of concrete is an important issue in the engineering practice. This holds for both the site practice and the prefabrication industry. A rapid strength development is of advantage in view of, for example, early demoulding and the application of prestress. A number of experiments on the effect of curing temperature on strength development and final strength has evidenced that high curing temperatures result in a lower ultimate strength. Strength reductions of 20 to 30% have been reported if curing takes place at temperatures of 50°C and there beyond (Rakel (1965), Kjellsen et al. (1991), Odler (1986)). Temperatures beyond 50°C may occur when rapid cements are used and in high strength concretes. Substantial strength reductions may affect the probability of cracking in an adverse way. Reliable predictions of the probability of cracking shall consider these strength reductions.

A practical tool for predicting the strength development is the maturity concept. This concept, however, is a predominantly phenomenological method and does not allow for the temperature-caused reduction of the ultimate strength. Neither does the more materials science-oriented "degree of hydration concept", in which the strength is assumed to be a function of the degree of hydration. In this paper a model will be presented in which the strength development and the effect of the curing temperature on it is explained on a microstructural basis.

Thermal Cracking in Concrete at Early Ages. Edited by R. Springenschmid. Published 1994 by E & FN Spon, 2–6 Boundary Row, London SE1 8HN, UK. ISBN: 0 419 18710 3.

2 Effect of temperature on microstructure and porosity of cement paste

2.1 Description of observed phenomena
Cement in contact with water starts to dissolve under formation of hydration products. These products, consisting of ettringite needles, calcium silicate hydrates (CSH) and calcium hydroxide crystals (CH), form a spatial network between the hydrating particles. This spatial network is gradually filled up with subsequently formed hydration products. There seems to be a distinct preference of the reaction products to precipitate in the close vicinity of the cement particles from which they are formed. The whole process can be considered as an "expansion process". With progress of hydration expanding particles make interparticle contacts. Between the expanding and interconnected particles a pore system remains in which the capillary water is accommodated.

If hydration proceeds isothermally the gel that forms around the hydrating cement particles is assumed to have a constant porosity, i.e. a constant density. Literature data on the effect of the reaction temperature on the microstructure and porosity is not free from contradictions. Many tests seem to have evidenced, however, that at elevated temperatures a denser gel is formed while the capillary porosity increases. Schematically the foregoing is shown in Fig. 1.

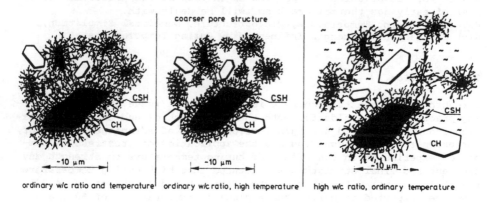

Fig. 1   Effect of temperature on microstructural development - Schematic representation (after van Breugel (1991)).

2.2 Strength considerations
Numerous test results have revealed a distinct correlation between porosity and strength. The lower the porosity the higher the strength. This seems evident, since a high porosity implies that less contacts, or smaller contact areas, exist between hydrating cement particles. Since high reaction temperatures result in a higher porosity, strength reductions observed at elevated temperatures seems to be explained.

For predicting the influence of temperature on strength development one should be able to quantify the effect of the curing temperature on the porosity or on the intensity of interparticle contact. The basic structure of a numerical model that is capable to quantify the temperature effects on microstructural development will be outlined in the next sections.

3 Numerical simulation of hydration and microstructural development

3.1 Background of the simulation model

In the computer-based simulation model called HYMOSTRUC, the acronym for HYdration, MOrphology and STRUCtural development, the effects of the particle size distribution and the chemical composition of the cement, the w/c ratio ω_0 and the reaction temperature T on the rate of hydration are dealt with. Unlike most previously proposed models the effect of physical interactions between hydrating cement particles on the rate of hydration of individual cement particles is modelled explicitly. In order to obtain a workable algorithm for the determination of the interaction between hydrating and expanding cement particles several assumptions and simplification had to be made. For a summary of them reference is made to Koenders et al. (1994). In view of modelling of the effect of temperature on microstructural development the ratio ν between the volumes of the reaction products and the anhydrous cement is assumed to be a function of the curing temperature. So $\nu = \nu(T)$, with decreasing values at elevated temperatures (see section 3.3).

3.2 Numerical modelling of expansion and interaction mechanisms

In the model the cement particles are considered to be distributed homogeneously in the cement-water system. An arbitrary particle is considered to be located in the centre of a cell "I_x^c" (Koenders et al. (1994)). The cell I_x^c is defined as a cubic space with a central particle with diameter x and further consists of $1/N_x$ times the original water volume and $1/N_x$ times the volume of all particles with diameter

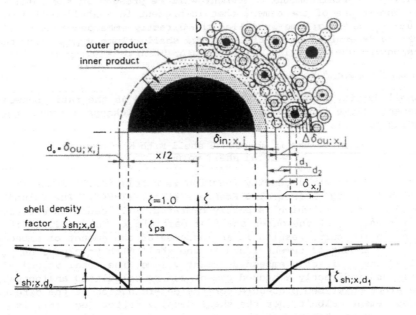

Fig. 2. Interaction mechanism for expanding particles (Breugel (1991)). Upper-left: free expansion; Upper-right: embedding of particles, several iteration steps. Bottom part: Shell density factor $\zeta_{sh;x,d}$. smaller

than x μm, N_x being the number of particles with diameter x μm in a certain paste volume.

As suggested in section 2.1 microstructural development can be considered as a process of the formation of contacts between expanding cement particles. An arbitrary stage of the expansion of a central particle x is shown schematically in the upper part of Fig. 2. At a certain time, say t_j, the degree of hydration of particle x is $\alpha_{x,j}$. For the corresponding penetration depth $\delta_{in;x,j}$ of the reaction front it holds:

$$\delta_{in;x,j} = (x/2) * [1 - \{1-\alpha_{x,j}\}^{1/3}] \tag{1}$$

The volume of outer product $v_{ou;x,j}$, i.e. the hydration products formed inside the original grain surface, of particle x which corresponds with the degree of hydration $\alpha_{x,j}$ follows from:

$$v_{ou;x,j} = (v-1) * \alpha_{x,j} * v_x \tag{2}$$

with v_x the volume of particle x in its anhydrous state. If no cement would be found in the spherical outer shell which surrounds the expanding central particle, the outer radius $R_{ou;x,j}$ of particle x would be:

$$R_{ou;x,j} = [\frac{v_{ou;x,j}}{4\pi/3} + (x/2)^3]^{1/3} \tag{3}$$

For the thickness $\delta_{ou;x,j}$ of the outer shell it would follow:

$$\delta_{ou;x,j} = R_{ou;x,j} - x/2 \tag{4}$$

In reality a certain amount of cement will be present in the outer shell. The volume of the cement that is present in a shell with thickness $\delta_{ou;x,j}$, which volume is called the directly embedded cement volume $v_{em;x,j}$, is determined by multiplying the shell volume $v_{ou;x,j}$ with the corresponding shell density factor $\zeta_{sh;x,d_0}$ (with $d_0 = \delta_{ou;x,j}$):

$$v_{em;x,j} = \zeta_{sh;x,d_0} * v_{ou;x,j} \tag{5}$$

The shell density factor $\zeta_{sh;x,d}$ has been defined as the ratio between the cement volume in the spherical shell with thickness d around the expanding particle and the total shell volume:

$$\zeta_{sh;x,d} = \frac{\text{cement volume in spherical shell with thickness d}}{\text{total shell volume}} \tag{6}$$

The course of the shell density function is schematically shown in the bottom part of Fig. 2 (see also Koenders et al. (1994)). The volume of the embedded cement, which is partly hydrated, will cause an additional expansion $\Delta\delta_{ou;x,j}$ of the outer shell of particle x. This additional expansion, on its turn, results in an increase of the amount of embedded cement. The amount of cement found in the shell with thickness $\Delta\delta_{ou;x,j}$ is called the indirectly embedded cement volume. The total volume of directly and indirectly embedded cement can be calculated analogously to the afore described procedure for the determination of the directly embedded cement, albeit that the shell density factor (eq. (6)) is a little higher now because of the meanwhile increased thickness of the outer shell (Fig. 2, bottom part: $\zeta_{sh;x,d_1} > \zeta_{sh;x,d_0}$). The increase of the amount of embedded cement, on its turn, causes another increase of the outer shell, etc. This expansion mechanism can be written in the form of a geometrical series. Algebraic evaluations result in workable expressions for, among other things, the amount of cement embedded in the outer shell of particle x. Addition of the amounts of cement embedded in the outer shells of all free particles in the system yields the amount of cement involved in the interaction process at time t_j.

3.3 Modelling of temperature effects on microstructural development

For quantification of the relationship between the curing temperature and the v-factor experimental test data of Bentur (1979) has been evaluated. In his tests on C_3S pastes Bentur has determined the capillary porosity of pastes as a function of the degree of hydration for isothermal curing temperatures of 4, 25 and 65°C (Fig. 3). The volume of the capillary porosity can be calculated with the equation:

$$V_{cp} = \frac{\omega_0}{\rho_w} - \frac{\alpha}{\rho_s} * (v(T)-1) \tag{7}$$

in which V_{cp} is the capillary porosity, ρ_s and ρ_w the specific mass of the solid phase and the water, respectively, ω_0 the w/c ratio and α the degree of hydration. In this equation the v-factor is supposed to be a function of the curing temperature. Its dependence on the curing temperature can be deduced from an evaluation of Bentur's experimental data. In this evaluation the following relationship could be established:

$$\tag{8}$$
$$v(T) = v_0 * \exp(-28\ 10^{-6} * T^2)$$

in which v_0 is the datum value of the v-factor which would hold for isothermal curing at 20°C. In accordance with literature data a value of $v_0 = 2.2$ is adopted. Eq. (8) is assumed to cover the temperature range from 0 to about 80°C.

A computation of the interparticle interaction, i.e. of the embedded cement volume, that accounts for the temperature dependency of the v-factor, will result in a lower volume of the embedded cement when hydration occurs at elevated temperatures.

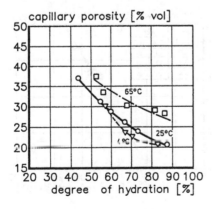

Fig. 3. Capillary porosity of C_3S pastes (Bentur (1979)).

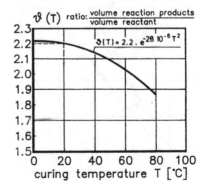

Fig. 4. v-factor as a function of temperature (van Breugel (1991)).

4 Microstructure and strength development

4.1 General Considerations
Strength has been considered to be function of the number of interparticle contact points or contact areas (van Breugel (1991)). The number
of contact points is supposed to be a universal strength parameter. The
number of interparticle contact points depends on the initial packing
density of the cement particles, i.e. the w/c ratio, the degree of hydration and the expansion coefficient v(T) (see Fig. 1). This implies
that strength can not be related to merely one of these parameters. For
example, the relationship between strength and degree of hydration is
not a universal one, but is strongly affected by the w/c ratio. This is
illustrated in Fig. 5a, where the strength of two mortars with wcr's
0.4 and 0.5 is presented as a function of the degree of hydration.

4.2   Strength as a function of the embedded cement volume
The number of contact points between hydrating cement particles can, in
a way, be calculated with HYMOSTRUC. As a first approximation it has
been assumed that the embedded cement volume, i.e. the volume of cement
that is involved in the formation of interparticle contacts, can be
considered as a measure for the number of interparticle contacts and
hence as a universal strength parameter. In Fig. 5b the measured compressive strength of mortar is presented graphically as a function of
the calculated embedded cement volume per unit mass of paste, i.e. v_{em}^{pa}.
The mortars were made with wcr's 0.4 and 0.5 and cured isothermally at
20, 32 and 44°C. The figure suggests that strength is unambiguously related to the embedded cement volume indeed and that this relationship
is hardly affected by the w/c ratio and the curing temperature.

4.3   Numerical simulation of temperature effects on strength
If it were true that the amount of embedded cement is a universal
strength parameter, it should be possible simulate the effect of the
curing temperature on the strength. This can be done by comparing the
amounts of embedded cement calculated for the same values of the degree

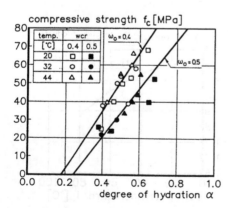
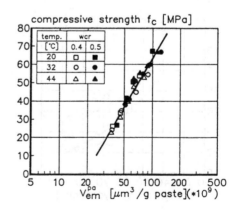

Fig. 5. Strength of semi-adiabatically cured mortars as a function of
the degree of hydration (a) and of the embedded cement volume v_{em}^{pa} (b).

of hydration but for different curing temperatures. Results of calculations of this kind are presented in Fig. 6. The ratio $\beta_{T/22}$ between the
amount of embedded cement calculated for curing temperatures of 30, 40,
50 and 65°C and the amount calculated for 22°C is presented with solid
curves as a function of the degree of hydration. These curves indicate
that isothermal curing at temperatures between 50...65°C may cause a
reduction of the amount of embedded cement up to 20 to 30%. Assuming an
unambiguous relationship between the amount of embedded cement and the
strength, this would also mean a strength reduction of the same order
of magnitude, viz. $\beta_{T/22} = v_{em}^{pa}(T)/v_{em}^{pa}(22°C) \approx f_c(T)/f_c(22°C)$.

4.4 Verification of temperature effects on strength development
Experimental evidence for a temperature-induced strength reduction has
been established in the work of, among others, Rakel and Kjellsen. Rakel (1965) has measured the 28-day compressive strength of specimen
cured at 22, 30, 40 and 50°C. The ratio between the strength reached
for curing at 30, 40 and 50°C and the strength at 22°C has been inserted in Fig. 6 with triangular points for the measured average degree of
hydration at 28 days of $\alpha \approx 74\%$ ($\alpha_{28} = 73...75\%$). There appears to be a
good agreement between these discrete points indicating the relative
strength and the theoretical strength reduction curves (solid lines).

   Kjellsen et al. (1991) has presented the compressive strength reached at curing temperatures of 5 and 50°C as a function of the degree of
hydration. The relative strength $f_c(50°)/f_c(5°)$ is inserted in Fig. 6
with solid squares. The temperature differential in Kjellsen's tests of
45°C makes these data comparable with the theoretical strength reduction curve $\beta_{65/22}$ which is based on a temperature differential of 43°C.
The fact that the data derived from Kjellsen's tests are found to be
in good agreement with the theoretical strength reduction curve $\beta_{65/22}$,
i.e. between $\beta_{50/22}$ and $\beta_{65/22}$, seems to confirm the reasonableness of
the assumed relationship between strength and amount of embedded cement.

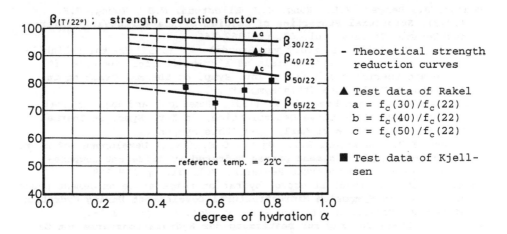

Fig. 6. Reduction factors $\beta_{T/22}$ for the influence of temperature on
strength development (van Breugel (1991), modified).

5 Discussion and concluding remarks

Curing at elevated temperature results in a reduction of the strength. The origin of this phenomenon could be explained on a microstructural basis. High temperatures result in a denser packing of the hydration products. This goes along with an increase of the capillary porosity, a corresponding reduction of the interparticle contacts and hence a reduction of the strength. Temperatures in mass concrete beyond 50°C may cause a strength reduction of 15 up to 30%. Strength reductions of this order of magnitude may well exceed the standard deviation of the tensile strength (van Breugel (1991), Eberhardt et al. (1994)) and may affect the cracking tendency substantially. For reliable predictions of the risk of cracking this temperature-induced strength reduction may, therefore, not be ignored.

Besides a lower strength and a corresponding increase of the probability of cracking a higher porosity will make the concrete more permeable and more susceptible to chemical and physical attack and hence less durable. This means that avoidance of early-age thermal cracking by redu-cing the curing temperature does not only improve the quality of the product in its early life, but also the long term durability.

With the computer-based programm HYMOSTRUC it appeared to be possible to simulate the microstructural background of macrostructural phenomena like temperature-induced strength reductions and the associated impact on the risk of thermal cracking. It is considered to be an example of the vast potential of computational materials science to bridge, bit by bit, the gap between theory and the engineering practice.

Acknowledgement

The author acknowledges with thanks the assistance of Mr. S. Lokhorst Msc for help with numerical evaluations of experimental data.

References

Bentur, A., Berger, R.L., Kung, J.H, Milestone, N.B., Young, J.F. (1979), **Structural Properties of Calcium Silicate Pastes**. J. American Ceramic Society, Vol. 62, pp. 362-366.

Eberhardt, M., Lokhorst, S.J., van Breugel, K. (1994) On the reliability of temperature differentials as a criterion for the risk of early-age thermal cracking, in **Int. Symp. on Thermal Cracking in Concrete at Early Ages** (this symposium).

Koenders, E.A.B., van Breugel, K. (1994) Numerical and Experimental Adiabatic Hydration Curve Determination, in **Int. Symp. on Thermal Cracking in Concrete at Early Ages** (this symposium)

Kjellsen, K.O., Detwiler, R.J., Gjørv, O.E. (1991) **Development of micro structures in plain cement pastes hydrated at different temperatures"**. Cement and Concrete Research, Vol. 21, pp 179-189.

Odler, I. et al. (1986) Effect of hydration temperature on cement paste structure, in **Symposium Microstructural Development During Hydration of Cement**. Materials Research Society, Vol. 85, pp. 33-38.

Rakel, K. (1965) **Beitrag zur Bestimmung der Hydratationswärme von Zement**, PhD, Aachen, 149 p.

Van Breugel K. (1991) **Simulation of hydration and formation of structure in hardening cement-based materials**, PhD, TU-Delft, 295p.

PART FOUR

MEASUREMENT OF THERMAL STRESSES IN THE LABORATORY

(Mesure en laboratoire
des contraintes d'origine thermique)

17 DEVELOPMENT OF THE CRACKING FRAME AND THE TEMPERATURE-STRESS TESTING MACHINE

R. SPRINGENSCHMID
Institute of Building Materials, Technical University of Munich, Germany
R. BREITENBÜCHER
Philipp Holzmann AG, Frankfurt, Germany
M. MANGOLD
Institute of Building Materials, Technical University of Munich, Germany

Abstract
The cracking frame and the temperature-stress testing machine were developed in order to measure the restraint stress in the young concrete for the temperature conditions as they occur on site. The cracking tendency of a concrete mix as well as the cracking sensitivity of a cement can be quantified in a cracking frame test. The deformation of the concrete is hindered to a large extent by the stiffness of the cracking frame. Therefore this testing device is able to simulate restraint conditions at early ages. The development of the Young's modulus E can be estimated from the very beginning of hydration. Tests at constant temperature allow the measurement of restraint stresses caused by early shrinkage and swelling.
Keywords: Testing Equipment, Cracking Frame, Temperature-Stress Testing Machine, Cracking Tendency, Cracking Sensitivity, Cement, Aggregates, Additives, Silica Fume, Fresh Concrete Temperature.

1 Requirement of New Laboratory Test Methods

Practical problems on site prompted the development of the cracking frame and the temperature-stress testing machine. The question of the stresses which arise in a concrete when
(1) the concrete temperature increases and
(2) the deformations are prevented
can be easily answered for hardened concrete or steel according to the equation $\sigma = E\ \alpha_T\ \Delta T$. However, this does not hold true for young concrete. The Young's modulus E increases from zero to a value of about 20 000 to 35 000 N/mm² during the first 24 hours. During the first day of hardening the concrete temperature increases even before stresses are produced. Generally valid mathematical formulae to describe the development of the Young's modulus and relaxation during the first 24 hours do not exist. To give realistic results such material laws should take into account e.g. the chemical composition of the cement (sulphate content, alkalies, C_3A content etc.).

Therefore the research was aimed at the development of new laboratory tests by which the "stress-response" of the young concrete could be measured if its thermal and nonthermal deformations were restrained. This enabled the investigation of the factors which influence thermal stresses in parameter studies. This is the necessary prerequisite to take measures against thermal cracking and to evaluate the effectiveness of these measures by comparative testing.

Thermal Cracking in Concrete at Early Ages. Edited by R. Springenschmid. Published 1994 by E & FN Spon, 2–6 Boundary Row, London SE1 8HN, UK. ISBN: 0 419 18710 3.

1.1 Crack Formation in Concrete Pavements

When the concrete pavement of the Salzburg-Vienna autobahn (Austria) was placed, many so-called "wild" transverse cracks occurred on the Mondsee construction site. The cracks occurred if the concrete cooled down by more than 15 K during the first night (e.g. due to a thunderstorm after a hot summer day) before joints were sawn. Adjacent to the Mondsee site was the construction site of Seewalchen. There the pavement was placed at the same time as on the Mondsee site. Though temperature conditions were the same, no wild cracks occurred on the Seewalchen site. Even after the joints were cut (one day after placement) it took several days until the pavement cracked underneath the cut. On the Seewalchen site the concrete was produced using different aggregates and a cement from a different plant. Did the aggregates, the cement or the curing method affect the cracking sensitivity of the concrete? How could the relevant factors be quantified? These questions could not be answered with the existing test methods.

1.2 Development of the Cracking-Frame

Obviously cracks occurred in the concrete pavement of the autobahn because the pavement could not shorten in the longitudinal direction during the cooling. It is well known that the friction between the concrete pavement and the underlaying basement is so high that most of the concrete pavement cannot move in the longitudinal direction. Laboratory experiments have to reproduce the fundamental strains and stresses which occur on the site as exactly as possible. In order to reproduce the tensile stresses which are effective in the longitudinal direction and the temperature of the concrete pavement as exactly as possible, a section of the pavement in form of a 1 metre long beam was built in the laboratory so that from the very beginning the concrete could neither shorten nor extend (fig. 1).

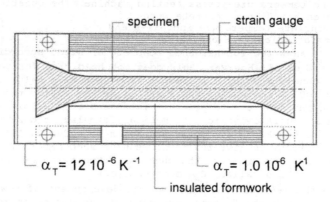

Fig. 1. The cracking frame

The ends of the beam were held by dovetails in two steel cross-heads. In order to keep the distance between the two cross-heads as constant as possible, they were connected with two steel longitudinal bars (steel and concrete approximately at equal cross-sections). A special steel (invar steel, code nr. 1.3912 according to DIN 17 007) with an extremely low thermal expansion coefficient ($\alpha_T = 1.0 \cdot 10^{-6}$ K^{-1}) was used, see RILEM Technical Recommendation (1993). The very small deformation of the steel bars was used to measure longitudinal stresses. The concrete cross-section had to be chosen quite small for practical reasons, primarily 10x10 cm², later 15x15 cm². The formwork had a thermal insulation in order to reproduce the temperature development in a thick cross-section.

The first experiments with the cracking frame were performed on concrete for road pavement in an air conditionned room of the Research Institute of the Austrian Cement Industry in Vienna. 18 hours after the placement of the concrete, the ambient air was cooled at 2 K/h. A transversal crack occurred in the beam, namely after cooling by 15 K, in agreement with the practical observations from the autobahn construction site.

1.3 Results from a Cracking Frame Test

The results of a cracking frame test can be judged by the second zero-stress temperature ($T_{z,2}$) and by the cracking temperature (T_c), fig. 2.

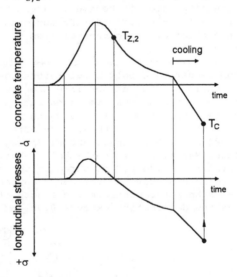

Fig. 2. Typical measurement using the cracking frame

The **second zero-stress temperature** is the temperature at which the concrete is stress free in the longitudinal direction under the conditions of the cracking frame test after the maximum temperature has occurred. The value of the second zero-stress temperature depends on the heat of hydration, the thermal expansion coefficient, the development of the Young's modulus and the relaxation of the concrete as well as on nonthermal effects such as chemical shrinkage or swelling of the cement.

The **cracking temperature** (i.e. the temperature of the concrete when it cracks) indicates the cracking tendency of a concrete mixture. A low cracking temperature means low cracking tendency. As well as the tensile strength of the concrete the same parameters as for the second zero-stress temperature are decisive for the cracking temperature.

In order to quantify the effect of the materials of the concrete, the concrete mix design and the properties of the fresh concrete on the cracking tendency, parameter studies with over 500 cracking frame tests were performed, see Breitenbücher (1989).

1.4 Development of the Temperature-Stress Testing Machine

For the construction of the 186 m high Zillergründl dam in Austria (1.37 million m³ of concrete) from 1983 to 1985, it was necessary to decide between the use of a blast furnace cement with a high slag content or an ordinary Portland cement with the addition of 33% fly ash on the site. The question proved to be controversial and was discussed by experts experienced in large dams built with a blast furnace cement as well as with an ordinary Portland cement and fly ash. It was considered whether the choice of the cement should depend only on its heat of hydration. With a higher slag content or a lower fineness of grinding, the 7 day heat of hydration could have been further reduced. In the course of the discussion it became clearer that the tensile stresses which are responsible for cracking could not be calculated from the heat of hydration of the cement or from the increase of the temperature in the concrete.

The concrete of massive structures (e.g. power plants or dams) reaches its maximum temperature only after several days. When the concrete is placed directly on the bedrock like the lower concrete layer of a dam, full strain restraint must be taken into account. In order to investigate in the laboratory this case which is of interest for theoretical reasons too, the first Temperature-Stress Testing machine (TSTM) was built 1984 by E. Gierlinger and R. Springenschmid at the Institute of Building Materials of the Technical University of Munich (fig. 3). As the strain is directly measured on the concrete, deformations of the testing machine have no influence on the measurements. A computer controlled step motor controls the length of the concrete specimen as soon as the deformation exceeds 0.001 mm.

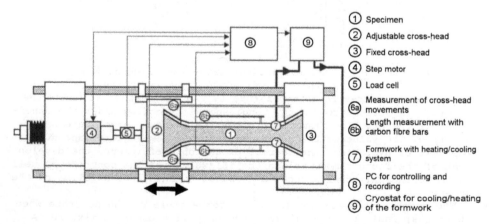

① Specimen
② Adjustable cross-head
③ Fixed cross-head
④ Step motor
⑤ Load cell
⑥a Measurement of cross-head movements
⑥b Length measurement with carbon fibre bars
⑦ Formwork with heating/cooling system
⑧ PC for controlling and recording
⑨ Cryostat for cooling/heating of the formwork

Fig. 3. The Temperature-Stress Testing machine (TSTM)

1.5 Results of Tests with the TST Machine
Fig. 4 shows the results from testing a concrete for a dam.

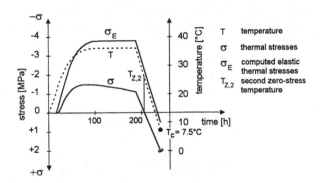

Fig. 4. Simulation of a 3 m thick concrete layer with the TST machine

The temperature development was calculated from a test with the adiabatical calorimeter, taking the heat transfer of a concrete layer of 3 m in thickness into consideration. After 96 hours the maximum temperature had been reached, the temperature was kept constant for 72 hours in order to measure the stress relaxation. Afterwards the beam was cooled at 0.5 K/h till a crack occurred at 7.5°C. The measured "stress response" of the concrete is shown in fig. 4. The elastic stress σ_E was calculated using the total temperature increase and the determined Young's modulus. The difference between σ_E and the measured stress σ is a measure for the relaxation.

2 Fundamental Knowledges from the Laboratory Experiments

2.1 Causes of the Thermal Stresses
The causes for the cracks due to the heat of hydration appear clearly from the results of the cracking frame tests. The concrete becomes its final dimensions not at the fresh concrete temperature but at the zero-stress temperature which is often considerably higher. During further cooling down to the ambient temperature, longitudinal stresses appear when the concrete cannot correspondingly shorten.

Therefore only the portion of the expansion due to the heat of hydration which is not converted into compressive stresses is unfavourable. It is mainly due to the heat of hydration which is released during the first hours when the Young's modulus is still low. The heat of hydration which is released after the first day does not give a measure of the cracking tendency. The period before the concrete reaches a high Young's modulus is decisive.

2.2 Effect of the Component Thickness on the Thermal Stresses
The stresses which develop in 40 cm and 1 m thick walls were compared using the TST machine, fig. 5. It appeared that the second zero-stress temperature is higher in thick walls than in thin walls. However this

difference is considerably smaller than the difference between the respective temperature increases. This is due to the higher temperature of the thicker wall, which takes place at a time where the Young's modulus is already high and where the relaxation is low. Therefore a substantially larger part of the restrained thermal extension is converted into compressive stresses.

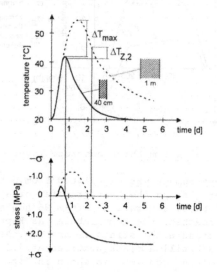

Fig. 5. 0.4 and 1 m thick concrete walls simulated in the TST machine

2.3 Comparison of Results with the Craking Frame and the TST Machine.

The cracking sensitivity of 2 cements has been determined by a series of tests with the cracking frame as well as with the TST machine. A standard concrete mix was chosen. The thermal conditions were the same in both cases. The concrete was artificially cooled after 4 days. The results are shown in table 1.

Table 1. Cracking temperature of two cements measured with the cracking frame and with the TST machine

Cracking temperature (°C)	Cement A	Cement B
TST machine	30	23
Cracking frame	21	13.5

From this follows that the cracking temperature of the cement determined with the TST machine is higher, because of the full restraint, than that determined with the cracking frame. Nevertheless the differences between the cracking temperatures of cement A and cement B are nearly the same. The reason for this is that the determining period is in the first 12 to 18 hours. During this period, the temperature curves and the degree of restraint were respectively nearly the same. When technological influences have to be studied, the results of the cracking frame tests can be used as an approximation applicable to the concrete of thick structures.

Based on new experiments a model was developed which allows the calculation of the stress curve corresponding to another degree of restraint from the stress curve measured with the cracking frame, see Mangold (1994). In this model the cracking frame is represented by an elastic spring which is firmly attached to the concrete beam (fig. 6). The stresses calculated in such a manner correlate very well with the stresses measured with the TST machine.

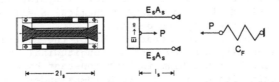

Fig. 6. Mechanical model of the cracking frame

3 Practical Application

3.1 Cracking Tendency of the Concrete Mix
Tests with the cracking frame are above all performed in order to determine the cracking tendency of concrete mixes in comparative test series. The lower the cracking temperature of the concrete in the cracking frame test, the lower is the cracking tendency of the concrete. The comparative tests are performed under well defined experimental conditions: the fresh concrete temperature is 20°C, the ambient air temperature is set to 20°C. The concrete hardens under semi-adiabatic conditions, in a thermally insulated formwork and water loss due to drying is prevented for the duration of the test. After 4 days the concrete is cooled at 1 K/h until it cracks.

3.2 Cracking Temperature of the Cement
In order to investigate the influence of the cement on the cracking tendency of the concrete, a standard concrete mix is used for the comparative tests. The composition for 1 m³ concrete is: cement 280 kg, fly ash 60 kg, water (and superplastisizer) 162 l, sand from the Main 646 kg and basalt chippings 1343 kg. Favourable cements with a low cracking tendency show with this standard concrete mix a concrete temperature of 10°C maximum. As well as for the cement, the influence of the aggregate, of the additives (e.g. silica fume) or of the fresh concrete temperature can be judged in comparative tests.

3.3 Isothermal Tests
With the cracking frame as well as with the TST machine the heat of hydration can be removed from the very beginning with suitable cooling so that the concrete hardens under nearly isothermal conditions. The stresses in the concrete which are measured in this cases are due to restrained nonthermal deformations. They are due to:
(1) The chemical shrinkage caused by the volume reduction of the hydration products of the cement, which is particularly pronounced for high strength concretes with silica fume (self-desiccation).
(2) The chemical swelling, obviously controlled by the sulphate phase of the cement.

4 Conclusions

(1) The cracking frame and the TSTM are used to measure the stresses which occur in young concrete under restraint. At the beginning of the test, the fresh concrete is placed in a formwork where dimensions are 15x15x ~150 cm³. During hydration, the temperatures and the stresses in the longitudinal direction are measured.

(2) The causes of the stresses are the heat of hydration, chemical shrinkage (self-desiccation) and/or chemical swelling due to a modification of the sulphate phase.

(3) In the cracking frame the deformations of the concrete are prevented to a large extent by two strong longitudinal steel bars. In the TSTM, the length of the beam is kept constant by a control device.

(4) In the cracking frame the concrete hardens under semi-adiabatic conditions and is artificially cooled in the standard test after it has reached the ambient temperature at an age of 4 days till it cracks.

(5) The lower the temperature of the concrete when it cracks (cracking temperature), the lower is the cracking tendency of the concrete.

(6) The influence of the cement (composition, fineness), the aggregate, the additives, the concrete mix design and the fresh concrete temperature on the cracking sensitivity of the concrete can be quantified in comparative tests with the cracking frame.

(7) The 7 day heat of hydration of the cement is not a criterion to estimate the thermal stresses or the cracking sensitivity of the concrete.

(8) As long as the concrete has a low Young's modulus (roughly during the first 12 to 18 hours) the temperature increase due to the heat of hydration leads to only low compressive stresses, which will be reduced by relaxation. Therefore the compressive stresses which develop for a temperature change of 1 K during the heating at the beginning are considerably lower than the tensile stresses which develop for the same temperature change during the later cooling.

(9) As soon as the Young's modulus of the concrete is high enough and the relaxation has reduced to a low value (after approximately 1 to 3 days) the major part of the heat of hydration is converted into compressive stresses. The heat of hydration which is released after 1 to 3 days has only a small effect on the cracking sensitivity of the cement.

5 References

Breitenbücher, R. (1989) Zwangsspannungen und Rißbildung infolge Hydratationswärme. Doctoral Thesis, Technical University of Munich.

Breitenbücher, R. and Mangold, M. (1994) Minimization of Thermal Cracking in Concrete Members at Early Ages, in **Thermal Cracking in Concrete at Early Ages** (editor R. Springenschmid), Chapman & Hall.

Mangold, M. (1994) The development of restraint and intrinsic stresses in concrete members during hydration, Doctoral Thesis, Technical University of Munich.

RILEM Technical Recommendation (1993) Testing of the Cracking Tendency of Concrete at Early Ages, 2nd Draft, December 1993.

18 INVESTIGATION OF CONCRETE BEHAVIOUR UNDER RESTRAINT WITH A TEMPERATURE-STRESS TEST MACHINE

G. THIELEN and W. HINTZEN
Forschungsinstitut der Zementindustrie, Düsseldorf, Germany

Abstract
The factors affecting crack formation in centrally restrained concrete components during the loss of the heat of hydration of the cement were investigated in a temperature-stress test machine. The important influencing factors are described. Concrete technology measures for reducing the heat development, and hence also the temperature rise in the component, are in conflict with the strength development necessary for cost-effective progress of work, for example in tunnel inner shells using in-situ concrete. Therefore in addition to the development of the heat of hydration the development of the early-age strength was also investigated on cylindrical specimens using normal commercial heat-insulated containers.
Keywords: Concrete Composition, Development of Strength, Elastic Modulus, Heat of Hydration, Relaxation, Restraint, Semi-adiabatic Calorimetry, Temperature-Stress Test Machine.

1 Introduction

The investigations reported in this paper aim at the evaluation of the main influences governing the behaviour of concrete at early age under restraint conditions. More detailed results of these investigations are reported in Thielen and Hintzen (1994). In addition to the experimental analysis of concrete behaviour under restraint conditions using a temperature-stress test machine, the development of the heat of hydration and the development of the early-age strength were investigated. The influence of intrinsic shrinkage, which exists especially for concretes with w/c-ratios below ≈0.50 is not regarded here, see Grube (1990).

2 Experimental analysis of concrete behaviour under restraint conditions

2.1 Temperature-stress test machine
A laboratory equipment for the measurement of thermal stresses in concrete at early age, a so called temperature-stress test machine, was built at the Research Institute of the Cement Industry, Germany (see Fig. 1). By means of a temperature controlled mould, which is placed in a horizontal compression-tension-testing machine, it is possible to

Thermal Cracking in Concrete at Early Ages. Edited by R. Springenschmid. Published 1994 by E & FN Spon, 2–6 Boundary Row, London SE1 8HN, UK. ISBN: 0 419 18710 3.

Fig. 1. Temperature-stress test machine

run any course of temperature in the hardening concrete beam with a cross section of 150x150 mm^2 and a length of 1200 mm. Thus the course of temperature development in any concrete member in practice can be simulated. The deformations of the concrete beam are measured with optical strain recorders not in direct contact with the test specimen. The precision is 1 μm over a measuring distance of 600 mm. The measurement starts with the setting of the concrete. A control program is able to vary the central restraint of the concrete beam between 0% and 100%. The stress history due to the temperature rise or decrease resp. is continually registered by means of a force transducer mounted at one head of the specimen against the testing frame. The evolution of the modulus of elasticity (secant modulus, measured up to an compressive strain of 25 μm/m) is measured seperately in order to differentiate the elastic and inelastic part of the deformation.

The heat liberation characteristic of the concretes under investigation was measured in a semi-adiabatic test using commercial heat-insulated containers, Grube and Hintzen (1993). This practical test method will be published by TC-119 TCE as a RILEM Technical Recommendation. The energy balance is made up in consideration of the heat development, the heat capacities of the system and the energy losses to the environment. The complete calculation of the hydration heat is done in terms of temperature changes, which means the property also relevant in real structural components. The sample holder used in the thermos vessel is also suitable for making cylindrical specimens for strength measurements, so that the development of strength esp. of early-age strength can be linked to the heat of hydration.

2.2 Description of the behaviour of concrete under restraint conditions as measured in the temperature-stress test machine

With the start of hardening, the thermal expansion of the concrete as it heats up leads to slowly increasing compressive stresses in the restrained component due to the increasing elastic modulus of the young concrete (see Fig. 2). Compared to the purely elastic behaviour the compressive stresses are reduced because of plastic deformations and stress relaxation, which show a considerable influence at this age (Fig. 2b, continuous line). The dotted line showing the ideal elastic course of stress has been calculated on the basis of the modulus of elasticity measured seperately as shown in Fig. 2c. The maximum compressive stress occurs shortly before the maximum of the component temperature T_{max} is reached. Because of the higher elastic modulus at this point the subsequent cooling leads to a rapid decrease in the compressive stress, so that the compressive stress is completely discharged only a few Kelvin below the maximum temperature, i.e. the stress becomes zero (so-called second zero stress temperature T_{02}, Springenschmid and Nischer (1973)). In the example of Fig. 2 the temperature difference $T_{max} - T_{02}$ is only about 8 % of the maximum temperature rise. Further cooling results in tensile loads which, on reaching the ultimate tensile strain (tensile strength) at the cooling temperature T_{crack}, leads to tensile cracking. The difference between T_{02} and T_{crack} characterizes the tensile strain capacity of the concrete for a given degree of restraint to movement. It amounts to 8 to 12 Kelvin with fully restricted movement (100% degree of restraint), which with a coefficient of thermal expansion of $\alpha_T = 1.2 \cdot 10^{-5}$ 1/K corresponds to a tensile strain at failure of 0.01 to 0.015%. In parallel tests on unrestrained specimens the influence of heating and cooling sequences on the coefficient of thermal expansion has been investigated. It could be demonstrated that a value of $\alpha_T = 1.2 \cdot 10^{-5}$ 1/K represents well the

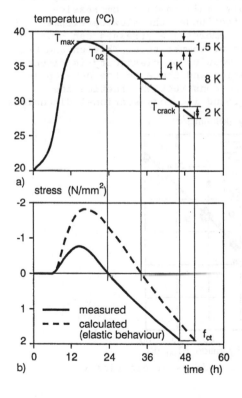

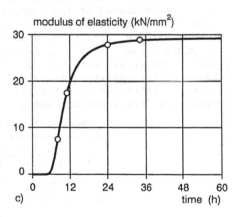

Fig. 2. Development with time of temperature rise (a), stress (b), and elastic modulus (c) during a trial with full restriction of movement

behaviour during the cooling phase of concrete with the silicious type
of aggregates used in these tests, Thielen and Hintzen (1994).

Because of plastic deformations and because of stress relaxation the
maximum compressive stress is much lower than the value which could be
expected for a relaxation-free, ideally elastic concrete (broken-line
stress curve in Fig. 2). The temperature differenz $T_{max} - T_{02}$ of the
relaxation-free, ideally elastic concrete is at least about 4 Kelvin
higher, but also reaches only about 30% of the total temperature rise.

The concrete also shows a relaxation capability in the tensile
loading after passing through the second zero stress temperature, which
can be seen from the somewhat flatter curve of the measured tensile
stress (stress curve shown by continuous line in Fig. 2) when compared
to the calculated elastic behaviour (broken-line stress curve). It
produces a somewhat lower increase in tensile stress during cooling,
but this compensates only slightly for the unfavourable influence of
the relaxation on the build-up of the compressive stress.

Such examinations of the influence of plastic deformations and
relaxation were carried out for large numbers of concretes under
investigation. As a result, it can be stated that although the
participation of plastic deformations and relaxation do influence the
stress development in the trials to a measurable extent they are of
very small importance in relation to the risk of cracking. This
applies to an even greater extent for slight differences in relaxation
behaviour of different concretes. This small influence is due to the
fact that the risk of cracking depends predominantly on the temperature
rise. The influence of the comparably small initial compressive
stresses is of minor importance. In fact under the slow development of
the modulus of elasticity only small compressive stresses can develop
even under a reasonably high increase of temperature, Thielen and Grube
(1990). The growth with time of the modulus of elasticity is directly
related to the progress of the hydration and hence to the development
of heat of hydration (see Fig. 3). In construction practice the
compressive "prestressing" which is found in cross-sectional dimensions

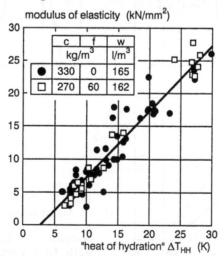

Fig. 3. Heat of hydration and modulus of elasticity

of about 40 cm is very largely meaningless because of its low value - it corresponds on average to 2 K.

2.3   Influence of restriction of movement (degree of restraint) on the development of the tensile stresses

The degree of restriction of movement exerts the decisive influence on the restraint stresses as they result from the loss of heat of hydration. It can be seen in Fig. 2a that with fully restricted movement the concrete under investigation can only be cooled by about 8 Kelvin below the second zero stress temperature T_{02}. At a coefficient of thermal expansion $\alpha_T = 1.2 \cdot 10^{-5}$ 1/K this corresponds, as mentioned in Section 2.2, to a tensile strain at failure ε_{ctu} of about 0.01%.

If the degree of restraint to movement is reduced, i.e. some of the tendency to deform can occur without stress (free from restraint), then there is a corresponding increase in the possible cooling before cracking occurs. Fig. 4 shows the measured deformation, designated as ε_{free}, which was able to occur without restraint during the cooling from T_{02} to T_{crack} in the trials and which corresponds to only elastic, i.e. non-rigid, restriction of the movement by the test frame. As this movement reduces the stress-generating proportion of the total movement, it is plotted in Fig. 4 with a negative sign. If the restraint-free movement is zero ($\varepsilon_{free} = 0$) then full movement restriction occurs. The slope of the straight line plotted corresponds to the measured coefficient of thermal expansion of the concretes under investigation. This straight line - notionally extended into the range above the rigid restraint - intersects the y-axis at the mean value of the tensile strain at failure ε_{ctu} of 0.009%. A reduction in the restriction of the movement to 50% doubles the possible cooling [$T_{02} - T_{crack}$], a reduction to 33% triples it. The proportion of deformation which was restricted during cooling under all support conditions up to cracking, i.e. which resulted in tensile stresses and led to cracking, was consistently approximately 8 Kelvin. The behaviour of concretes with the same coefficient of thermal expansion and with the same tensile strain at

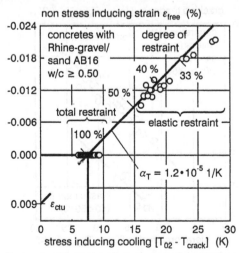

Fig. 4. Degree of restraint and tolerated cooling

Table 1. Summary of parameter variations

Cements	portland cements PZ 35F, PZ 35L blastfurnace slag cement HOZ 35L		
Aggregate	Rhine gravel sand		
Concrete composition	1	2	3
Cement content (kg/m³)	330	300	270
Fly ash content (kg/m³)	0	0	60
Water content (l/m³)	165	180	162
Fresh concrete temperature (°C)	12 to 25		
Ambient temperature (°C)	12 and 20		

failure can be expected to show a direct relationship between the cracking temperature and the maximum temperature rise above the ambient temperature. This relationship is plotted in Fig. 5 for the concretes under investigation. In spite of the wide variation in parameters shown in Table 1 (type of cement, concrete composition, ambient temperature and fresh concrete temperature) the trials confirm the long-known fact that the cracking tendency of a concrete component under external restraint conditions increases approximately in proportion to the highest temperature reached, Breitenbücher (1989). For the concretes under investigation with $\alpha_T \approx 1.2 \cdot 10^{-5}$ 1/K and $\varepsilon_{ctu} \approx 0.009\%$ the crack occurs at approximately 10 Kelvin for full restraint to movement, and at approximately 20 Kelvin below the maximum temperature for a degree of restraint of 40 to 50%. For concretes with lower coefficients of thermal expansion these spacings become larger for the same tensile strain at failure. To reduce the risk of cracking

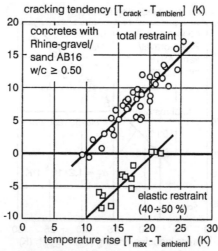

Fig. 5. temperature rise and tendency to crack

it is therefore primarily important to reduce the degree of restraint.
For a given degree of restraint those concretes with smaller
temperature rise and low thermal expansion are preferable to further
reduce the risk of cracking.

Generally the latter objective conflicts on economic grounds with
the required progress of construction work. Take the example of tunnel
constructions where a cycle time of 24 hours can only be maintained if
the formwork can be removed from the component after only 9 to 16
hours. This means that by this time the concrete must have sufficient
compressive strength to be able to support the load due to its own dead
weight safely and without cracks caused by insufficient strength.
Release strengths of approximately 3 to 6 N/mm^2 are normally required
for this in practice.

3 Relationship between early-age strength and heat of hydration

In addition to the investigations into the development of the heat of
hydration and the behaviour of the concretes under restraint, the
compressive strengths of 100/200 cylinders were measured at the age of
12 hours on test pieces which had also been exposed to the temperature
cycle of a component approximately 40 cm thick. In Fig. 6 the
compressive strength at the age of 12 hours is plotted against the heat
of hydration developed up to this age expressed as the temperature
change in the concrete (intrinsic temperature rise); this covers 6
different cements, two differing concrete compositions (1 and 3 in
Table 1), two ambient temperatures (12, 20°C) and fresh concrete
temperatures of 12 to 25°C. A relatively strong correlation is
apparent between the heat development and the compressive strength at
the same age of 12 h. Concretes with the same cement lie both above
and below the fitted curve which has been drawn in. The early-age
strength defined here in accordance with practical demoulding
requirements as 12 hours strength increases roughly in proportion to

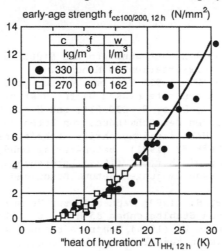

Fig. 6. Heat of hydration and early-age strength

the square of the heat of hydration. The two concrete compositions under investigation have no detectable influence on the heat - strength relationship. This is because the lower heat development caused by a reduction in the cement content by approximately 20% with a simultaneous increase in w/c-ratio from 0.5 to 0.6 led to a decrease in the early-age strength of approximately 40%. So the required early-age strength e.g. at 12 h is the main aspect for the choice of adequate concrete composition for this use. It corresponds automatically to the lowest temperature rise (cracking temperature) possible, if the needed strength is not essentially exceeded.

4 Conclusions

The behaviour of early-age concrete under restraint conditions is on principle governed by the degree of restraint and the temperature rise of concrete during hydration. This paper shows that the influence of possible compressive stresses during the temperature rise on the risk of cracking is very small as compared to the tensile stress developing during cooling. The temperature rise depends on the heat development in concrete as it results from the heat of hydration of the cement and is strongly influenced by the concrete constituents and mix proportions.

Further influences on the risk of cracking are the thermal expansion of concrete - primary governed by the aggregates, see Thielen and Hintzen (1994) - and the tensile deformation capacity of concrete. The latter is basically independent of possible parameter variations in the concrete composition and amounts for most concretes to the very small limit of about 0.01%. Important for the practical optimization of concrete mix design for concrete components under restraint conditions is not only the limitation of the temperature rise but also the early-age strength development. Therefore the correlation between these two characteristics limits the parameter range in which optimal selections for the concrete mix design can be found, Thielen and Hintzen (1994).

5 References

Breitenbücher, R. (1989) Zwangspannungen und Rißbildung infolge Hydratationswärme, Dissertation TU München.

Grube, H. und Hintzen, W. (1993) Test method for predicting the temperature rise in concrete caused by the heat of hydration of the cement. **Beton** 43, H. 5, S. 230/234, und H. 6, S. 292/295.

Grube, H. (1990) Ursachen des Schwindens von Beton und Auswirkungen auf Betonbauteile, Habilitationsschrift, Schriftenreihe der Zementindustrie, Heft 52 (1991), Beton-Verlag GmbH, Düsseldorf.

Springenschmid, R. und Nischer, P. (1973) Untersuchungen über die Ursache von Querrissen im jungen Beton. **Beton-und Stahlbetonbau** 68, H. 9, S. 221/226.

Thielen, G. und Grube, H. (1990) Maßnahmen zur Vermeidung von Rissen im Beton. **Beton- und Stahlbetonbau** 85 H. 6, S. 161/167.

Thielen, G. und Hintzen, W. (1994) Concrete technology measures for avoiding cracks in tunnel inner shell concrete. **Beton** 44, H. 9 u. H. 10.

19 DETERMINATION OF RESTRAINT STRESSES AND OF MATERIAL PROPERTIES DURING HYDRATION OF CONCRETE WITH THE TEMPERATURE-STRESS TESTING MACHINE

K. SCHÖPPEL
Ingenieurbüro, Munich, Germany (formerly at Institute of Building Materials)
M. PLANNERER and R. SPRINGENSCHMID
Institute of Building Materials, Technical University of Munich, Germany

Abstract
The temperature-stress testing machine was developed to measure continously the longitudinal stresses in a concrete beam specimen during early hardening under full restraint. The temperature conditions for the specimen can be controlled isothermal, semiadiabatic or adiabatic. Young´s modulus and relaxation of the hardening concrete can also be estimated. The degree of restraint can be set variable so conditions like they occur in concrete structures could be simulated realistically. Further the influence of non-thermal deformations due to chemical shrinkage and swelling on the development of thermal stresses can be estimated.
Keywords: Testing Equipment, Heat of Hydration, Chemical Shrinkage and Swelling, Young´s modulus, Relaxation.

1 Introduction

The first experimental investigations of the cracking tendency of young concrete took place 50 years ago. The different laboratory methods of testing stresses due to restraint which are used all over the world are described by Springenschmid and Adam (1980). Rings or beams were widely used as specimens for the determination of restraint stresses see Schrage et al. (1989). The most important advantage of the beam specimens is the ability to measure compressive and tensile stresses under uni-axial conditions. To investigate the restrained stresses due to hydration heat, beam specimens are more suitable than ring specimens because the temperature distribution over the cross-section is almost constant. Furthermore also compressive stresses are easier to control.

At the Institute of Building Materials the cracking frame was improved and the first temperature-stress testing machine was developed in 1983 (Springenschmid et al. (1984)) and used for experimental determination of the restraint stresses see Breitenbücher (1989). The difference between these two testing devices is the degree of restraint. The degree of restraint can be modified with the temperature-stress testing machine as opposed to the cracking frame where the degree of restraint depends on the stiffness of the steel frame and of the concrete.

Thermal Cracking in Concrete at Early Ages. Edited by R. Springenschmid. Published 1994 by E & FN Spon, 2–6 Boundary Row, London SE1 8HN, UK. ISBN: 0 419 18710 3.

2 Temperature-Stress Testing Machine

Fig. 1 shows systematically the construction of the machine. The concrete is freshly casted into the framework of the testing machine. The length of the beam is approximately 1500 mm and the cross-section is 150 x 150 mm². In contrast to the cracking frame, one cross-head is adjustable the other is rigidly connected to the massive longitudinal steel bars. The ends of the beam are fixed to the cross-heads. The

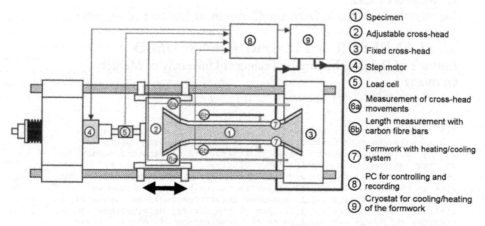

① Specimen

② Adjustable cross-head

③ Fixed cross-head

④ Step motor

⑤ Load cell

⑥ₐ Measurement of cross-head movements

⑥ᵦ Length measurement with carbon fibre bars

⑦ Formwork with heating/cooling system

⑧ PC for controlling and recording

⑨ Cryostat for cooling/heating of the formwork

Fig. 1: Schematic construction of the temperature-stress testing machine

formwork can be cooled or heated so it is possible to simulate temperature developments as they occur in structural members of any thickness. On the lateral sides of the concrete beam the longitudinal deformations of the concrete are registered continuously with an accuracy of 0.1 µm. Since the concrete has no sufficient stiffness in the first few hours, the relative movement between the cross-heads is continuously registered by the outer measuring system (see 6a in Fig. 1). As soon as the concrete has sufficient stiffness, i.e. when stress initially occurs, the lateral and upper formwork are pulled back 5 mm. The inner measuring system is applicated on steel bars embedded in the concrete (see 6b in Fig. 1). To control the adjustable cross-head, only the inner measuring system is used. The adjustable cross-head is moved against the change of length if the mean relative movement between the embedded steel bars is greater than 1 µm. The base length of this measuring system is 500 mm. Hence, the restraint of the longitudinal deformation is controlled by the movements of the cross-head. The force due to the restraint is registered continuously by a load cell.

In table 1 an overview is given of possible methods to determine the material properties and restraint stresses depending on the degree of restraint and on the temperature conditions.

Using a variable degree of restraint in the temperature-stress testing machine the stresses in concrete structural members can be determined realistically. If the concrete hardens under isothermal conditions the influence of non-thermal effects due to chemical shrinkage and/or swelling can be estimated.

Table 1. Methods for the determination of material properties and restraint stresses depending on the degree of restraint and on the temperature conditions.

Temperature condition during hardening	Restraining conditions	
	0%	variable 0% < δ < 100 %
Semiadiabatic Simulation of structural members with different thickness	Thermal and non-thermal unrestrained deformations	Thermal and non-thermal restrained stresses Development of Young´s modulus Tensile strength and failure strain
Constant temperature (isothermal)	Non thermal unre-strained deformations	Non-thermal restrained stresses Relaxation of tensile/compressive stresses Development of Young´s modulus Tensile strength and failure strain

3   Tests with the Temperature-Stress Testing Machine

The concrete is casted into the polyethylene-lined formwork and compacted with an internal vibrator. The concrete is then wrapped with plastic foil to prevent it drying out.

3.1 Control of Temperature

In the case of isothermal tests (constant temperature), the formwork is cooled in the first few hours to remove the heat of hydration and to keep a constant temperature in the specimen.

In order to simulate the temperature development in structural members, the adiabatic temperature development must be determined in an adiabatic calorimeter. A finite elements program is then used to calculate the temperature distribution in concrete members of different thickness and for different environment conditions. The formwork of the testing machine is then heated according to the calculated temperature curve.

3.2 Determination of Thermal Stresses

The thermal stresses in the longitudinal direction are measured either under full restraint (100 %) or under variable restraint depending on the development of stiffness. For instance at full restraint the relative displacement is shifted back to zero. Every controlling step of the cross-head causes restraint forces which are continuously measured by a load cell. The single controlling steps and the resulting stresses are schematically shown in Fig. 2 see Schöppel (1993). Fig. 3 shows measured restraint stresses due to semiadiabatic and to isothermal temperature conditions for one typical concrete mixture see Schöppel (1994).

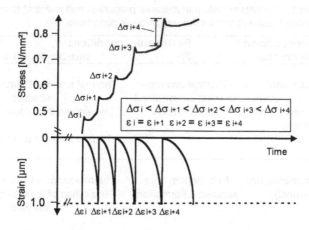

Fig. 2: Schematic stress-strain relationship due
to the controlling steps in the TST machine.

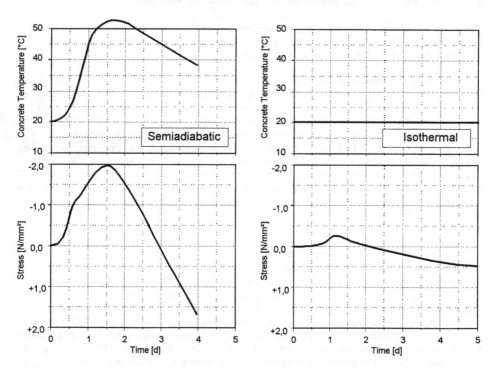

Fig. 3: Longitudinal restraint stresses under semiadiabatic (1 m thick
wall) and isothermal conditions for one specific concrete composition.
Cement content: 340 Kg/m^3, ω = 0,5 , OPC 35 F, plant D

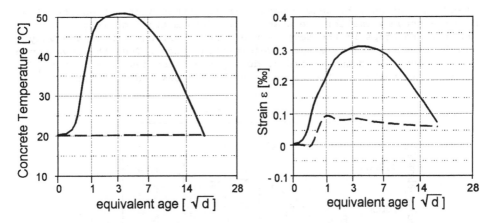

Fig. 4: Unrestrained deformations under semiadiabatic and isothermal
conditions for one specific concrete composition
Cement content: 340 Kg/m³, ω = 0,5 , OPC 35 F, plant D

3.3 Determination of the Unrestrained Strains

In order to determine unrestrained strains the same testing procedure
is applicable, but with a differently controlled cross-head. Not the
relative movement of the cross-head, but the restraint force is shifted
back to zero as soon as the restraint force exceeds 250 N (i.e.
equivalent to 0,01 N/mm² concrete stress).

In Fig. 4 examples of unrestrained longitudinal deformations of the
same concrete are given under semiadiabatic and under isothermal
conditions.

4 Determination of the Young´s modulus with the Temperature-Stress
   Testing Machine and Comparisons with Other Methods

During the adjustment of the length of the specimen the longitudinal
force will be measured (see Fig. 2). For every single controlling step
the Young´s modulus can be determined. The Young´s modulus is
calculated from the re-deformation and the corresponding force. The re-
deformation is in case of "full" restraint constant at 0.002 ‰.

The Young´s modulus determined in the temperature-stress testing
machine is equivalent to the tangential Young´s modulus at the origin
of the stress-strain curve. As shown in Fig. 5 the tangential Young´s
modulus as determined in the temperature stress testing machine is
somewhat higher than the static (secant) Young´s modulus according to
the german standard test (DIN 1048). Ultrasonic testing gives a even
higher Young´s modulus see Schöppel (1993).

The difference between these test methods decreases continuously
during the first few days. These differences are mainly affected by the
large viscose and visco-elastic deformations of young concrete. With
increasing strength the part of visco-elastic deformations decreases
compared to the elastic deformations so that the difference between the

standard test and the determination in the temperature-stress testing machine becomes smaller. Because the re-deformations during the controlling steps in the temperature-stress testing machine are very small (0,002 ‰) mainly elastic deformations occur. Thus the development of the Young´s modulus is faster during the first few days if the concrete is tested in the temperature-stress testing machine.

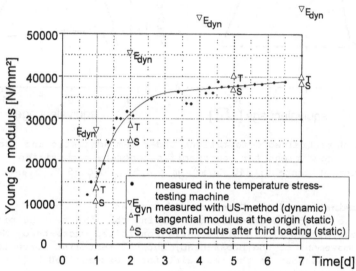

Fig. 5: Development of the Young´s modulus depending on different test methods

5 Computation of the Elastic Restraint Stresses under Semiadiabatic Conditions - Relaxation and other Effects.

By means of the developments of temperature and of the Young´s modulus the elastic restraint stresses $\sigma(t)_{elast.}$ can be computed as a function of time as follows.

$$\sigma(t)_{elast.} = \int \alpha_T \cdot dT/dt \cdot E(t) \, dt \qquad (1)$$

with α_T coefficient of thermal expansion
$E(t)$ Young´s modulus at time t
dT/dt temperature change with respect to time.

The calculated elastic restraint stress $\sigma(t)_{elast.}$ differs strongly from the measured stresses $\sigma(t)_{meas.}$ due to the high relaxation $\psi(t)$ and because of non-thermal stresses caused by chemical shrinkage and swelling. The following formula describes the development of restraint thermal stresses in young concrete including the non-thermal effects see Schöppel (1993).

$$\sigma(t)_{meas.} = \int (\alpha_T \cdot dT/dt + d\varepsilon_{S,Sw}/dt) \, E(t) \, [1 - \psi(t)] \, dt \qquad (2)$$

with $\psi(t)$ relaxation at time t
$d\varepsilon_{S,Sw}/dt$ non-thermal deformations due to chemical shrinkage (s) and/or swelling (sw) with respect to time.

Fig. 6 shows schematically the elastic stress curve calculated on the basis of the development of concrete temperature, the coefficient of thermal expansion and the Young's modulus and the measured stress development of young concrete. For comparison, the calculated elastic stress for hardened concrete with E = 40000 N/mm² is also shown.

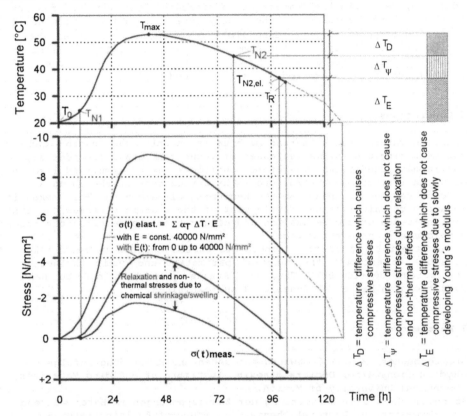

Fig. 6: Schematic development of temperature in a 1m thick concrete wall and corresponding stresses under full restraint – Comparison of calculated and measured stresses.

In young concrete with growing Young's modulus only a small compressive stress occurs at the maximum temperature. If the same temperature rise would be applied to hardened concrete the compressive stress would be as much as 9 N/mm². Thus in young hardening concrete more than roughly 50 % of the temperature deformations could not result in compressive stress due to the slowly increasing Young's modulus.

The difference between the calculated elastic stress curve for young concrete with the measured growing Young's modulus and the measured stress curve depends on the stress relaxation and on non-thermal effects.

Only a small part of the total temperature rise causes compressive stresses in concrete. This part is equivalent to the temperature difference $\Delta T_D = T_{max} - T_{N2}$ and usually does not exceed 20 % of the total temperature rise (for a 1 m thick structure). The remaining temperature rise does not lead to compressive stresses because the relaxation is high and the Young´s modulus develops slowly. The difference $\Delta T_E = T_{N2,el.} - T_0$, where T_0 is the temperature of fresh concrete, describes the temperature difference which does not results in compressive stress due to the slowly developing Young´s modulus. The remaining temperature difference $\Delta T\psi = T_{N2} - T_{N2,el.}$ describes the effect of the relaxtion on the compressive stress generation. Also non-thermal effects superimpose so in case of chemical shrinkage the compressive stresses are additionally reduced.

6 Conclusions

The temperature-stress testing machine is an universal testing machine for measuring the thermal stresses of young concrete also under any restraining conditions. Therefore conditions can be simulated like they occur in real concrete structures. The development of the Young´s modulus can be measured from the beginning of hardening. This makes it possible to calculate the development of the elastic stresses, so that the influence of the relaxation and of non-themal effects on the stress development can be estimated.

The flexibility of the temperature-stress testing machine makes it possible to estimate the material parameters and their influences on the stress development. Thus the stress generation due to heat of hydration can be better understood and explained.

7 References

Breitenbücher, R. (1989) Zwangsspannungen und Rißbildung infolge Hydratationswärme, **Doctoral thesis**, Institute of Building Materials, Technical University of Munich.

Schöppel, K. (1993) Entwicklung der Zwangspannungen im Beton während der Hydratation, **Doctoral thesis**, Institute of Building Materials, Technical University of Munich.

Schöppel, K. (1994) The effect of thermal deformations, chemical shrinkage and swelling on restraint stresses in concrete from the beginning of hydration, in: **Proceedings of the international symposium, Thermal cracking of concrete at early ages**, Munich, October, 10-12, 1994.

Schrage, I.and Breitenbücher,R. (1989) Experimentelle Ermittlung lastunabhängiger Spannungen in jungem Beton in **Betonwerk + Fertigteil-Technik**, H.11, pp.48-55.

Springenschmid, R. and Adam, G. (1980) Mechanical Properties of Set concrete at Early Ages. Test Methods, in **Matériaux et Constructions**, Vol. 13, No. 77 pp. 391 -399.

Springenschmid, R. Gierlinger, E. and Kiernozycki, W. (1985) Thermal stress in mass concrete: A new testing method and the influence of different cements, in **Proceedings of the 15th Intern. Congress for large dams**, Lausanne, R4, pp.57-72.

20 MATERIAL CHARACTERIZATION OF YOUNG CONCRETE TO PREDICT THERMAL STRESSES

A. NAGY and S. THELANDERSSON
Department of Structural Engineering, Lund University, Lund, Sweden

Abstract
This paper is concerned with thermal stresses in concrete at early ages generated by the heat of hydration. It is shown that thermal stresses can be estimated successfully using a very simple constitutive model describing the behaviour of the young concrete. Regarding the development of thermal stresses the most important material characteristic is the early age stiffness of the concrete. Effects like creep and plastic behaviour are of minor importance by comparison. To determine concrete stiffness a simple non-destructive method is used. The method is based on measurements of the dynamic resonance frequency and damping in a concrete prism using a FFT-analyser. Comparison with static tests shows that the method gives a reliable measure of the stiffness development of young concrete. The advantage is that the method enables measurement of progressive changes in the stiffness of the concrete without damaging the specimen. The method can easily be used to characterize concrete used in practice.
Keywords: Dynamic Resonance Frequency, Elastic Modulus, Heat of Hydration, Modelling of Thermal Stress.

1 Introduction

It is normally not an easy task to assess the risk of cracking in concrete due to thermal stresses induced by heat of hydration. The problem consists essentially of two parts. The first one is to predict the time-dependent temperature distribution in the concrete structure during the hydration phase. Utilizing modern computer programs, this can be done with relatively high precision, considering the actual conditions during concrete production on site, Dahlblom (1990), Jonasson (1988), Munch-Petersen et al (1992) . The second part of the problem is to estimate the thermal stresses developed during hydration, as a consequence of a given time-dependent temperature field within the structure.

Traditionally, crack control in practice has often been implemented by limiting the maximum temperature rise and/or the maximum temperature difference within the structure during hydration. This method is not very reliable, however, and may in many cases lead to erroneous results, implying that cracking may occur even if the temperature requirements are fulfilled or that expensive, unnecessary measures are undertaken during construction, Emborg and Bernander (1993).

Thermal Cracking in Concrete at Early Ages. Edited by R. Springenschmid. Published 1994 by E & FN Spon, 2–6 Boundary Row, London SE1 8HN, UK. ISBN: 0 419 18710 3.

Stresses induced by temperature change depend to a great extent also on the geometry and mechanical boundary conditions of the structure. Another important factor is the strong age dependence of the mechanical properties of young concrete. This dependence may be quite different for different mixes and cement types. The thermal stresses developed in the concrete are highly sensitive to the rate at which certain properties change with age. It is generally advantageous if the growth of stiffness in the fresh concrete is as rapid as possible, Nagy and Thelandersson (1992). On the other hand the heat release due to hydration should be as late as possible. During recent years, sophisticated analysis methods have been developed, which make it possible to simulate the evolution of thermally induced stresses in a concrete structure from the time of casting and onwards, e. g. Emborg (1989), Dahlblom (1992). These programs can also represent the boundary conditions in the structure in a fairly realistic manner, thus enabling a correct estimation of the degree of constraint exerted on the concrete.

Such methods of stress analysis may be very valuable tools in the design and plan-ning of the construction work for heavy concrete structures where the requirements with respect to crack control are severe (bridges, tunnels, dams, etc.). One problem with many of the existing computer codes, however, is that extensive information is required about various concrete properties and their age dependence, and this information is usually not available in practice. Therefore, it is desirable to identify which properties are most decisive for the development of thermal stresses in young concrete and to find simple methods to determine these properties for a given concrete.

In this paper a simplified constitutive model is used to predict thermal stresses in young concrete. The input material parameters required by the model are kept to a mini-mum. The validity of the model is tested against experimental data available in the litera-ture. The most important material input for the model is the development of material stiffness ("elastic modulus") with age. A simple non-destructive method to determine this property already from the age of 5 hours is presented.

2   Simplified model

The model has been formulated assuming uniaxial stress and restraint conditions, for re-inforced and plain concrete. The applications considered in this paper only refer to plain concrete. The incremental stress-strain relation is given by:

$$\Delta\sigma_c = E_c(t)[\Delta\varepsilon_c - \alpha_c\Delta T]$$ (1)

where $\Delta\sigma$ and $\Delta\varepsilon$ are stress and strain increments (index c for concrete), ΔT is the temperature increment, α is the coefficient of thermal expansion and E is the elastic modulus. E_c for concrete is assumed to be age dependent. The coefficient of thermal expansion α_c is assumed to be independent of age but with different values α_{ce} and α_{cc} during expansion ($\Delta T>0$) and contraction ($\Delta T<0$), respectively. Δu_c is incremental displacements in the concrete. Aging is described by the well-known concept of equivalent maturity time t_e.

The constitutive equations in combination with standard kinematical and equilibrium relations were transformed into finite element (FEM) relations and implemented as additional subroutines in a FEM program. More detailed presentation of the model is made in Nagy and Thelandersson (1992).

Equation (1b) gives a very simple description of the mechanical response of the young concrete. It is assumed that some creep is implicitly included in the elastic modulus $E_c(t_e)$. The "material stiffness" E_c may be determined from creep tests performed on young concrete, where the load is applied at different ages. Fig. 1 shows an example of such tests performed by Emborg (1989). The delayed response after loading (creep) is quite small compared with the initial deformation. The main feature is the strong age dependence of the initial response. Estimation of an effective value of E_c, including some creep effects, is indicated in Fig. 1, where $J(t_e, t'_e)$ is the strain at the time t_e corresponding to a unit load applied at time t'_e. A set of creep tests for different loading ages thus determines the function $E_c(t_e)$.

It should be pointed out that the young concrete is non-elastic also for low stresses, Kasai et. al. (1974). Some part of the strain is irreversible if the material is loaded and then immediately unloaded. This behaviour is most significant at very early age. At the time when actual unloading usually takes place in practical structures the approximation of elastic behaviour inherent in Eqs. (1) is reasonable.

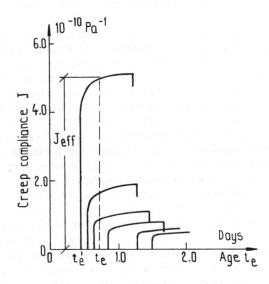

Fig. 1. Creep in young concrete with loading applied at different ages. An effective material stiffness is estimated from $E_c = 1/J_{eff}$, where $J_{eff} = J (t'_e + 5h, t'_e)$ and t'_e is the age of loading in the creep test. Based on tests by Emborg (1989).

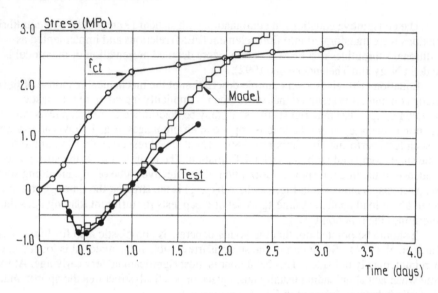

Fig. 2. Stress development in fully restrained concrete prism from the time of casting. The predicted growth of tensile strength f_{ct} with age is also shown.

Fig. 2 shows the experimentally obtained restraint stress developed in a plain concrete prism exposed to a prescribed temperature history starting at 23°C, increasing to a maximum of 55°C at the time 0.45 days, and then slowly cooling to 25°C at the time 6 days, Emborg (1989). The prism was restrained from axial movement from the time of 0.23 days. This test was simulated with the above mentioned simplified model and the development of elastic modulus with age was estimated from creep tests performed on the same concrete, see Fig.1. The temperature variation and the restraint conditions were the same as in the test. The result is shown in Fig. 2, where it can be observed that the simplified model gives surprisingly good conformity with the test results. Similar comparisons for other cases, also with another type of cement, have shown that the simplified model can predict the experimental thermal stresses in a satisfactory manner.

By using the simulation model it is found that the computed thermal stresses are very sensitive to the age dependence of the material stiffness. The effect of a change in age dependence for E_c on the stress development in a constrained concrete prism is illustrated in Fig. 3. The curve marked normal corresponds to a simulation with $E_c(t_e)$ determined from direct measurements, cf. Fig. 1. The curve marked rapid is defined as $E_{c,rapid}(t_e) = E_c(t_e + 0.3$ days$)$. The temperature development, the boundary conditions as well as all other material parameters are the same in both cases. An earlier development of E_c gives a significant increase in compressive stress during the heating phase. This is an advantage since the tensile stress during the subsequent cooling period becomes much smaller. A cement which can give an early growth of stiffness and a delayed release of heat is therefore advantageous with respect to thermal cracking. Another conclusion is that the age dependence of the material stiffness must be accurately known in order to predict thermal stresses in a reliable way.

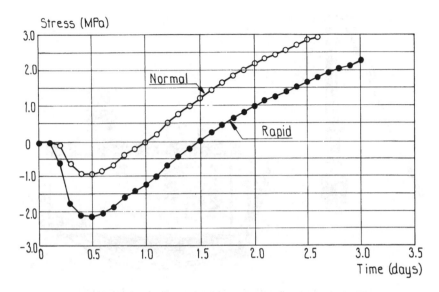

Fig. 3. Influence of early growth of material stiffness on thermal stresses in fully restrained concrete prism.

3 Non-destructive measurement of material stiffness for young concrete

In contrast to the static stiffness the dynamic E-modulus is easily tested even at early ages by means of the resonance frequency method. The test is accomplished at very low stresses and short periods which makes it refer to almost purely elastic effects, unaffected by creep. For very young concrete, however, the result may be affected by significant material damping due to plastic behaviour also at very low load levels.

The test set up with the FFT-analyser (HP3582 Spectrum Analyser, frequency range 0-25 Hz) is shown in Fig 4. The measurements were made on concrete prisms with the dimensions 40x40x160 mm (the maximum aggregate size is 8 mm). The concrete prism is placed freely on a support covered with a pad of expanded polyurethane. The basic idea is to permit the specimen to vibrate as a free beam. An oscillator generates a dynamic load at the top of the specimen and the response is registered through a pick up attached to the specimen near the oscillator. The output signal from the pick up is registered by the FFT-analyser, which displays the so called mechanical mobility (velocity divided by input force) as a function of frequency. The mechanical mobility can be interpreted as a measure of the "willingness to be set in motion" of the material. The first peak in the mobility curve is the fundamental resonance frequency f. Besides the resonance frequency, the damping coefficient η of the material is also determined by the analyser. The damping coefficient characterises the degree of departure of the material from perfect elasticity. A more detailed presentation of the dynamic test method is given in Nagy (1994).

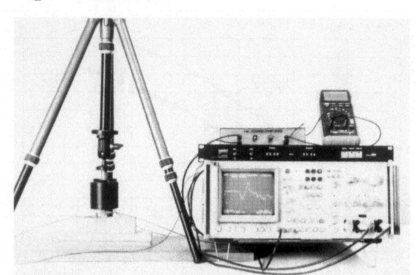

Fig. 4. Test set-up with the Spectrum Analyser HP3582.

The dynamic modulus E_{dyn} is calculated according to the ASTM-formula $E_{dyn} = C \cdot W \, f^2$, where W is the weight of the specimen (kg), f is the resonance frequency at transverse vibration (Hz) and C (m^{-1}) is a factor depending on the dimensions of the specimen and Poisson's ratio of the material.

The dynamic modulus for a concrete made of slow hardening cement (Degerhamn standard) during the first 7 days is shown in Fig. 5. The corresponding static modulus was measured at selected times on concrete cylinders from the same mix. These results are also shown in Fig. 5. There is obviously some scatter in the measured static moduli, partly due to experimental difficulties. It is also hard to find a consistent definition of the static modulus for concrete at very young age, in which case the stress-strain relation is non-linear already at low stress levels. The ratio between the dynamic and static moduli for hardened concrete is typically of the order 1.25, which is reflected in the results.

The fact that the dynamic measuring procedure gives rapid results with significantly smaller effort, makes it convenient to have an empirical model which uses only dynamic parameters to obtain the static E-modulus. It was found that the following formula is suitable for such a conversion:

$$E_{stat} = \frac{E_{dyn}}{1 + \eta^a} \tag{2}$$

where a is an empirical factor and η is the damping coefficient.

Simultaneous tests of dynamic and static moduli have been performed for two different concretes each with different types of cement. In both cases it was found that a value of a=0.35 gives the best agreement for both mixes. The static E-modulus estimated with Eq. (2) is compared with the measured static modulus in Fig. 5. The predictive capacity of the dynamic method can be regarded as acceptable. Similar conclusions could be drawn from tests with the other concrete mix.

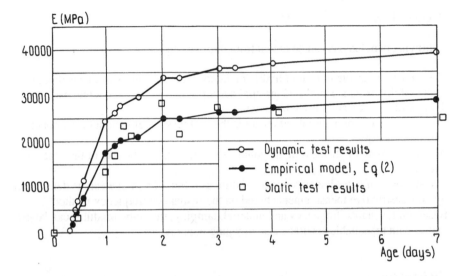

Fig. 5. Evolution of material stiffness measured by dynamic and static methods.

4   Discussion

A simplified constitutive model was used to predict thermal stresses in young concrete. Although the input parameters required by the model is kept to a minimum, the simplified model has a good capability to predict thermal stresses in concrete under restraint. The results indicate that the proposed material description is sufficiently accurate for thermal stress analysis in practical applications, where the information available for a more sophisticated analysis usually is incomplete. There may be situations, however, where the description is too simple to capture the behaviour in an adequate way. Therefore, it is desirable to compare results obtained with the simplified model for some typical cases with the results obtained with more sophisticated computer programs available today.

The most important material characteristic governing thermal stress development during hydration is the age dependent material stiffness. Therefore, it should be of great practical value to be able to determine the progressive changes in material stiffness in a simple way with some non-destructive method. A promising way to do this is by resonance frequency measurement, which can be performed on a small beam with relatively inexpensive equipment. The results so far indicate that the dynamic properties obtained from such measurements can be used to determine the evolution of static modulus of the concrete by a simple formula. Further verifications are needed, however, before the method can be widely used for practical purposes.

5 Conclusions

The following conclusions can be drawn from this investigation:

1. Thermal stresses due to heat of hydration in concrete structures can be estimated successfully using a simplified material description of young concrete partially disregarding creep and plastic behaviour.
2. The most important material parameter is the stiffness development of the concrete during hardening.
3. An early growth of stiffness in the young concrete reduces the risk of thermal cracking.
4. The stiffness of a particular concrete mixture as a function of age can be determined by non-destructive measurements based on the resonance frequency method.
5. Based on resonance frequency and material damping, the static modulus can be estimated with reasonable accuracy also in young concrete.

6 References

Dahlblom, O. (1990) HACON-T - A Program for Simulation of Temperature in Hardening Concrete. R,D&D-report, Serial number U 1990/31, Swedish State Power Board, Vällingby, Sweden.

Dahlblom, O. (1992) HACON-S - A Program for Simulation of Stress in Hardening Concrete. R,D&D-report, Serial number H1992/3, Swedish State Power Board, Vällingby, Sweden.

Emborg, M., Bernander, S. (1993) Concrete with no Temperature Cracks Due to Hydration Process. Proc. of the Nordic Concrete Research Meeting, Gothenburg, pp 227-229.

Emborg M. (1989) Thermal Stresses in Concrete Structures at Early Age. Doctoral Thesis 1989:73D, Division of Structural Engineering, Luleå University of Technology, 285pp.

Haugaard, M., Berrig, A., Frederiksen J. O. (1993) Curing Technology. A 2-dimensional Simulation Program, Proc. of the Nordic Concrete Research Meeting, Gothenburg, pp 222-224.

Jonasson, J-E. (1988) HETT-A Computer Program for The Calculation of Strength, Equivalent Hydration and Temperature. (in Swedish), Swedish Cement and Concrete Research Institute, Stockholm.

Kasai, Y., et.al. (1974) Tensile Properties of Early Age Concrete (The Plastic and Elastic Strain and The Extensibility). Proceedings of the 1974 Symposia on Mechanical Behaviour of Materials. Kyoto, August the 21- 24th, 1974, pp. 433-441.

Nagy, A., Thelandersson, S. (1992) Modelling Thermal Effects in Young Concrete. Nordic Concrete Research, March 1992.

Nagy, A. (1994) Cracking in Concrete Structures due to Early Thermal Deformation. To be published. Dep. of Structural Engineering, Lund University, Sweden.

PART FIVE

MEASUREMENT OF THERMAL STRESSES IN SITU

(Mesure in situ des contraintes
d'origine thermique

21 THERMAL STRESS IN FULL SIZE RC BOX CULVERT

T. MISHIMA
Maeda Corporation, Tokyo, Japan
H. UMEHARA
Nagoya Institute of Technology, Nagoya, Japan
M. YAMADA and M. NAKAMURA
Nagoya Expressway Public Corporation, Nagoya, Japan

Abstract
In this paper, we discuss our instrumental observation on construction site of a RC box culvert with crack inducers and our quasi 3 dimensional thermal stress analysis. It consists of 2 dimensional heat diffusion analysis and plane stress analysis. By comparing both results, we verify that our method has sufficient accuracy to forecast the temperature and thermal stress characteristics of RC structures. Then, we go on to calculate the spacing and crack width induced by crack inducers using the smeared crack model. Finally, we demonstrate that the quasi 3-dimensional analysis can be an effective tool to design thermal crack control of RC wall type structures.
Keywords: Crack Inducer, Spacing of Cracks, Stress Meter, Structural Members (Box Culvert), Thermal Stress Analysis, Width of Cracks.

1 Foreword

Following the increasing number of constructions by rapid operation of mass concrete placing with advanced concrete pumping methods, thermal stress caused by the heat of the hydration of cement has become an important factor, even to wall type structures in terms of quality.

Generally it is not necessary to prevent all cracking in wall type structures, since the occurrence of certain cracks is allowed in design. However, it is necessary to control the location of cracking and the width of cracks to ensure the durability of the structure. Thus the provision of crack inducers and control of reinforcement is becoming a popular measure to deal with thermal stresses. Especially with the successful performance of waterproofing condition, crack inducers are quite effective in coping with thermal stress without impairing the durability of the structure.

This paper reports the results of the instrumental observations and the quasi 3 dimensional thermal stress analysis conducted for a reinforced concrete box culvert in which crack inducers were provided. The concrete temperature, thermal stress and strain of the side walls and the width of cracks are principally discussed in this paper.

Thermal Cracking in Concrete at Early Ages. Edited by R. Springenschmid. Published 1994 by E & FN Spon, 2–6 Boundary Row, London SE1 8HN, UK. ISBN: 0 419 18710 3.

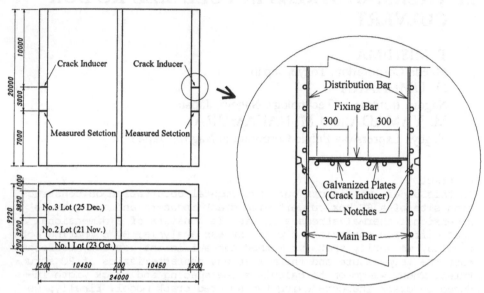

Fig.1. RC Box Culvert   **Fig.2. Crack Inducer**

2 Instrumental Observations

2.1 RC Box Culvert Structure

The structure in which instrumentation was installed was a reinforced
concrete box culvert as shown in Fig. 1. Concreting to the box was
carried out firstly to the base slab, secondly to the lower part of
the side walls and lastly to the upper part of the side walls together with the roof slab. The distribution reinforcement of D13 bars
which influence crack width in the concrete were of 150mm spacing
outside and of 300mm spacing inside of the walls.

Although two types of concrete mixes with different slumps (8cm
and 12cm) were used respectively for the walls, no significant difference was recognized in stress hysteresis and appearance of cracks.
The following discussions refers to the wall with the 8 cm slump concrete.

2.2 Crack Inducers

A crack inducer was placed in the middle of each side wall. Due to
limited experience, a standard method of crack inducer installation
has not yet been established. In this case we used galvanized steel
plates type as shown in Fig. 2. 50 % of the decreasing ratio of sectional area was adopted to ensure complete inducement, since 35 % had
been reported to be not very effective from the past experience.

2.3 Items and Location of Instrumental Observations

Since it is empirically known that thermal cracking usually occurs in
the walls in the case of box culvert structures, instrumental obser-

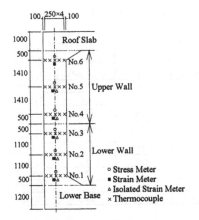

Fig.3. Measured Section

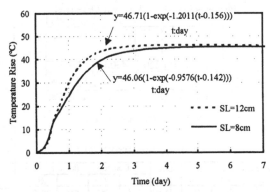

Fig.4. Adiabatic Temp. Rise Test

vations were carried out mainly in the walls with exception of four temperature reading points in the base slab. Table 1 lists the parameters measured and the instruments used to carry out the measurement.

It is known that when cracking takes place in close proximity to the being measured section, the reliability of data tends to diminish. Therefore the measured sections were set up where cracking would be least expected. The locations of the instrumental observations are indicated on Fig. 3. The strain of the reinforcement and the opening of induced crack were measured at the crack inducer sections.

2.4 Material Tests

Material tests were conducted to study hardening properties of concrete in terms of compressive strength, splitting tensile strength and elastic modulus. The concrete mix and the results of material tests are shown on Table 2 and 3 respectively. The heat generation properties were examined in an adiabatic temperature rise test. Fig. 4 shows the test results that were estimated by the least square method. The thermal expansion coefficient of the concrete was obtained by a strain meter which was placed in the stress isolation box, installed in the center of each wall. An average value of $10.9 \times 10^{-6}/°C$ for all the walls was taken as the thermal expansion coefficient for the following examination.

Table 1. Measured Item

		Points in a Section	Number of Section	Total Points	Type of Sensor
Wall	Concrete Stress	3	4	12	Effective Stress Meter
	Concrete Strain	3	4	12	Strain Meter
	Expansion Coef.	1	4	4	Isolated Strain Meter
	Base Temp.	2	2	4	Thermocouple
	Side Wall Temp.	15	4	60	Thermocouple
	Air Temp.	1	2	2	Thermocouple
Joint	Steel Strain	6	4	24	Strain gauge
	Joint opening	3	4	12	Opening Meter
Total				130	

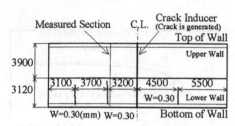

Fig.5. Crack Pattern

Table 2. Concrete Mix

Slump	Air	W/C	S/A	Composition. (kg/m³)			
(cm)	(%)	(%)	(%)	W	C	S	G
8±2.5	4±1	56	45.5	157	281	835	1039

Table 3. Material Tests

Age (day)	Compressive Strength (kgf/cm²)	Tensile Strength (kgf/cm²)	Young Modulus (kgf/cm²)
0.5	23	2.9	81000
1.0	55	6.4	119000
2.0	111	11.4	190000
3.0	126	14.4	207000
5.0	189	19.5	216000
7.0	215	20.0	219000
10.0	238	21.3	237000
14.0	253	22.7	248000
21.0	266	25.7	268000
28.0	280	28.5	285000

3 Results of the tests and analysis

3.1 Occurrence of cracks

The cracks that occurred in the side wall are shown in Fig. 5. A couple of cracks were found at the lower part of the wall, which did not run through the crack inducer. This indicates that a single inducer in the whole wall was not sufficient to induce all cracks. The minimum crack spacing was approximately equal to the height of a concrete lift. Thus it is concluded that spacing of crack inducers also needs to be the same as the height of concrete lift.

3.2 Quasi Three Dimensional Thermal Stress Analysis

In wall type structures, generally, the temperature gradient predominates in the transverse section and the thermal stress predominates in the longitudinal direction. Therefore in the case where the thermal stress analysis is carried out in two dimensions, the thermal analysis needs to be carried out in the normal section to the longitudinal direction and the stress analysis needs to be carried out in the longitudinal section. However, since these two sections perpendicularly intersect, 2 dimensional analysis can not be applied. Usually the CP-Method, which is the Japanese original method proposed by the Japanese Concrete Institute(JCI) and now is recommended by the Japanese standard specification, is used for this kind of structure. The CP-Method is an excellent method with shorter calculation time, however it has a shortcoming in that the treatment of the degree of restraint is indirect.

We therefore adopted a quasi three dimensional stress analysis (QTD analysis) that conducts thermal stress analysis on two different planes by assuming that the average temperature taken in the transverse section from the results of the temperature analysis distributes evenly in the longitudinal direction. The reliability of this analysis is examined here by comparing to the results of the instrumental observations and CP-Method. The concept of QTD analysis is schematically shown on Fig. 6.

Table 4. Condition of Thermal Analysis

		Soil	Lower Wall	Upper Wall
Thermal Conductivity	(kcal/·m·h·°C)	0.92	2.3	2.3
Specific Heat	(kcal/kg·°C)	0.4	0.23	0.23
Weight per Unit Vol.	(kg/m³)	1600	2300	2300
Heat Transfer Coef.	(kcal/m²·h·°C)	10	8	6
Air Temp.	(°C)	Monthly Average Temp.		

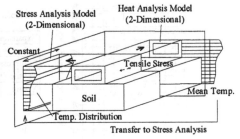

Fig.6. Quasi 3-Dimensional Analysis

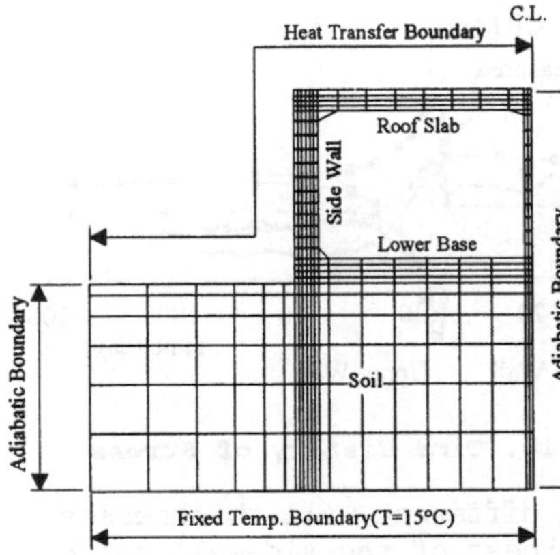

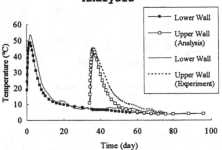

Fig.7. Model of Thermal Analysis

Fig.8. Time History of Temp.

3.3 Result of Thermal Analysis

The model for the thermal analysis is shown in Fig. 7. Taking advantage of its symmetry, only the left half was used for the analysis. The conditions of the analysis are shown in Table 4.

Since the air circulation of this structure was poor due to semi-underground construction and wooden formwork was used, the heat transfer coefficient is considered to be smaller than in the normal case. As a matter of fact, when the normal value of coefficient of 10 kcal/m²·h·°C is used, speed of the temperature fall tends to be faster than the measured results. In order to correct this, we reduced the value of heat transfer coefficient.

Fig. 8 shows the comparison results of the analysis and the observations. By adjusting the heat transfer coefficient the analysis values agree fairly well with those of the observations with a marginal difference. Although this difference can be reduced by further adjustment, the coefficient becomes remote from normal. Therefore we consider that no further adjustment is desirable.

3.4 Result of Stress Analysis

(1)Analysis Model
In QTD analysis, stress analysis is conducted in the longitudinal direction with a different FEM mesh from that of the thermal analysis. The analytical model used is shown in Fig. 9. In this analysis

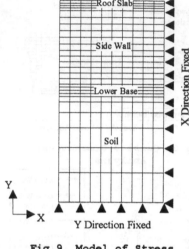

Fig.9. Model of Stress Analysis

Table 5. Condition of Stress Analysis

Expansion Coef.	(1/°C)	0.0000109
Poisson's Coef.		0.17
Compressive Strength	(kgf/cm²)	fc=158log M-167
Tensile Strength	(kgf/cm²)	ft=0.248fc^0.812
Young Modulus	(kgf/cm²)	Ec=(108.2log M-27.8)×10³

M:Maturity(=30×Effective Age)

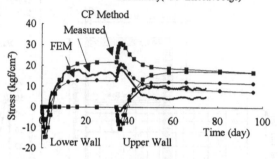

Fig.10. Time History of Stress

the differences of stiffness due to the different wall thicknesses are to be considered. Although the thickness of the sidewall can be regarded as thickness for analysis, however, the effective thickness of the soil portion, base slab and roof slab can not be simply determined. In this analysis, with various studies, the effective thickness is determined as four times that of the sidewall thickness.

Table 5 shows the assumptions of the analysis. The concrete hardening properties are based on the results of the material tests. Considering that the effect of creep is small, the concrete is assumed to be an elastic body. Since the ground, on which this structure is situated, is of sand, its elastic modulus is considered to be minimal. In this analysis, the elastic modulus of the ground was taken to be nearly zero. In the CP-Method analysis, the equivalent restraint conditions to the QTD analysis were used for comparison.

(2) Comparison with CP Method and Experiments

Fig. 10 shows the results of this analysis compared to those of measured observations and the CP-Method analysis. Generally both methods are fairly agreeable with the observed results. Since both methods are of similar degree of accuracy, it is difficult to determine which one is superior.

Fig. 11 shows the time history of the minimum crack index. The minimum crack index of the lower wall had been 1.2 before the upper wall was concreted and became less than 1.0 after concreting. The crack index of the upper wall was 1.3. The probability of crack occurrence, which was figured out from the crack index, is 60% for the lower wall and 20% for the upper wall, according to the Japanese specifications (Fig. 12). These figures are agreeable with the actual occurrence of cracks.

(3) Spacing of Crack Inducers

In CP-Method analysis, where the restraint ratio is assumed to be zero, the effect of structure length i.e. the effect of the L/H ratio

can not be taken into account, because the effect of L/H ratio is
included in the restraint ratio. Therefore, if the CP method is applied to analyze this kind of structure, generated stress would remain constant regardless of the spacing of crack inducers. For this
reason, it will be very difficult to determine the spacing of crack
inducers by the CP-Method where the restraint ratio is very small.

On the other hand, if the analysis is carried out with the spacing of crack inducers being approximately taken as structure length,
the QTD analysis is able to reflect the effect of the L/H ratio.
Therefore, QTD analysis makes it possible to determine spacing of
inducers analytically even if the restraint ratio is small, which
makes QTD analysis more advantageous than CP Method.

Fig. 13 indicates the relation between numbers of crack inducers
and crack indexes. When the limit of crack index is set at 1.5, the
number of inducers required become three. This corresponds with the
fact that three to four cracks actually occurred in the side wall.
For the above reason, this analytical method is confirmed to be of
sufficient accuracy for determining the spacing of crack inducers.

3.5 Analysis for Width of Cracks by Smeared Crack Model

(1) Analysis Model
In thermal cracks, because of the crack spacing is very large, the
smeared crack model is considered to be inapplicable. However, even
with the smeared crack model it is possible to analyze the width of
thermal cracks by limiting the cracking elements.

Fig. 14 represents a model of smeared crack that is recommended
by JCI. With the reinforcement being represented by truss element,
and by setting the element length of cracking element into the
equivalent length of bond slip zone, the bond action of the reinforcement can be evaluated reasonably. In this analysis, 20 cm of the
equivalent length of bond slip zone was used as recommended by JCI.
(2) Comparison with Observation Results
Fig. 15 shows the comparison of crack width. Since the observed values vary significantly in each wall, the accuracy of the instrumentation may be questionable. Bearing this in mind, however, the average
of the observed widths of the induced cracks in both walls was nearly
same as the analyzed results. Furthermore, analyzed widths are corresponding to the results of the naked eyes observations. While further
examinations are required, there is a possibility that this analytical method is able to predict crack widths with practicable accuracy.

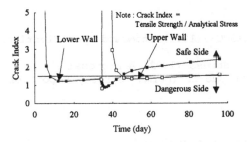

Fig.11. Time History of Crack Index

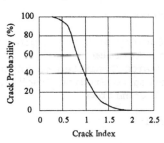

Fig.12. Crack Probability

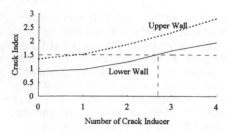

Fig.13. Effect of Crack Inducer

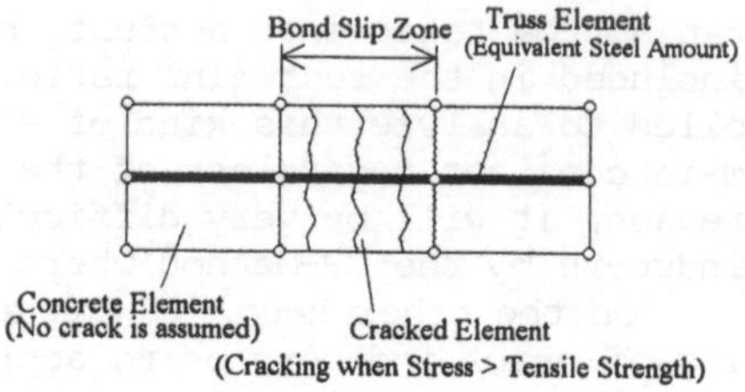

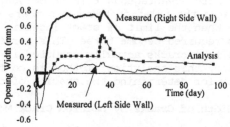

Fig.14. Model of Crack Element

Fig.15. Time History of Crack Width

4 Conclusion

Instrumental observations were carried out for the box culvert structure to collect measurement data of the thermal stress of a wall type
structure installed with crack inducers. QTD analysis was also conducted to examine its applicability by comparison with the observed
results.

The conclusions are summarized below.
1) In examining the observed results, it is confirmed that observation results of temperature and stress are reliable.
2) Comparing between the CP method and the QTD analysis, both have a
similar accuracy, while the QTD analysis is superior in modeling restraint conditions and is able to take into account the L/H ratio.
3) The QTD analysis has proved to be useful in predicting the probability of cracking and in determining the spacing of crack inducers.
4) Crack analysis using a smeared crack model has proved to have a
possibility in predicting crack widths with further examinations to
be carried out in future.

5 References

Japan Society of Civil Engineers(JSCE)(1991) **Standard Specifications
for Concrete**
JCI Research Committee on Thermal Stress (1992) A Proposal about a
Method to Predict Crack Width due to Thermal Stress, **Report of
JCI Committee**

6 Appendix

Unit Conversion Factor : 1 kgf/cm² = 0.098 MPa

22 THERMAL CRACKING IN WALL OF PRESTRESSED CONCRETE EGG-SHAPED DIGESTER

T. YOSHIOKA
Oriental Construction Co. Ltd, Tochigi, Japan
S. OHTANI
Oriental Construction Co. Ltd, Tokyo, Japan
R. SATO
University of Utsunomiya, Tochigi, Japan

Abstract
When egg-shaped digesters (ESD) are constructed, thermal cracks often
develop due to heat of hydration. Because of very complicated external
and internal restraints and boundary conditions Finite Element Method
(FEM) is the most applicable to analyze mechanical behavior of ESDs. On
the other hand effect of creep is indispensable for analysis of thermal
stresses due to heat of hydration generated by cement.
In this study a simplified "step-by-step" method in accordance with
Japanese Standard is adopted to incorporate in FEM for axi-symmetrical
structures. In this method reduced Young's modulus is used considering
effect of creep of concrete at early ages.
Computed temperatures and thermal stresses are compared with experimental results measured in an ESD under construction. Computed tensile
stresses agree comparatively well with experimental results.
Keywords: Numerical Simulation, Stress Meter, Structural Members (Axi-
symmetrical).

1 Introduction

Since 1983 prestressed concrete egg-shaped sludge digesters (ESD),
which have the advantage of sludge digestion and require little maintenance in service, have been constructed in Japan. ESDs are often constructed on very soft ground such as reclaimed land at the sea. In
order to assure the safety during severe earthquakes ESDs are normally
supported on pile foundations, which require massive foundation rings
to connect ESDs to pile foundations(refer to Fig.1).

As the ESD is normally concreted circumferentially in several layers, heat of hydration and restraints due to pre-existing rigid concrete could cause cracks, which develop vertically. These cracks surely
impair water-tightness and durability of the concrete structures, while
a suitable way has not been established to avoid these cracks. However,
few investigations have been performed on the prediction of thermal
cracks of such structures.

In the present study thermal stresses were computed by Finite Element Method (FEM). In order to take the effect of creep of concrete
into account "step-by-step" method and "effective Young's modulus",
which are proposed in "Standard Specification for Design and Construction of Concrete Structures" (Japan Society of Civil Engineering (JSCE)
1991), were adopted.

Thermal Cracking in Concrete at Early Ages. Edited by R. Springenschmid. Published 1994
by E & FN Spon, 2–6 Boundary Row, London SE1 8HN, UK. ISBN: 0 419 18710 3.

At the same time circumferential thermal stresses and temperatures were measured to know the mechanism of thermal stresses in in situ structures

The validity of the present method was discussed based on comparison of experimental and computed results.

2 Construction procedure and measurement in situ

2.1 Construction procedure

The vertical section of the ESD, whose content is 4,000 m³, is shown in Fig.1. The ESD was vertically divided in four concrete sections up to the equator. The first joint, which is not dealt with in this paper, is located at the top of the lower conical cone. The second joint is located in the massive foundation ring at the height of 1.0 m above the ground and the third one is located in the curved wall at the height of about 4.2 m above the ground. Concrete sections, joints and concrete lifts are indicated in Fig.3. The second lift was concreted 28 days after the first lift. Concrete mix proportion is shown in Table 1. The

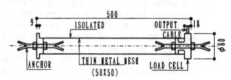

Fig.2 Stress meter

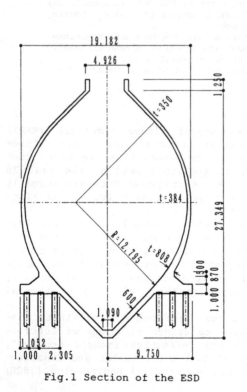

Fig.1 Section of the ESD

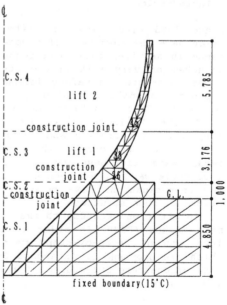

Fig.3 Mesh of FEM

Table 1 Concrete mix proportion

Max. size of coarse aggre- gate	Slump	Air cont.	Water- cement ratio (W/C)	Sand aggre- gate ratio (s/a)	Unit content				
					Water W	Cement C	Fine aggre- gate s	Coarse aggre- gate G	Admix- ture
(mm)	(cm)	(%)	(%)	(%)	kg/m3				
25	8.0	-	38	41.1	149	392	720	433	5.88

Cement:OPC,   Mix proportion strength:350MPa
Admixture:superplasticizer

ESD has been covered with wooden formworks, whose thickness is 27 mm, for more than two weeks after placing concrete and special curing such as moist or temperature controlled curing was not carried out.

2.2 Measurement in situ
Thermal stresses and temperatures were measured in situ with stress meters and thermo-couples, respectively, which were circumferentially embedded at the same locations. Fig.2 shows details of the available stress meter.

Measuring points are indicated in Fig.3 as node No. 26, 33 and 45 which correspond with node No.s for stress analysis.

3 Numerical computation

3.1 Thermal analysis
Two dimensional thermal analysis for radical cross section was carried out with a computer program supplied by Japan Concrete Institute (JCI). This analysis for heat transfer and conduction is based on Fourier's law. Mesh used for finite element analysis is shown in Fig.3. The coefficient of heat transfer for the formwork used was decided on referring to JSCE Standard with slight modification considering the thickness of the wooden panels. Material properties and other conditions for computation are summed up in Table 2.

Adiabatic temperature rise was estimated in compliance with JSCE Standard as follows;

$$Q(t) = Q_\infty(1 - e^{-\gamma t}) \tag{1}$$

where
Q_∞:the ultimate adiabatic temperature rise to be determined by testing,
γ:the constant on rate of temperature rise to be determined by testing,
t:age, in days,
$Q(t)$:the adiabatic temperature rise at an age of t, in ℃.
For lack of testing Q_∞ and γ for ordinary Portland cement (OPC) were determined at placing temperature 25 ℃ as follows;
$Q_\infty = 0.11 \cdot C + 12.5 = 0.11 \times 390 + 12.5 = 55.4$
$\gamma = 3.9 \cdot C + 0.151 = 3.9 \times 390 + 0.151 = 1.672$
where C:unit cement content

Table 2 Material properties and other conditions

Item	Unit		Remarks
Thermal conductivity	W/mK	2.33	concrete & soil
Specific heat	J/kgK	921	concrete & soil
Density	kg/m3	2,500	concrete & soil
Coefficient of heat transfer	W/m²K	5.82	wood formwork
		11.63	exposed concrete surface
Ambient temperature	°C	22	for lift 1
		18	for lift 2
Concrete temperature at placing	°C	27	for lift 1
		25	for lift 2

3.2 Thermal stress analysis

Thermal stress analysis was conducted with FEM in which axi-symmetrical solid elements were used. The mesh shown in Fig.3 was also applied in thermal stress analysis excluding the elements for soil. The ESD is externally restrained by piles in both radial and vertical directions, which were modeled by elastic springs.

The effective Young's modulus, which is different from the so-called effective modulus of elasticity, may be empirically formulated so that it can be applicable for step-by-step analysis, was determined in compliance with JSCE Standard as follows;

$$E_e(t) = \psi(t) \times 1.5 \times 10^3 \sqrt{10 f_c'(t)} \tag{2}$$

where

$E_e(t)$: effective Young's modulus at an age of t days, in MPa,

$f_c'(t)$: compressive strength of concrete at an age of t days, in MPa,

$\psi(t)$: correction factor of Young's modulus considering the effect of creep under temperature rise,

:0.73 and 1.0 for up to three days and after five days, respectively.

Estimated compressive and tensile strengths are obtained according to JSCE Standard as follows;

$$f_c'(t) = \frac{t}{a + bt} f_c'(91) \tag{3}$$

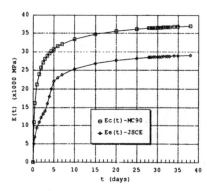

Fig.4 Young's modulus

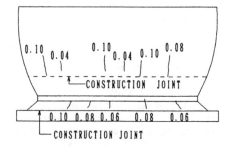

Fig.5 Cracks

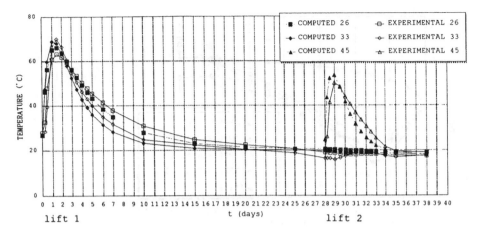

Fig.6 Temperatures

$$f_t(t) = c\sqrt{10f_c'(t)} \tag{4}$$

where

$f_t(t)$: tensile strength of concrete at an age of t days, in MPa,

$f_c'(91)$: compressive strength of concrete at an age of 91 days,

a,b : 4.5 and 0.95, respectively, for OPC,

c : 1.4 in standard practice.

The computed results of effective Young's modulus (Ee(t)-JSCE) are shown in Fig.4 compared with Young's modulus in accordance with CEB/FIP Model Code 90(Ec(t)-MC90). The temperature rise in the jth time interval was determined as the difference between temperature at the beginning t_j and that at the end t_{j+1} of the jth time interval. Thermal stresses developed during the jth time interval was computed using effective Young's modulus at the middle of the jth time interval in accordance with Eq.(2).

4 Discussion of results

Visible cracks observed on outer surface are illustrated in Fig.5, in which crack widths are also shown. Cracks could have developed upward from the construction joint or just above the joint. Crack widths are ranging from 0.04 mm to 0.1 mm.

Computed results of temperatures are shown in Fig.6 compared with experimental ones. Fig.6 shows that computed temperatures agree approximately with experimental results at all measuring points. While experimental results show some delay of generation of hydration heat at very early age before one day, computed results do not express the phenomenon. This is due to that Eq.(1) for adiabatic temperature rise does not consider the delay.

Regarding peak temperatures computed ones are slightly higher than experimental ones at nodes of 26 and 45 and almost the same at node 33.

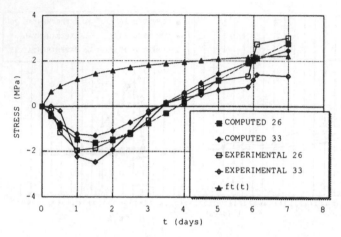

Fig.7 Thermal stresses for the first seven days (lift 1)

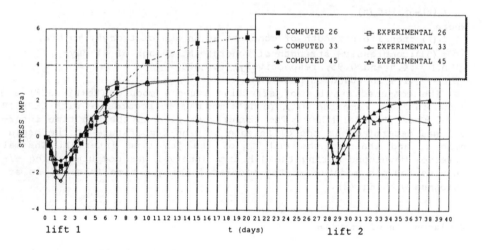

Fig.8 Thermal stresses for the whole period

After the peak temperatures computed temperatures decrease faster than experimental ones. This could be due to incorrect estimation of heat conductivity and coefficient of heat transfer.

Computed thermal stresses for the first lift are shown in Fig.7 compared with experimental results for the first seven days before cracking with estimated tensile strength of concrete $f_t(t)$ according to Eq.(4). Fig.8 shows them for the whole period of measurement. From these figures computed stresses qualitatively agree well with experimental results at all measuring points.

In the first lift cracks occurred first at the age of six days considering sudden change of stresses at the age. In the second lift cracks probably developed this between four and five days because of the stop of increment of stress. Assuming that thermal stresses were precisely measured, cracks developed first below the thermal stress of 2.0 MPa at an age of four to six days. Nevertheless computed tensile strength of concrete reaches 2.0 MPa at an age of four days. This fact means that actual tensile strength at early ages under variable sustained stress could be less than the computed one.

In Fig.7 some differences can be found between computed and experimental results. Computed results can not express the behavior at very early ages before one day and underestimate the peak stresses. Considering tensile stresses after peak stresses, those in experimental results grow more rapidly than those in computed ones. However the growth of tensile stresses in experimental results becomes less steep thereafter. On the other hand tensile stresses increase steadily in computed results. This could depend on the difference between actual creep behavior and behavior given by the present model.

It is also indicated in Fig.7 that comparison between computed thermal stresses and estimated tensile strength of concrete predicts the age at cracking with satisfactory accuracy, while computed tensile stresses overestimate thermal stresses as described before.

Although farther study is requested to estimate more accurate thermal stresses, the present method could be useful to predict thermal cracking in the practical design works.

5 Conclusion

Thermal analysis for radial cross section of an ESD was carried out by FEM. Using these temperatures, cracking in the wall of ESD was investigated based on thermal stresses computed by FEM in which both step-by-step method proposed by JCI and reduction of Young's modulus due to effect of creep adopted in JSCE Standard were taken into account. Consequently following conclusions were obtained in the present study;
1) Computed peak temperatures agree comparatively well with experimental results. However, some difference was recognized in decreasing velocity of temperature between experimental and computed results. It is important to determine suitable coefficient of heat transfer considering material and thickness of formworks.
2) Present method predicted the age of cracking with satisfactory accuracy. However, maximum compressive stresses were underestimated and tensile stresses overestimated by 50% at stabilized temperature. In order to gain more accurate computation results it is necessary to apply more accurate creep model as well as creep analysis.

References

Deguchi,H. et al (1993) A Prestressed Concrete Egg-shaped Digestion Tank in Western KYOTO. **Proc. of FIP Symposium '93**,Vol.1,pp.627-634.

Japan Concrete Institute (1985) The method of calculation of thermal stress in massive concrete structures and its personal computer program.

Japan Concrete Institute (1986) **Recommendations for Control of Cracking in Massive Concrete**.

Japan Society of Civil Engineering (1991) **Standard Specification for Design and Construction of Concrete Structures**.

Kiyomiya,O. (1991) Recent Measurement Technology on Concrete. **Concrete J**.Vol.29,No.5,May 1991,pp.5-14.

Miyazaki,N. et al (1991) **Analysis of Thermal Stress, Creep and Heat Transfer with Finite Element Method**. Science-sha Co., LTD.

Nagataki,S. Sato.R. (1988) Recent Trend of Control of Thermal Cracking in Massive Concrete. **Concrete J**.,Vol.26,No.5,May 1988,pp.4-12.

23 STUDY OF EXTERNAL RESTRAINT OF MASS CONCRETE

M. ISHIKAWA
Tokyu Construction Co. Ltd, Institute of Technology, Tokyo, Japan
T. TANABE
Department of Civil Engineering, University of Nagoya, Nagoya,
Japan

Abstract
This study aims at gaining a detailed understanding of the
external restraint mechanism of mass concrete during the
period of its hardening. For this purpose a thermal stress
experiment was carried out by following two factors which
have a influence upon the effects of the external
restraint as parameters; the shape of concrete which can
be explained by L/H and the bonding condition of a joint.
This paper describes the actualities of the complicated
external restraint through the carrying out of a numerical
study by the finite element method(F.E.M.).
Keywords: Numerical Simulation, Mass Concrete, Restraint.

1 Introduction

It is necessary that the values of the stresses,strains,
and displacement obtained from an analysis corresponds
closely to those obtained measurement. For the analytical
estimate of the thermal stress for mass concrete the
evaluation of the external restraint is extremely
important. It is well known that the effects of the
external restraint, in general, changes due to: 1) the
shape of the concrete indicated by length (L) and height
(H), 2) the bonding characteristics of a joint and, 3) the
stiffness ratio of placed concrete to bedrock or already
placed concrete. However, in order to closely simulate the
behavior of mass concrete, more detailed data needs should
be obtained on the restraint. For example, 1) relative
displacement at a joint. 2) Cleavage occurs on a joint
owing 'to the thermal deformation of placed concrete.
Therefore, in this study, the detailed external restraint
mechanism was implemented to FEM program, and an
analytical investigation were made on the thermal
deformation behavior of the specimen.[1] Through full
consideration of the deformation as well as the
temperature, stress and strain comparison were made
between the experimental result and the analytical result.

Thermal Cracking in Concrete at Early Ages. Edited by R. Springenschmid. Published 1994
by E & FN Spon, 2–6 Boundary Row, London SE1 8HN, UK. ISBN: 0 419 18710 3.

2 Outline of test specimens

The five specimens M1 ~ M5 which has tested in from 1985 to 1987, shown in Fig. 1, were subjected to the analysis.

Each of these five specimens is composed of a restraining body and a restrained body. The behavior of the restrained body during its hardening of the case of concrete of the upper part of the restraining body being joined to that of the restrained body is subject to the analysis which will be described later. The mix proportion of concrete for each specimen is equal. Table 1 shows the mix proportion. The value of L/H of these specimen is 15. The shape and the dimensions of M1, M2 and M3 are the same. The restrained body of M1 which is made of plain concrete and placed carefully on the top face of restraining body after sand blasting. M2 has a special sliding mechanism with the value of 0.1 for the coefficient of friction (a static coefficient of friction obtained from the actual measurement by pushing the restrained body using a jack after the completion of the test) at its joint to allow free thermal deformation of the restrained body as much as possible.

With regard to M3, the restraining body and the restrained body are united using by reinforcement. The value of L/H for the restrained body of M4 and M5, both of which are made of plain concrete, are 5, and 2.5 respectivily. The condition of the joint for the restrained body for M4 and M5 is equivalent to that for M1.

These five specimens can be classified into the following two groups: M1, M2 and M3 in which the bonding state of a joint is used as a parameter and M1, M4 and M5 whose shapes(L/H) are used as a parameter.

Table 1. Mix Proportion of Concrete

Nominal strength (kg/cm²)	Slump (cm)	W/C (%)	s/a (%)	Air content (%)	Quantity of material per unit volume (kg/m³)			
					C	W	S	G
240	9+1	58.9	49.7	4	280	165	913	948

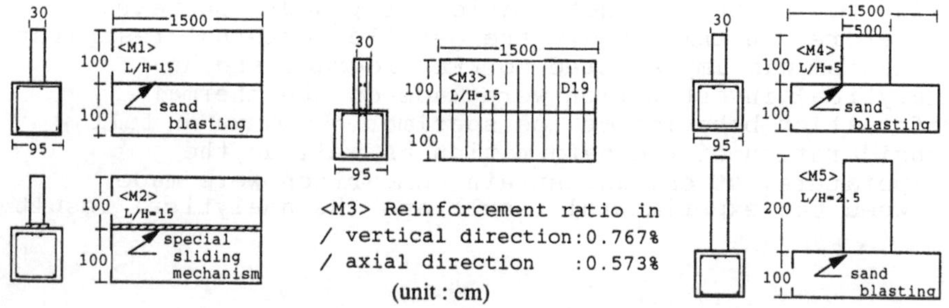

<M3> Reinforcement ratio in
/ vertical direction:0.767%
/ axial direction :0.573%
(unit : cm)

Fig.1. Outline of experimant specimens

3 Analytical method

3.1 Quasi-Three Dimensional Stress Analytical Method
As for wall structures a stress analysis in a three
dimensional field must be made if the thermal field can be
regarded as a two dimensional area. However, the three
dimensional analysis which is carried out in general is
unsuitable for a parametric study due to the large degree
of freedom. Therefore, an analysis was made on the basis
of the following assumption so that the variation of the
stress of a wall structure shown in Fig.2a in the wall
thickness direction can be taken into account.

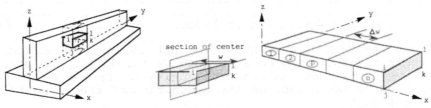

a. Wall structure   b. Element (i, j, k, l)   c. Subelemnts

Fig.2.Quasi-Three Dimensional stress analysis method

The strain distribution is set to be constant in the x
- direction at an arbitrary point of the (z, y)
coordinate. As one of the assumptions the thermal
distribution in the thickness direction for the element
(i, j, k,l) in the case of this element being taken out
(See fig.2b) must be symmetrically arbitrary toward the
central surface.

When the element (i,j,k,l) is divided into subelements
(1-n) as shown in Fig.2c, the load $\{F_P\}$ acting on the node
of the (z, y) plane of an arbitrary subelement of p can be
explained by the following equation, assuming that the
thermal distribution for the subelement is set at $T_P(z, y)$.

$$\{F_p\} = \Delta W \int_A [B]^T [D] \, \alpha_c \, T_p \, (z,y) \, dz \, dy \tag{1}$$

where, ΔW: Width of element p, [B]:Strain displacement
matrix, [D]: Materials matrix, c: Thermal expansion
coefficient of concrete

Therefore, the equivalent nodal force of $\{F\}^e$ can be
explained in the following equation:

$$\{F\}^e = \sum_{p=1}^{n} F_p \tag{2}$$

The value for equation (2) indicates the equivalent
external nodal loads which is used in a two dimensional
stress analysis. In the case of displacement being
obtained from a Hooke's low, the stress of the $\{\sigma_p\}$ for

subelement P can be obtained based on the values for the
displacement of $\{U\}^e$ at the node of the element (i,j,k,l)
as below:

$$\{\sigma_p\} = [D]\ [B]\ \{U\}^e - [D]\ \alpha_c\ \Delta T\ (z,\ y) \tag{3}$$

3.2 Modeling of spring for discontinuous element

When virtual springs with the spring coefficients of K_h and
K_v are inserted into the discontinuous field surface which
is assumed to develop to the width of Δv and the slip of
Δu both in the horizontal and vertical directions, shown
in fig.3, the following equation can be formed in the
relationship between the stress of the σ_z' τ_{yz}' acting on
the element and the expansion of Δv, Δu.

$$\sigma_z' = K_v\Delta v, \qquad \tau_{yz}' = K_h\Delta u \tag{4}$$

In the case of converting Δv and Δu into the amount of
strain the following equation can be formed for the stress
of σ_y' , σ_z' and τ_{yz}' and the strain E_y, E_z and σ_{yz}
induced by the stress on the assumption of the element's
height of h.

$$\begin{pmatrix}\varepsilon_y\\ \varepsilon_z\\ \gamma_{yz}\end{pmatrix} = \begin{bmatrix} 1/E & -\nu/E & 0 \\ -\nu/E & 1/E+1/h/Kv & 0 \\ 0 & 0 & 2(1+\nu)/E+1/h/Kh \end{bmatrix}\begin{pmatrix}\sigma_y'\\ \sigma_z'\\ \tau_{yz}'\end{pmatrix} \tag{5}$$

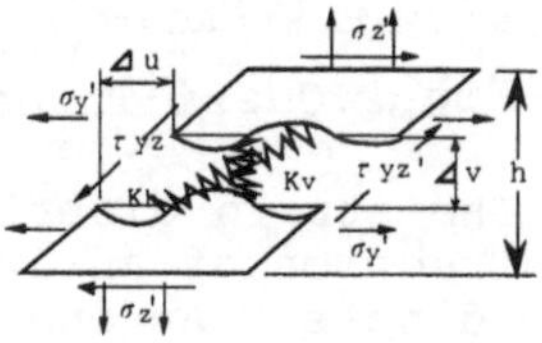

Fig.3. Spring
model for
discontinuity

4 Analytical results

4.1 Results of analyses using bonding state as parameters for specimens of M1,M2, and M3

Numerical analyses of the specimen M1 ~ M5 were carried
out by the finite element method through the employment of
the quasi-three dimensional analytical method and through
the use of the discontinuous element possessing the
virtual springs of K_h and K_v. Figs.4 and 5 illustrate the
comparison between the analytical result and measured
result for the stress, strain and deformation all of which
were obtained by using the aforementioned spring
coefficients.

The variation of the analytical result for M1, M2 and
M3 obtained using the bonding state of the joint as a
parameter depending on the application of the spring
coefficient will be described. At the same time the
process necessary to identify the spring coefficient,
which is roughly equivalent to the value obtained from the
actual measurement, will be explained.

Stress, Total strain (Longitudinal direction of the specimen)

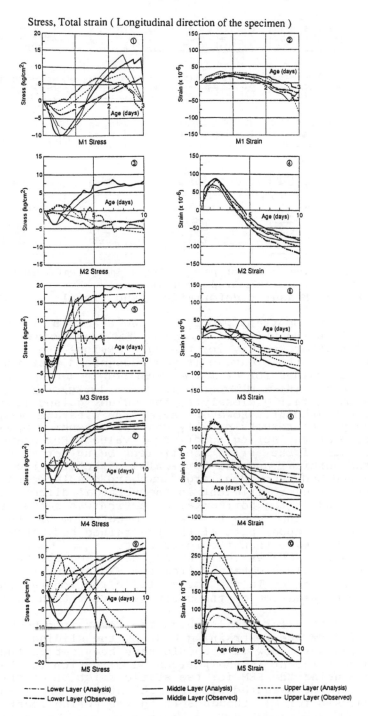

Fig.4. Comparison for stress and strain between measurement and FEM

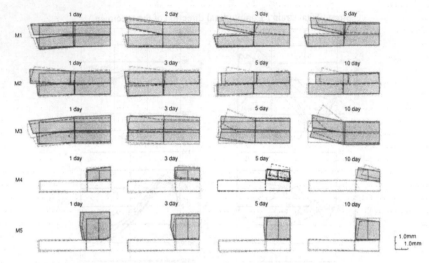

Solid line : prototype   Net : Analytical results
Dotted line : Measured deformation
Fig.5. Measured deformation and result of FEM

It was an important problem in the analysis of M1 how
to establish the value of Kh and Kv for the cleavage
occurring after one day for the age. Before the
occurrence of the cleavage it was recognized that the
analytical values for Kh and Kv obtained by using somewhat
higher values (Kh = 105kg/cm^2/cm, (Kv = 106kg/cm^2/cm,)
corresponded fairly closely to the values gained from the
actual measurement. However, after the cleavage occurred
the analytical results greatly changed depending upon the
variation of the spring coefficient value. For example,
in the case of the spring coefficient being set at 0 in
both the vertical and the horizontal directions the
restrained body showed a large downward displacement.
When the spring coefficient in the vertical direction was
set at 0 and in the horizontal direction was set at the
same value as that used before the occurrence of the
cleavage, the upper part of the restrained body showed a
greater shrinkage than the lower part.
Fig. 6 illustrates the finally identified spring
coefficient values obtained from the FEM analysis. The
spring coefficient in the horizontal direction was changed
with the propagation of the cleavage rather than by the
age.
The question for M2, in which the restrained body is
isolated from the restraining body by the sliding
mechanism, was, `how much should have been set as the
spring coefficient in the horizontal direction in the
analysis using the FEM for the static friction coefficient
of 0.1 for the measured sliding mechanism?' The result of
the analysis for M1, though setting the spring coefficient

at 1/10 of the K_h before the cleavage, corresponded closely
to the results obtained from the actual measurement.   An
analysis of M3 using the FEM was carried out through the
employment of the same spring coefficient value as that
used before the occurrence of the cleavage of M1.   As a
result, the cracks of the restrained body which were
observed on the third date for the age could be
analytically explained. The analytical result of the
restrained body possessing the age before the occurrence
of cracks corresponded quite closely to the measurement
result.

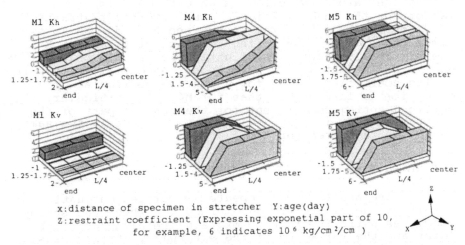

x:distance of specimen in stretcher Y:age(day)
Z:restraint coefficient (Expressing exponetial part of 10,
         for example, 6 indicates 10^6 kg/cm^2/cm)

Fig.6. Spring coefficients identified in FEM

4.2 Results of analyses using L/H as parameters for specimens of M4 and M5

The problem in the identification of the spring
coefficient for M4 and M5 was in being unable to take into
account the variation of the spring coefficient in the
same manner for M1.
   The following three rules exist in the arrangement of
the identification process for the spring coefficient in
the analysis of M1.
1)   The vertical spring coefficient value is set at 0 when
the condition attains to the occurrence of the cleavage.
2)   The horizontal stiffness is gradually reduced when the
cleavage occurs.
3)   Fundamentally there is no time variation for the
spring coefficient.
   However, when trying to identify the spring coefficient
for M4 and M5 under the same rules as those for M1, the
analytical result corresponding to the measured behavior
was not recognized. In an extreme case, the analytical
result was found to be different from the measured result
in the generating tendency of the stress.
   As the final conclusion, the spring coefficient values

were changed in a point of time at the following three stages: (I) temperature rise, (II) temperature drop and (III) steady temperature.

The location where the vertical spring coefficient values is small, displacement of restrained body tends to upward, otherwise the values is large, displacement tends to downward.

The deformation of the restrained body, which can be presumed from the change of the spring coefficient values, corresponds quite closely to the deformation which was observed. Considering from this point of view it was shown that the thermal deformation behavior of concrete during the hardening period is an extremely important factor for the investigation of the external restraint.

Since the height of the restrained body for M5 is two times that for M4, the spring coefficient value for M5 in the vertical direction is slightly larger than that for M4 due to the effect of the self-load. According to this the spring coefficient value for M5 in the horizontal direction is larger than that for M4. In particular, the vertical spring coefficient for the central part of the restrained body in stage (I) is not released in the same manner as that seen for M4.

5 Conclusion

1) The analysis was repeatedly carried out until the result of the finite element method, in which the spring model was built in order to explain the cleavage of the joint and the variation of the external restraint caused by the relative displacement, corresponded to the values for the stress, strain and displacement which were observed for each of the specimens. Then, each value for L/H and the spring coefficient of the joint according to the bonding state were identified.

2) From the study of the identified spring coefficient it was found that in the case of L/H being 15 (M1,M2,M3), the spring coefficient value changes with the occurrence of cracks. On the other hand, in the case of the value for L/H being 5 or 2.5 (M4, M5), the result obtained from the actual measurement can not completely be explained unless the spring coefficient is changed in a point of time. Therefore, within the scope of this paper, it can be considered that there is a possibility of the mechanism for the external restraint which is more complicated in the case of the value for L/H being small than in the case of it being large.

6 Reference

Ishikawa,M. Maeda,T. Nishioka,T. and Tanabe,T.(1989) An experimental study on thermal stress and thermal deformation of massive concrete. JSCE 408/v-11 pp.121-130

PART SIX

INFLUENCE OF CONSTITUENTS AND COMPOSITION OF CONCRETE ON CRACKING SENSITIVITY

(Influence des constituants
et de la composition du
béton sur la fissurabilité)

24 THE EFFECT OF SLAG ON THERMAL CRACKING IN CONCRETE

M.D.A. THOMAS
Cement & Concrete Studies Ltd, Milton, Ontario, Canada
P.K. MUKHERJEE
Ontario Hydro Technologies, Toronto, Canada

Abstract
This paper reviews problems of high temperature rises (>45°C above ambient) and thermal cracking encountered in the field with concrete foundation pads containing 30% to 35% slag and reports the results of a laboratory study to evaluate the effect of slag on thermal cracking. The heat development of concretes with a range of slag levels was determined in the laboratory by semi-adiabatic calorimetry. In addition, the flexural strength and tensile strain capacity of companion specimens were determined at 1, 3 and 7 days. The results show that slag is effective in reducing both the initial rate of heat development and the maximum temperature rise. However, it is shown that the benefit of reduced thermal stress is offset by a concomitant reduction in the tensile strength and strain capacity of slag concrete at early ages. It is concluded that the partial replacement of Portland cement with the slag used in this study may not be effective in mitigating thermal cracking.
Keywords: Cracking, Effect of Slag on Strength, Slag, Development of Temperature, Tensile Strain Capacity.

1 Introduction

In recent years, Ontario Hydro has experienced problems with early-age cracking in reinforced concrete transmission tower foundation pads despite the precautionary use of slag to control thermal cracking. An examination of 173 accessible foundation pads showed the cracking to vary from none to extensive (i.e., crack widths greater than 0.25mm over a large area of the exposed pad). A total of 49 foundation pads were classified as being extensively cracked. These pads contained between 350 to 400 kg/m^3 of cementitious material, which comprised Type 10 cement with 25 to 35% slag in 48 cases and Type 20 cement in one case. In some cases, the temperature rise in the slag concrete was in excess of 45°C above ambient, this value being higher than the predicted temperature rise for a comparable concrete without slag. It was considered that relatively high placing temperatures and high average ambient temperatures would have contributed to the high temperature rise within the foundation pads. Such high temperatures may also have accelerated the reaction of the slag and thereby its contribution to the temperature rise. Another factor that may have contributed to thermal cracking is the high temperature gradients generated in the pads; temperature differences between the mid-point and 75mm to 100mm from the surface of the pad were observed to be as high as 20°C after only 48 hours in some cases.

Thermal Cracking in Concrete at Early Ages. Edited by R. Springenschmid. Published 1994 by E & FN Spon, 2–6 Boundary Row, London SE1 8HN, UK. ISBN: 0 419 18710 3.

It was also considered that the slower strength development of slag concrete may have contributed to thermal cracking. The reduced tensile strength and strain capacity of slag concretes at early ages lowers the ability of the concrete to resist thermal stresses and strains. In order to evaluate the effect of slag on thermal cracking, a laboratory study was initiated to determine the effect of slag on the autogenous temperature rise and tensile strength and strain capacity at early ages.

2 Experimental details

Five concrete mixes were prepared using Type 40 cement or Type 10 cement blended with slag, at slag contents of 0%, 25%, 35% and 50%; concrete mix proportions are given in Table 1. The mix proportions were selected to have similar cement contents and water/cement ratios as those concrete mixes that had previously exhibited thermal cracking in transmission tower footings.

The semi-adiabatic temperature rise for each concrete mix was determined by casting a concrete cylinder (150 X 300 mm), incorporating a temperature probe, into a 160-litre thermally insulated drum. The output from the temperature probe was recorded and later analyzed to determine the heat of hydration for the cementitious materials in the concrete (the insulated drum having been calibrated for heat loss). In addition, the heat release characteristics of cements and cement/slag combinations (25, 35 and 50% slag) were determined by conduction calorimetry using pastes with water/cement ratio of 0.5.

A further three cylinders (150 mm x 300 mm) were cast from each mix for compressive strength determination and six prisms (90 mm x 100 mm x 400 mm) for flexural strength tests. The specimens for strength tests were demoulded at 1 day and stored in a fog room at 23°C until test.

The prisms were tested for flexural strength under third point loading. Prior to test, a foil strain gauge (50 mm long) was attached to the tensile surface at the centre of the test specimen. Loading was applied at 0.28 MPa (outer fibre stress) per minute.

Table 1. Concrete Mix Proportions

	Type 10	Type 40	Type 10/ 25% Slag	Type 10/ 35% Slag	Type 10/ 50% Slag
Cement	350	354	263	229	175
Slag	0	0	88	123	175
Water/Cement	0.4	0.4	0.4	0.4	0.4
Coarse Agg.	1080	1092	1081	1086	1079
Fine Agg.	760	768	760	764	759

All proportions are in kg/m^3. A water reducing admixture and an air-entraining admixture were used to achieve slump values from 60 mm to 100 mm and air content from 6% to 7%.

3 Results

Compressive and flexural strength results are given in Table 2. The absolute tensile strain capacity (i.e. strain at failure) was difficult to determine from the load-strain curves in many cases. Consequently, the strain at near failure, defined as the strain at 90% of the ultimate strength, was taken from the curves and the mean values are given in Table 2.

Table 2. Results from compressive and flexural strength tests.

	Test Age (days)	Compressive strength (MPa)	Flexural Strength (MPa)	Strain at 90% Flexural Strength (microstrain)
Type 10	1	-	3.9	82
	3	-	4.6	91
	7	-	4.8	107
	28	32.9/31.2	-	-
Type 40	1	-	2.4	62
	3	-	3.7	81
	7	-	4.0	75
	28	38.9/32.3	-	-
Type 10/ 25% slag	1	-	3.2	66
	3	-	3.9	79
	7	-	4.4	81
	28	37.0	-	-
Type 10/ 35% slag	1	-	2.8	62
	3	-	3.8	80
	7	-	4.5	81
	28	38.6	-	-
Type 10/ 50% slag	1	-	2.0	56
	3	-	3.1	68
	7	-	4.0	74
	28	39.3	-	-

These results show a general trend of decreasing early-age tensile strength and tensile strain capacity with increasing slag content. The concrete containing Type 40 cement also exhibits reduced strength and strain capacity compared with the concrete with Type 10 cement. The tensile strength increases with age and the effect is more marked for concretes with high levels of slag. The strain capacity increases with age but to a lesser extent than the strength, indicating an increase in elastic modulus with age. After 7 days curing the strain capacity of the slag concretes is still considerably lower than that of the plain Type 10 concrete.

The autogenous temperature rise measured for the five concrete mixes is shown in Figure 1. The concrete with Type 10 cement (no slag) exhibits the most rapid temperature rise and the highest peak temperature (50°C). The effect of slag is to delay the peak temperature and reduce its magnitude, especially at the higher slag level. The concrete with Type 40 cement exhibits similar behaviour to the concrete with moderate levels of slag. After reaching a peak, the temperature of the Type 10 concrete decreases rapidly and after just over 2 days the Type 10 cement concrete has the lowest temperature.

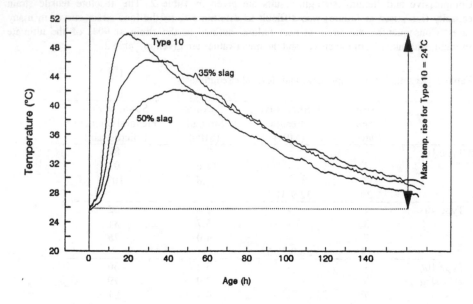

Fig. 1. Autogenous temperature rise of concrete cylinders

The cumulative heat of hydration, normalized to heat per unit mass of cementitious material, for these concretes is shown in Figure 2. Although there are marked differences in the rate of heat evolution at early ages, only the concrete with 50% slag has a significantly lower heat of hydration after 7 days.

The cumulative heat development curves of the different cements and cement/slag combinations measured using conduction calorimetry are shown in Figure 3. The heat of hydration determined on these cements by conduction calorimetry are consistently higher than those determined on the concrete samples under semi-adiabatic conditions. In addition, conduction calorimetry indicates substantial differences between cement types, the Type 10 having the highest 7 day heat of hydration.

4 Discussion

The higher strength and equal heat of hydration of the Type 40 cement concrete compared with the Type 10 cement concrete is unusual. The 7-day heat of hydration for the Type 10 cement as determined by conduction calorimetry (332 J/g) is normal for ordinary Portland cement (Neville, 1987). Conversely, the result for the Type 40 cement was higher than that usually observed for low-heat Portland cement and was only just below the limit specified by CSA A5 (275 J/g at 7 days). The compound composition (Bogue) of this Type 40 sample shows relatively high C_3S

and C3A and low C2S contents for a low-heat Portland cement, although the sample complies with the compound requirements of CSA A5 for low-heat Portland cements (i.e., C3A<5.5%).

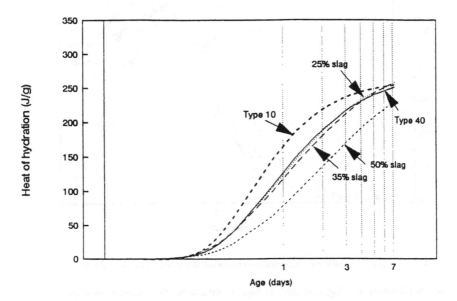

Fig. 2. Heat of hydration - from autogenous temperature rise of concretes.

For a given blend of cementitious materials, the heat of hydration determined from the autogenous temperature rise measured for concrete under semi-adiabatic conditions is lower than that determined by conduction calorimetry using cement paste samples. These differences might be due to the inherent differences in the test methods and materials used. The different water/cementitious material ratios (0.5 for cement in conduction calorimetry and 0.4 for concrete in semi-adiabatic test) will undoubtedly lead to disparities in hydration rates, lower ratios resulting in a reduced rate and degree of hydration (Lea, 1970).

Conduction calorimetry tests on cement samples indicated that partial replacement of Type 10 cement with slag has a large effect on the 7-day heat of hydration, whereas analysis of data from semi-adiabatic tests on concrete show comparatively little difference in the heat released during 7 days. Conduction calorimetry is an isothermal test and any effects of temperature rise on hydration rates are eliminated. Conversely, in the semi-adiabatic test, the temperature of the thermally-insulated concrete test sample increases as hydration proceeds (with only a small amount of heat loss). The higher temperature accelerates hydration reactions and the effect is more marked for slag than Portland cement such that, at sufficiently high temperatures the heat released from slag blends may exceed that of Type 10 Portland cement (Coole, 1988; Roy and Idorn, 1982). Thus slag is less effective in reducing heat evolution under semi-adiabatic conditions that are representative of mass concrete.

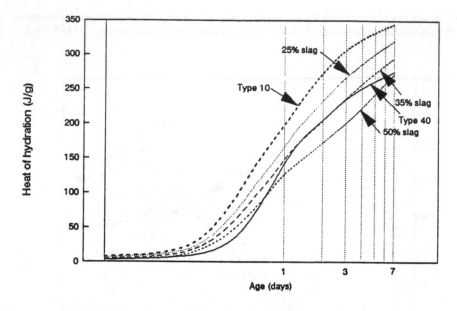

Fig. 3. Heat of hydration - conduction calorimetry tests on cement pastes

Despite the acceleration of slag reactions at high temperature, the incorporation of slag was effective in reducing the maximum autogenous temperature rise in concrete. The maximum temperature rises from Figure 1 for the different concretes are as follows:-

Type 10	24°C
Type 40	22°C
Type 10 + 25% slag	21°C
Type 10 + 35% slag	20°C
Type 10 + 50% slag	16°C

These results indicate that the temperature rise in a large pour may be reduced by the partial replacement of Type 10 cement with the slag and placing temperatures used in this study. This will lead to smaller volume changes on cooling and, consequently, smaller stresses if the volume change is restrained.

However, the results from the flexural tests indicate that the tensile strength and tensile strain capacity of concrete at early ages generally decrease as the slag content increases. This means that slag concretes are less able to withstand thermal stresses at early ages. The reduced tensile strength may be expected to partially or wholly offset the benefits of reduced temperature rise in slag concrete.

An indication of the risk of thermal cracking in slag concrete compared to plain Type 10 cement concrete can be made by a consideration of the data obtained in this study. If it is assumed that all the compressive stresses generated as the concrete temperature increases are relieved by creep and that cooling of the concrete occurs rapidly thereby permitting no relaxation of tensile stresses by creep (i.e. worst case), cracking will occur when the restrained thermal strain exceeds the tensile strain capacity of the concrete (Harrison, 1981), i.e.:

$$\text{when} \qquad \alpha.\Delta\theta.R > \varepsilon_t \qquad\qquad (1)$$

where α = coefficient of thermal expansion
$\Delta\theta$ = maximum temperature rise
R = restraint factor
ε_t = tensile strain capacity

The restraint factor is a measure of the degree of external restraint and is dependent on the dimensions of the pour and the nature of the external restraint; typical values of restraint factor are given in reference (Bamforth, 1982).

Although this is an oversimplification of the mechanisms controlling thermal cracking, it serves to compare the risk of cracking in concrete for concretes with different slag levels. The restraint factor will not be affected by cement type, and the coefficient of thermal expansion is largely a function of the aggregate type. Hence, for concrete with the same aggregate, the risk of thermal cracking is proportional to the maximum temperature rise during hydration and inversely proportional to the tensile strain capacity of the concrete. The ratio of maximum temperature rise (T_{max}) to tensile strain (ε_1, ε_3 and ε_7 = strain capacity at 1, 3 and 7 days) for the five concretes used in this study are as follows:

Concrete mix	T_{max}/ε_1	T_{max}/ε_3	T_{max}/ε_7
Type 10	0.29	0.26	0.22
Type 40	0.32	0.27	0.29
Type 10 + 25% slag	0.31	0.27	0.26
Type 10 + 35% slag	0.32	0.25	0.24
Type 10 + 50% slag	0.29	0.23	0.22

The risk of thermal cracking increases as this ratio increases. The range of values for a particular concrete reflects the change in tensile strain capacity with age from 1 to 7 days. It can be seen that the slag content has little influence on this ratio as the reduction in temperature is offset by a concomitant reduction in tensile strain capacity. This implies that the slag may not be effective in limiting thermal cracking.

This approach makes no allowances for stress relaxation due to creep or the rate of temperature rise, both of which may be expected to be affected by the presence of slag. However, the results emphasize that reducing the temperature rise in concrete through the use of supplementary cementing materials may not be a sufficient precaution against the occurrence of thermal cracking without consideration of other controlling factors.

5 Conclusions

1. The slag used in this study was effective in reducing the rate of heat evolution and the maximum temperature rise in concrete when blended with a Type 10 cement.

2. The tensile strength and tensile strain close to failure (at 90% of strength) for concrete up to 7 days old decreased with increasing slag content.

3. Based on a simplified approach to evaluate the effect of slag on the thermal cracking of concrete, it is concluded that the benefits of reduced temperature rise in slag concrete may be offset by the lower strength and strain capacity at early ages. This also appears to be the case for the concrete made using Type 40 cement.

Acknowledgement

Permission to publish this paper has been given by the President of Ontario Hydro Technologies.

References

P.B. Bamforth, **Early-age Thermal Cracking in Concrete**, Institute of Concrete Technology Report, Slough, (1982).

M.J. Coole, **Magazine of Concrete Research**, 40(144), pp 152-158, (1988).

T.A. Harrison, **Early-age Thermal Crack Control in Concrete**, CIRIA Report 91, Construction Industry Research and Information Association, London, (1981).

F.M. Lea, **The Chemistry of Cement** (3rd Edition), p. 294, Chemical Publishing Company Inc., New York, (1970).

A.M. Neville and J.J. Brooks, **Concrete Technology**, p. 26, Longman Scientific & Technical, London, (1987).

D.M. Roy and G.M. Idorn, **ACI Journal**, 79(6), pp 444-457, (1982).

25 MINIMIZATION OF THERMAL CRACKING IN CONCRETE MEMBERS AT EARLY AGES

R. BREITENBÜCHER
Philipp Holzmann AG, Frankfurt, Germany
M. MANGOLD
Institute of Building Materials, Technical University of Munich, Germany

Abstract
If the deformations due to temperature changes or shrinkage are prevented in concrete members, restraint stresses are caused which can
result in cracking. The heat of hydration is thus of importance for
thick members. To reduce the risk of cracking, besides constructional
measures the concrete mix must be optimized. For this purpose the
influence of technological parameters on thermal cracking in concrete
at early ages was studied in numerous laboratory tests with the cracking-frame. With this test device the restraint stresses in a concrete
specimen can be measured continuously from the beginning of hydration,
whereby the concrete hardens under the same temperature conditions as
a member of about 50 cm in thickness. The cracking tendency of different concretes can be estimated by the cracking temperature.
Keywords: Cracking Frame, Restraint Stresses, Cracking Tendency,
Cement, Aggregates, Fresh Concrete Temperature.

1 Introduction

In concrete members cracking often occurs already a few days or a few
weeks after pouring without the action of any external force. Such
cracks are caused by restraint stresses which are formed when the
deformations due to temperature changes or shrinkage are restrained.
Especially in thick members, the heat of hydration is thus particulary
important. This not only has to be taken into consideration for mass
concrete but also for members between about 30 to 80 cm in thickness,
e.g. in tunnel linings (see Springenschmid and Breitenbücher (1985),
Breitenbücher (1989)). In such cases the maximum temperature is
reached about one day after pouring. The concrete then requires about
one week to cool down to the ambient temperature. If the restraint
(tensile) stresses reach the tensile strength, which is still low,
cracking occurs.
   The risk of such thermal cracking can partly be reduced by constructive measures. For this purpose limited dimensions of the single
sections are useful as well as the minimization of friction, e.g. by
the use of foils between the new concrete member and the subbase.
By reinforcement only the crack width can be limited, cracking itself
cannot be prevented by this measure. Thermal cracking is mainly caused
by technological parameters. It therefore appears to be more appropriate to minimize the restraint stresses by technological measures
from the very start, so that cracks are not produced at all rather
than only limiting the crack width by reinforcement.

Thermal Cracking in Concrete at Early Ages. Edited by R. Springenschmid. Published 1994
by E & FN Spon, 2–6 Boundary Row, London SE1 8HN, UK. ISBN: 0 419 18710 3.

For a long time such measures were limited to the use of low-heat cements. However, by this means only the temperature rise is lowered. To reduce the risk of cracking, consideration must also be given to the development of the modulus of elasticity and the relaxation, which also influence the restraint stresses, and the development of the tensile strength (Springenschmid and Breitenbücher (1986)).

2 Restraint Stresses in Concrete Members at Early Ages - Tests with the Cracking Frame

Before the above-mentioned technological measures can be taken the parameters which influence the restraint stresses must be studied. It must be taken into account, that the E-modulus and relaxation change continuously over a wide range especially at this early age. For this reason a calculation of thermal stresses is very difficult with only low accuracy. Also it is not possible to determine restraint stresses **in situ** directly by measuring the deformation, because these stresses are caused by the inhibition of deformations. If nevertheless deformations are measured, they are not restrained and so cannot result in restraint stresses.

The continuous change in the concrete properties results in a non-simultaneous development of the temperature and thermal stress. Because of these difficulties, comparative tests in the laboratory are necessary for studying restraint stresses and thermal cracking in concrete at early ages. For this purpose a special test device, the so called cracking frame (Fig. 1), was developed at the Institute of Building Materials at the Technical University of Munich (see e.g. Breitenbücher (1989)).

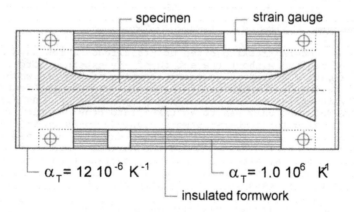

Fig. 1. The Cracking Frame.

This device consists of a closed frame with two cross-heads and two massive steel bars (diameter 100 mm). The fresh concrete is placed directly into the formwork of the frame and is compacted by vibration. Due to this, the deformations are restrained from the beginning in the longitudinal direction.

Since the formwork is built with a thermally insulating material (d=50 mm), the concrete specimen with a cross-section of only 150 mm x 150 mm heats up without any external heating in the same way as a member of approximately 50 cm in thickness.

The deformations of the specimen are not prevented completely but by at least 75%: the steel bars of the frame deform to a small extent due to the reaction force, which is of the same order as the restraint force in the specimen, only with an opposite sign. Thereby the specimen is able to deform in the same way. By continuously measuring the deformations of the steel bars with strain-gauges and multipliying by their stiffness, the force and the restraint stresses in the concrete can be determined at all times.

A typical result of such a test is shown in Fig. 2. Soon after pouring the temperature increases due to the onset of hydration and the corresponding heat development. At this stage no compressive stresses are caused, since the concrete can still be deformed plastically. After about 3 to 6 h compressive stresses are obtained due to an increasing E-modulus and further heating. However, these stresses are still reduced by the relatively high relaxation.

Thus the maximum in compressive stresses is obtained a few hours earlier than the temperature maximum. Having reached the maximum temperature the concrete temperature falls and the remaining compressive stresses decrease rapidly, because the relaxation is still high and the modulus of elasticity is already higher than during the first heating.

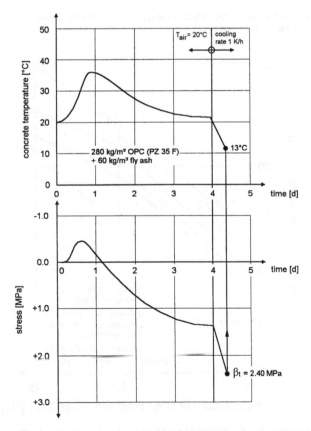

Fig. 2. Temperature and stress curves of concrete tested in the cracking frame

Due to this drop in temperature, the concrete reverts to a stress-free condition at a temperature which is far above the initial temperature and only a few degrees below the temperature maximum. The concrete temperature at this time is called **zero stress temperature, T_z** (Springenschmid and Nischer (1973)). On further cooling relatively high tensile stresses develop because the E-modulus has increased and the relaxation is smaller than in the heating phase.

After about 4 days the ambient temperature is normally reached again in the specimen. If the concrete has not yet cracked by this time, the specimen is further cooled down artificially at 1 K h^{-1} to obtain a crack. When the tensile stresses exceed the tensile strength, cracking occurs. The specimen temperature at the time of cracking is defined as the cracking temperature of the concrete. It characterizes the cracking tendency due to heat of hydration (see e.g. Breitenbücher (1989)): the lower the determined cracking temperature, the less the cracking tendency of concrete and the less the risk of cracking on site. In repetitive tests the results differed by less than 1.5 K and 0.1 MPa, showing a high reproducibility of the results.

3 Test Results

In extensive tests various technological factors have been investigated in the cracking frame.

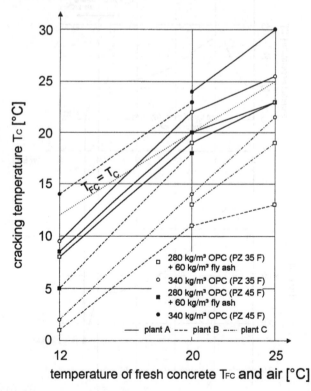

Fig.3. Influence of the temperature of fresh concrete on the cracking temperature

3.1 As expected, the **temperature of the fresh concrete** was found to be of great importance regarding thermal cracking. By lowering the temperature of the fresh concrete from 25°C to 12°C (i.e. by 13 K) the cracking temperature could be decreased by about 15 to 19 K (Fig. 3).

This large improvement owing to a low temperature of the fresh concrete can be attributed to a smaller temperature increase during hydration and an increase in the tensile strength. The last point can be explained by a stronger bonding of the hydration products (CSH) at low hardening temperatures, which was also found in other investigations (see e.g. Verbeck and Helmuth (1968), Wischers (1961)).

In field work, the absolute cracking temperature is decisive for the estimation of the risk of cracking. Thus the most effective way of lowering the risk of cracking is to cool the fresh concrete to a suitably low temperature.

3.2 **Aggregates,** especially coarse aggregates, influence the cracking tendency of a concrete essentially by their coefficient of thermal expansion and their roughness. In concretes with quartzite aggregates, which have a coefficient of thermal expansion 50% higher than e.g. those with limestone or basalt, approximately 50% higher restraint stresses were observed in the cracking frame for almost the same temperature curve; in concretes with quartzite aggregates the tensile strength was exceeded at a 6 to 9 K higher cracking temperature than in concretes with limestone or basalt.

3.3 A high **cement content** or a high **cement strength** class leads to more heating and therefore to a relatively high cracking temperature. Corresponding to this the cracking temperature decreased with increasing water/cement ratio in the normal range of about 0.4 to 0.7. If the water/cement ratio is much higher than about 0.7 the cracking temperature increases again because then the tensile strength of the concrete is very much lowered (Breitenbücher (1989)). When the cement content of the concrete tested was lower than 340 kg m^{-3} fly-ash was added to the mix to obtain sufficient workability. The fly-ash itself did not influence appreciably the cracking tendency of the concrete, because in the testing period, i.e. the first 4 to 5 days, the pozzolanic fly-ash does not strongly react and can be considered here as inert material, fig.4.

3.4 **Slag cements** are normally used as **low-heat cements** for mass concrete. Such cements lead to a relatively small temperature rise during hydration and also to small tensile stresses. However, in some concretes with slag cements a higher cracking temperature was found than in concretes using OPC from the plant with the same clinker (Springenschmid, Breitenbücher, Kussmann (1987)). This is related to a very slow developement in the tensile strength of the concrete.

3.5 **Cements of the same type and strength class, but from different plants,** i.e. with different chemical composition and/or fineness, showed a very different behaviour (Fig. 5). In 17 concretes made with PZ 35 F (OPC) the cracking temperatures were found to range from 6°C to 22°C (Breitenbücher (1989)). Although the temperature curves of these concretes are different, the high variation in the cracking temperature can only be partly attributed to this. The difference also points to additional non-thermally induced expansions of various magnitudes. For this behaviour the chemical composition and the fineness are of importance. It could be proved, that cements with low alkali content (K$_2$O, Na$_2$O), high sulphate content related to the C$_3$A-content and not too high fineness lead to low cracking temperatures. In view of these results it could be expected, that a higher sulphate content (i.e. additional adding of gypsum) could decrease the temperature. This, however, is true only for cements with low-alkali cements (Breitenbücher (1991)). In high-alkali cements an increased sulphate content does not improve the resistance against thermal cracking.

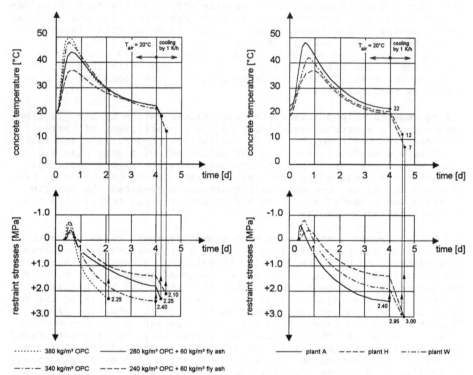

fig.4. Temperature and stress curves of concretes with different cement contents OPC (PZ 35 F).

fig. 5. Temperature and stress curves of concretes with cement OPC (PZ 35F) from different plants.

3.6 However, when not only gypsum or anhydrite, but calciumsulpho-aluminates are added as **expansive additives** to the concrete mix (such are already used e.g. in Japan, USA or Italy) the compressive stresses were found to be much higher than in concretes without such additives, although the temperature behaviour was essentially unchanged. In this way the thermally induced tensile stresses were minimized or fully compensated by the compressive stresses due to prevented expansion, and the cracking temperature could be lowered strongly.

3.7 Concretes containing **air-entraining agents** have a lower modulus of elasticity so that the restraint stresses are reduced. Furthermore the ultimate strain (strain capacity) of air-entrained concrete is higher than in normal concrete. It was found in the cracking tests that by adding air-entraining agents (air content approximately 3 to 6 Vol.-%) the cracking temperature could be reduced by about 5 K.

4 Experiences on Site

The use of suitable concrete constituents and mix compositions with a low crack susceptibility resulted in a considerable reduction of cracking in practice, see Springenschmid, Breitenbücher, Mangold (1994)

5 Conclusions

In mass concrete but also in members with a thickness of only about 30 to 80 cm the heat of hydration is of particular importance regarding stresses and thermal cracking. Up to now low-heat cements were normally used to prevent thermal cracking in such members.

Using these cements, only the temperature rise during hydration can be lowered. Important parameters like E-modulus, relaxation and tensile strength remain unconsidered.

To study the influence of the technological parameters on thermal cracking, different concretes were investigated in the cracking frame. Although not all the problems in connection with this have yet been solved, it is already possible to give the following guidelines for construction practice:

- The decisive criteria for lowering the risk of cracking are a low heat of hydration and also - to put it simply - the conversion of a considerable proportion of the restrained deformation into compressive stresses.
  The E-modulus and the relaxation together with their development are responsible for the latter. Thus during cooling only small tensile stresses are caused. Furthermore the strength must develop sufficiently quickly.
- The restraint stresses in concretes at early ages are difficult to calculate with sufficient accuracy because the E-modulus and the relaxation are changing continuously.
- The tendency to thermal cracking in concretes at early ages can be assessed by the cracking temperature determined in the cracking frame: the lower the cracking temperature, the less the cracking tendency.
- By cooling the fresh concrete, the tendency to thermal cracking can be reduced to a great extent.
- Low-heat cements do indeed lead to a small temperature increase during hydration; in some cases, however, the E-modulus and the tensile strength of the concrete develops so slowly that cracking occurs at a relatively high temperature.
- There are some favourable cements which swell slightly during hydration. In this case the thermally induced tensile stresses are compensated to some extent and the cracking tendency is lowered. In the same way expansive additives improve the concrete behaviour.
- Due to a higher coefficient of thermal expansion the restraint stresses and also the tendency to thermal cracking of concretes containing quartzite are much higher than those of concretes with limestone or basalt.
- A high early concrete strength increases the risk of cracking. By an adequate optimization of the concrete mix regarding the temperature of the fresh concrete, it is possible, however, to obtain a sufficient early strength as well as a low tendency to thermal cracking.

6 References

Breitenbücher,R. (1989) **Zwangsspannungen und Rissbildung infolge Hydratationswärme**, Doctoral thesis, Technical University of Munich.

Breitenbücher, R. (1991) Reißneigung von jungem Beton - Einfluß des Zements. **Proceedings of 11th ibausil Symposium HAB**, Weimar 1991, Vol.1, pp. 70-85.

Springenschmid, R. and Nischer, P. (1973) Untersuchungen über die Ursache von Querrissen im jungen Beton, **Beton und Stahlbetonbau** 68 (1973), no. 9, pp. 221-226.

Springenschmid, R. and Breitenbücher, R. (1985) Über die Ursache und das Vermeiden von Rissen im Beton von Tunnelauskleidungen, **Felsbau** 3 (1985) no. 4, pp. 212-218.

Springenschmid, R. and Breitenbücher, R. (1986) Are low-heat cements the most favourable cements for the prevention of cracks due to heat of hydration?, **Concr. Precasting Plant Technol.** 52 (1986) no. 11, pp. 704-711.

Springenschmid, R., Breitenbücher, R., Kussmann, W. (1987) Cracking tendency of concretes with blastfurance slag cements due to outflow heat of hydration, **Concr. Precasting Plant Technol.** 53 (1987) no. 12, pp.817-821.

Springenschmid, R., Breitenbücher, R., Mangold, M. (1994) Practical Experience with Concrete Technological Measures to Avoid Cracking, **Proceedings from Rilem International Symposium on "Thermal Cracking in Concrete at Early Ages"**, Munich, October 10-12, 1994, (to be published).

Verbeck, G.J. and Helmuth, R.H. (1968) Structure and physical properties of cement paste, **Proceedings of 5th International Symposium on Chemistry of Cement**, Tokyo, 1968, Part III, pp. 1-32.

Wischers, G.(1961) **Einfluß einer Temperaturänderung auf die Festigkeit von Zementstein und Zementmörtel mit Zuschlagstoffen verschiedener Wärmedehnung,** Doctoral thesis, RWTH Aachen 1961.

26 THE EFFECT OF THERMAL DEFORMATION, CHEMICAL SHRINKAGE AND SWELLING ON RESTRAINT STRESSES IN CONCRETE AT EARLY AGES

K. SCHÖPPEL
Ingenieurbüro, Munich, Germany (formerly at Institute of Building Materials)
R. SPRINGENSCHMID
Institute of Building Materials, Technical University of Munich, Germany

Abstract
In the Temperature-Stress Testing machine the stress of concrete specimens under full restraint caused by thermal expansion and chemical shrinkage and swelling have been measured. Concretes made with different cements were investigated under isothermal (12, 20 and 30°C) and semi-adiabatic conditions with complete or without any restraint. Besides the thermal deformation, the nonthermal deformation can decisively influence the development of the restraint stresses in concrete at an early age. The development of the nonthermal deformation was mainly affected by temperature, cement content and type and composition of the cements.
Keywords: Chemical (Autogenous) Shrinkage, Swelling, Thermal and Nonthermal Deformations, Restraint Stresses.

1 Introduction

During the early hydration of concrete the thermal deformation superposes on volume changes due to nonthermal deformations. Thermal deformation is caused by the heat of hydration or external temperature effects and the resulting change in temperature of the concrete element. Nonthermal deformation is caused by shrinkage and swelling of the concrete. For concrete at an early age between three types of shrinkage can be distinguished, see Grube (1991)

(1) **Plastic** or **early shrinkage** which occur as a result of capillary force in the plastic non hardened concrete due to loss of water
(2) **Chemical shrinkage** which occurs because the volume of the reaction products is less than the sum of the volumes of the original constituents, i.e. the water and the anhydrous cement. If water cannot penetrate from outside into the concrete internal self-desiccation occurs and causes a contraction of the concrete volume. This occurrence is also called **autogenous shrinkage.** (High strength concretes have a high autogenous shrinkage.)
(3) **Drying shrinkage** of the hardening concrete by loss of moisture.

Type 1 and 3 of shrinkage will not be dealt with because in the present investigations loss of moisture was prevented.

Swelling of the concrete can be caused by the absorbtion of water and/or the chemical reaction of the sulphate components of the cement (**chemical swelling**), see Nolting (1989).

Thermal Cracking in Concrete at Early Ages. Edited by R. Springenschmid. Published 1994 by E & FN Spon, 2–6 Boundary Row, London SE1 8HN, UK. ISBN: 0 419 18710 3.

If deformations are prevented restraint stresses occur which cause cracking when the tensile strength of the concrete has been reached. Up to now the cracking sensitivity of a concrete mixture can be determined only experimentally. Numerous investigations have been carried out with the cracking frame at the Institute of Building Materials of the Technical University of Munich, see Breitenbücher (1989), Springenschmid and Breitenbücher (1990, 1991) and Springenschmid et al. (1994). On the basis of these investigations it is possible to assess the effect of technological measures on the cracking sensitivity in situ. It is well known that the cracking sensitivity of a concrete mixture is mainly influenced by the fresh concrete temperature, the cement content, the type and the chemical composition of the cements and the coefficient of thermal expansion of the aggregate.

2 Investigations with the Temperature-Stress Testing Machine

With the cracking frame it is not possible to distinguish between the different types of causes for deformation and consequently assess their influence on the development of the restraint stress. Therefore a special testing machine, the Temperature-Stress Testing machine (TST machine, see fig. 1), was developed which allows measurements of the longitudinal deformation without restraint as well as the longitudinal stress under full restraint for any temperature development of the concrete specimen, see Schöppel (1993) and Schöppel et al. (1994)

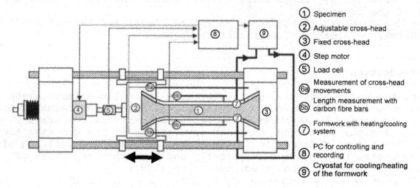

① Specimen
② Adjustable cross-head
③ Fixed cross-head
④ Step motor
⑤ Load cell
⑥ⓐ Measurement of cross-head movements
⑥ⓑ Length measurement with carbon fibre bars
⑦ Formwork with heating/cooling system
⑧ PC for controlling and recording
⑨ Cryostat for cooling/heating of the formwork

Fig. 1. Temperature-Stress Testing machine (TST machine)

The concrete is placed and compacted in the TST machine. The tests described in this paper were carried out either with **full restraint** or **without restraint**. The temperature was either kept constant (**isothermal conditions**) or **semi-adiabatic** conditions applied. In the semi-adiabatic tests the temperature conditions of one meter thick concrete elements were simulated. Based on the results of adiabatic tests, the temperature increase due to the hydration heat was calculated. The concrete temperature in the TST machine was controlled according to the calculated temperature development. The investigations were limited to the influence of the following factors on the development of the restraint stresses
- Fresh concrete temperature,
- Cement content and type
- Chemical composition of the cements.
   The concrete components and mixtures listed in table 1 were used.

Table 1: Concrete components and concrete mixtures

Cement	ordinary Portland cement 35 F (CEM I 32,5 R) or blast-furnace slag cement 35 L (CEM III A 32,5)
Aggregates	morainal sand and gravel (mainly limestone) from the area of Munich
Cement content	340 kg/m³ or 280 kg/m³ and 60 kg/m³ fly ash
Water content	170 kg/m³

Commercially available Portland cements from five different German plants were used. The cement data are compiled in table 2.

Table 2: Characteristic data for the Portland cements (OPC) used.

cement	C$_3$A %	C$_3$S %	C$_2$S %	C$_4$AF %	SO$_3$ %	soluble K$_2$O %	soluble Na$_2$O %	soluble Alk %	d' μm	n
OPC A	11.62	46.68	24.18	6.75	2.77	0.75	0.11	0.60	28.0	0.97
OPC B	10.93	53.67	21.64	6.41	2.53	0.39	0.05	0.31	27.2	0.91
OPC C	10.96	50.92	21.65	7.80	3.10	0.86	0.08	0.65	25.1	0.96
OPC D	10.01	54.09	18.91	7.87	3.19	0.89	0.03	0.62	24.1	0.91
OPC E	10.92	52.35	18.79	9.24	2.81	1.47	0.01	0.98	19.7	1.02

soluble Alk = soluble alkali equivalent (= soluble Na$_2$O + 0.658 soluble K$_2$O)
d'= position parameter in the RRSB granulometry diagramm, see DIN 66145 (1976)
n = slope in the RRSB granulometry diagramm, see DIN 66145 (1976

3 Free Deformation Due to Temperature and Nonthermal Effects

Figure 2 shows the longitudinal deformation as a function of temperature during hydration and figures 3 and 4 show the deformation as a function of the equivalent age according to Freiesleben et al. (1977). Results are shown for two concretes of the same composition, but made with Portland cement 35 F from two different plants. OPC D has a relatively high SO$_3$ content with regard to the fineness of the cement and a low content of soluble alkalis , OPC E has a relatively low SO$_3$ content and a high content of soluble alkalis.

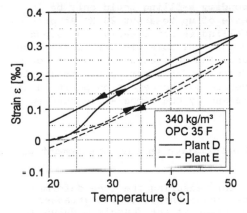

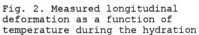

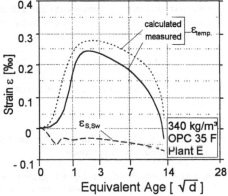

Fig. 2. Measured longitudinal deformation as a function of temperature during the hydration

Fig. 3. Measured longitudinal deformation, calculated thermal and nonthermal deformations at early ages

Up to a temperature increase of about 4 to 5 K or up to an age of 10 hours, the length increase was small although larger deformations should occur due to the large thermal expansion of the fresh concrete. The coefficient of thermal expansion of fresh concrete is 8 to 10 times higher than for hardened concrete. It decreases during the first ten hours of hydration and then remains constant, see ACI Commitee 517 (1980). In the first 10 hours, the plastic and autogenous shrinkage combine with the temperature deformation to produce a reduction of the total deformation. After this stage the longitudinal expansion of the concretes due to temperature increased clearly, but not linearly as expected for a constant coefficient of thermal expansion. On returning to the initial temperature, an expansion (OPC D) or a contraction (OPC E) remained. Similar results were obtained in the investigations of Emborg (1989) where the concretes showed different, but always positive residual deformations. The non-linearity of the curves and the differences between the cements were caused by different nonthermal deformations which occur simultaneously with the thermal deformations.

To determine the nonthermal deformations, the curves for the thermal deformation were calculated using the assumption that the coefficient of the thermal expansion is constant after about 10 hours, fig. 3. The magnitude of a_T was calculated according Dettling (1961) at $7.8 \ 10^{-6} \ K^{-1}$. It was assumed, that the difference between the measured and the calculated thermal deformations is equal to the nonthermal deformation. The curves for the nonthermal deformations $\varepsilon_{s,sw}$ are shown in figures 3 and 4. The nonthermal deformation components are defined in accordance with their cause and their time of occurrence as primary autogenous shrinkage and chemical swelling, secondary autogenous shrinkage, see Buil and Baron (1980) and Ziegeldorf et al. (1982) secondary swelling and tertiary autogenous shrinkage (Shrinkage due to drying did not occur because the specimens were protected against moisture loss by carefully wrapping with PE sheet). According to the investigations of Springenschmid and Breitenbücher (1989) and Nolting (1989), the **primary chemical swelling** is caused by surplus sulphate which was not consumed during the reaction between water and cement at the beginning of hydration. The **secondary swelling** could only be observed during a temperature increase of up to about 23 K. At present, the cause of the secondary swelling is not clearly understood. This swelling could be caused by an internal moisture transport in gel pores due to the temperature increase. The autogenous shrinkage continues in the tertiary state. The degree and the period of the single deformation types were mainly influenced by the cement composition and type, the fresh concrete temperature and by the temperature development.

4 Effect of the Different Types of Deformation on the Development of the Restraint Stresses

The same temperature development as for measuring the free deformations (see chapter 3) was used to determine the restraint stresses under full restraint. Also the development of the Young's modulus at early age was determined during the same test in the TST machine, fig. 4.

The measurements show that the **primary autogenous shrinkage** caused as little stress as did the large coefficient of thermal expansion at

the beginning of the hydration. This is because the concrete was not
able to built up stresses due to its plasticity. Only after the
Young's modulus started to increase the concrete could build up compressive stresses during the subsequent temperature increase. Additional compressive stresses were caused by the **primary chemical swelling**. This swelling corresponds, depending on the cement, to an estimated temperature increase of about 2.5 to 8 K. The short-term intensive **secondary autogenous shrinkage** which occurred after the primary
chemical swelling caused the compressive stress curve to flatten
although the temperature increased continuously during this period.
The following **secondary swelling** corresponds to an estimated temperature increase of about 1.5 to 5 K. This swelling continued to some
extent during the first hours of the cooling stage. Because the
Young's modulus had already reached a relatively high value, the
reduction of the compressive stress due to the cooling was retarded by
the secondary swelling or was strengthened by the autogenous shrinkage. The tensile stress which developed on further cooling was partly
strengthened by the **tertiary autogenous shrinkage**. This intensified
the risk of cracking.

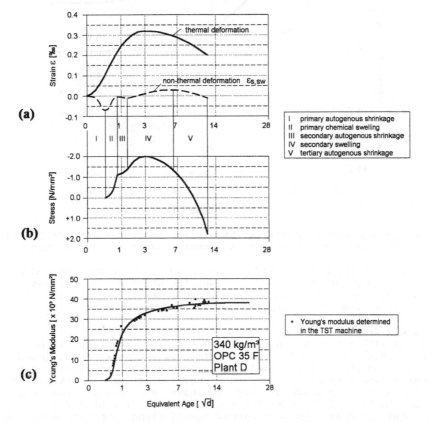

Fig. 4. a) Calculated thermal and nonthermal deformation based on the
measurement of the total deformation of an unrestrained concrete
specimen. Development of stress b) and Young's modulus c) for the same
concrete mix under full restraint in the TST machine

5 Technological Influences on the Development of Nonthermal Deformation and Restraint Stresses (Schöppel 1993)

5.1 Fresh Concrete Temperature
Concretes with fresh concrete temperatures of 12, 20, and 30°C were investigated under isothermal, fig. 5, and semi-adiabatic conditions. The longitudinal stresses were continuously recorded. Under isothermal conditions, the concretes with low temperatures showed longer and, in some cases greater restraint compressive stresses due to a larger chemical swelling than the warmer concretes. Also the cooler concretes had smaller tensile stresses due to a smaller secondary autogenous shrinkage.

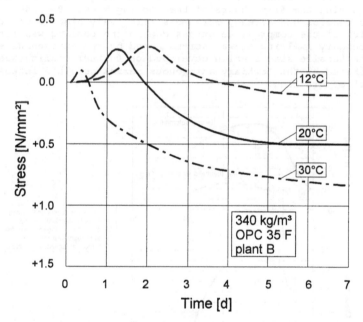

Fig. 5. Development of restraint stress for concretes with different fresh temperatures under isothermal conditions in the TST machine

Additional tests under semi-adiabatic conditions confirmed that the cooler concretes showed a chemical swelling which was larger and lasted longer as well as an autogenous shrinkage which was smaller, especially during the cooling stage.

5.2 Cement Content
Concretes with a cement content of 340 kg/m^3 and concretes with a cement content of 280 kg/m^3 and 60 kg/m^3 fly ash were investigated under semi-adiabatic conditions. The concrete with the smaller cement content showed a smaller temperature increase. It was remarkable that the concretes which contained fly ash showed a late autogenous shrinkage in the first four days. Therefore in spite of significantly lower temperature increase, in some cases, the cracking temperature of the concretes with lower cement content was only negligibly lower, sometimes even higher than as for the comparable concretes with higher cement content. Low cracking temperatures indicate low cracking sensitivity.

5.3 Type of Cement

The investigations under isothermal and semi-adiabatic conditions were performed using concretes made with cements from different plants. Each plant supplied a Portland cement and a blast-furnace slag cement. Similar to the concretes with fly ash, a large autogenous shrinkage occurred in the concrete with slag cement during the first 5 days.

5.4 Chemical Composition and Granulometric Grading of the Cements

Concretes with OPC 35 F plant A to E (table 2) were investigated under isothermal conditions. Concretes with cements of a high sulphate content showed larger and longer lasting compressive stresses due to a larger and longer swelling. Concrete made with cement E with a large alkali content exhibited large tensile stresses due to a large autogenous shrinkage, see fig. 6.

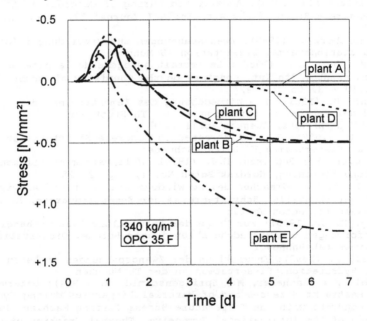

Fig. 6. Development of restraint stress for concretes with 5 different Portland cements 35 F under isothermal conditions

As well as the different nonthermal deformations, different temperature developments and Young's moduli resulted from the different composition and granulometric grading of the cements. For example at the age of 3 days the Young's modulus varied between 34 000 and 44 000 N/mm². For most of these concretes the different effects on the development of the restrained compressive stresses cancelled each other out.

6 Summary

During the early stages of hydration, the temperature deformation due to the heat of hydration, as well as nonthermal volume changes occur. These factors can decisively influence the development of the restraint stresses. The degree and development of nonthermal volume changes depend strongly on temperature. This means that under isothermal conditions differences in primary autogenous shrinkage and swelling and secondary autogenous shrinkage can be observed. However, the results of isothermal measurements can not be easily transferred to concretes which harden under semi-adiabatic conditions. Regarding the temperature development of a one meter thick element the nonthermal volume changes affect the development of restraint stresses mainly in the compressive stage, but to some extent also in the tensile stage. The latter is mainly applicable to concretes containing cement with fly ash or blast-furnace slag cement.

7 References

ACI Commitee 517 (1980): Accelerated Curing of Concrete at Atmospheric Pressure - State of the Art. In: **ACI Journal 77**, H. 6, pp. 429 - 448

Breitenbücher, R. (1989): Zwangspannungen und Rißbildung infolge Hydratationswärme. Dissertation TU München

Buil, M.; Baron, J. (1980): Le retrait autogène de la pâte de ciment durcissante. **7th Int. Congress on the Chemistry of Cement**, Paris, pp. VI - 37 to VI - 42

Dettling, H. (1961): Die Wärmedehnung des Zementsteins, der Gesteine und der Betone. Dissertation an der TH Stuttgart

DIN 66145 RRSB-Netz, Ausgabe April 1976

Emborg, M. (1989): Thermal Stresses in Concrete Structures at Early Ages. Dissertation TU Luleå (Schweden)

Freiesleben, H.; Pedersen, E.J. (1977): Maleinstrument til kontrol af betons haerdning, **Nordisk Beton** No. 1, pp. 21-25

Grube, H. (1991): Ursachen des Schwindens von Beton und Auswirkungen auf Betonbauteile. **Schriftenreihe der Zementindustrie**, H. 52, Düsseldorf: Beton-Verlag

Nolting, H. E. (1989): Zur Frage der Entwicklung lastunabhängiger Verformungen und Wärmedehnzahlen junger Betone. Dissertation Universität Hannover

Schöppel, K. (1993): Entwicklung der Zwangspannungen im Beton während der Hydratation. Dissertation an der TU München

Schöppel, K.; Plannerer, M.; Springenschmid, R. (1994): Determination of Restraint Stresses and of Material Properties during Hydration of Concrete with the Temperature-Stress Testing Machine, **Proceedings of the International Symposium, Thermal Cracking of Concrete at Early Ages**, Munich, October, 10-12

Springenschmid, R; Breitenbücher, R. (1989): Schlußbericht zum Forschungsvorhaben "Einfluß der Zementeigenschaften und Zusatzmittel auf die Reißneigung von Beton". Baustoffinstitut TU München

Springenschmid, R.; Breitenbücher, R. (1991): Cement with low crack-susceptibility. In: **Advances in Cementitious Materials, Ceramic Transactions**, The American Ceramic Society 16, pp. 701 - 713 (Proceedings of Conference 1990 in Gaithersburg)

Springenschmid, R.; Breitenbücher, R.; Mangold, M. (1994): Development of the Cracking Frame and Temperature-Stress Testing Machine, **Proceedings of the International Symposium, Thermal Cracking of Concrete at Early Ages**, Munich, October, 10-12

Ziegeldorf, S.; Müller, H. S.; Plöhn, J.; Hilsdorf, H. K. (1982): Autogenous Shrinkage and Crack Formation in Young Concrete. **Int. Conf. on Concrete at Early Ages**, Paris, Vol. I, pp. 83 - 88

27 EFFECT OF AUTOGENOUS SHRINKAGE ON SELF STRESS IN HARDENING CONCRETE

E. TAZAWA
Civil Engineering, Hiroshima University, Higashi-hiroshima, Japan
Y. MATSUOKA
Technical Research Institute, Taisei Corporation, Yokohama, Japan
S. MIYAZAWA
Civil Engineering, Hiroshima University, Higashi-hiroshima, Japan
S. OKAMOTO
Technical Research Institute, Taisei Corporation, Yokohama, Japan

Abstract
It has been thought that thermal cracking in mass concrete
at early age was initiated by self stress caused by thermal
deformation. From recent study, however, it is suggested
that autogenous shrinkage could be an another important
factor to dominate so-called thermal cracking of mass
concrete. In this study, it was experimentally investigated whether the principle of superposition was applicable to
the thermal strain and the autogenous strain or how the
self stress caused by autogenous shrinkage was superimposed
to the thermal self stress. For these purposes, measurement of autogenous shrinkage simultaneously occurring with
thermal deformation and cracking simulation test by restraining apparatus were conducted. From the results of
these tests, it was concluded that temperature strain and
autogenous shrinkage strain are the two main causes of self
stress generation in early age concrete, and the importance
of the two factors is dependent upon type of cementitious
material and water cement ratio of concrete.
Keywords: Autogeneous Shrinkage, Cracking, Creep, Heat of
Hydration, Superposition.

1 Introduction

Prevention of early age cracking associated with temperature change is very important particularly for massive
members of reinforced concrete. This problem is becoming
more important since richer mix and higher strength is
preferably specified. Mechanism of thermal cracking,
however, is not yet understood well. The main reason seems
to be the fact that self stress is generated during hydration process of cement simultaneously with drastic change
in microscopic structure and macroscopic properties of
concrete. Recently, a large amount of autogenous shrinkage
of concrete is reported for a certain type of concrete and
evaluation of the new factor with respect to thermal cracking problem is considered to be one of the most urgent
subjects. In this paper, effect of autogenous shrinkage on

Thermal Cracking in Concrete at Early Ages. Edited by R. Springenschmid. Published 1994
by E & FN Spon, 2–6 Boundary Row, London SE1 8HN, UK. ISBN: 0 419 18710 3.

thermal stress generation was experimentally investigated.

2 Test apparatus

Apparatus for cracking test consist of 4 restraining pipes
of stainless steel (A_s = 2.13 cm^2 for each pipe) and two
plates rigidly fixed to the pipes with nuts (Fig. 1).
Sample concrete is cast in a mould between the plates.
Pipes are heat insulated and water of constant temperature
is circulated to keep the length of the pipes.   Cross
section of concrete was 10 x 10 cm and length of test
section, both ends of which are tapered as shown 125 mm in
length, was 50 cm.   In order to reduce restraint foamed
polystyrene was used as the material of mould for the test
section.   An electric strain gage is buried at the center
of the specimen and the thermo-couples were used for temperature measurement.   Polyethylene sheet was covered on
top of concrete to prevent evaporation of water.

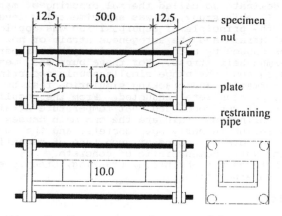

Fig. 1   Apparatus for cracking test

3 Test procedures

3.1 Measurement of autogenous shrinkage
Autogenous shrinkage was measured at three different temperatures, namely at 20 °C, 40 °C and 60 °C.   Mix proportion is shown in Table 1.   Normal portland cement replaced
with 10 % of silica fume and low heat cement containing 56
% of belite were used for mix 1 and 2 respectively.   Cracking test was also conducted for the same mix proportion.

3.2 Cracking test
Cracking test was started just after concrete was cast in
the form.   Temperature of the room, where the apparatus was
placed, was controlled so as to simulate temperature of

Table 1 Mix proportion

Mix No.	Type of cement	W/C (%)	s/a (%)	Unit content (kg/m^3)							Slump (cm)	Slump flow (cm–cm)	Air (%)	Cracking test
				W	C	SF	BS	S	G	Ad				
1	NPC	20	34	160	720	80	–	502	1012	12.0	23	45–42	2.6	○
2	LHC	30	37	160	533	–	–	633	1120	1.60	7	– – –	1.9	○
3	NPC	30	37	160	533	–	–	633	1120	1.07	5	– – –	2.0	–
4	NPC	30	37	160	160	–	373	610	1102	1.60	23	50–48	0.7	–
5	NPC	50	43	185	370	–	–	738	1016	3.70	7	– – –	4.1	–

NPC : Normal Portland Cement, LHC : Low Heat Cement, SF : Silica Fume, BS : Blast furnace Slag

history of a referred point of mass concrete. In this paper temperature history at a point 1.5 meter from top surface was selected. If cracking did not occur after 14 days, temperatures of pipes were raised and the specimen was cut.

Thermal stress generated in a specimen was measured by electric strain gages attached to the restraining pipes. Calibration of pipes was made in advance.

Temperature and strain of restrained specimen and free specimen, which were placed side by side in temperature control room, were automatically recorded at a fixed interval.

Compressive strength, tensile strength and modulus of elasticity were measured at peak point of temperature and at the time of cracking.

4 Test results

4.1 Autogenous shrinkage of concrete at different temperatures

From Fig. 2, the following items can be pointed out.

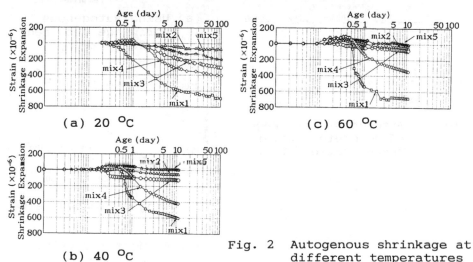

(a) 20 °C

(b) 40 °C

(c) 60 °C

Fig. 2 Autogenous shrinkage at different temperatures

1 Autogenous shrinkage of concrete is changed with curing temperature. At higher temperature shrinkage tends to occur at an earlier age.

2 The amount of autogenous shrinkage for the same water cement ratio is varied with type of cementitious material used. Autogenous shrinkage is low for low heat cement. Replacement of blast furnace slag increased autogenous shrinkage.

3 Autogenous shrinkage is increased with decreasing water cement ratio. For 20 % of water cement ratio, about 700×10^{-6} of shrinkage was observed for each curing temperature.

All these observations coincide with the previous data reported on cement paste.

4.2 Estimation of autogenous shrinkage under varying temperature by maturity

As an example is shown in Fig. 3, autogenous shrinkage versus temperature curves observed for three different curing temperatures, roughly approach to single curve when shrinkage is plotted against maturity in stead of age. So, autogenous shrinkage observed at constant temperature was

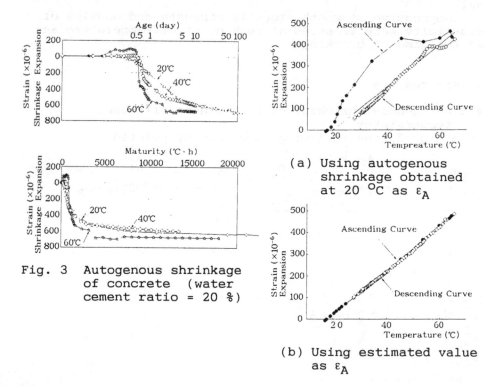

Fig. 3 Autogenous shrinkage of concrete (water cement ratio = 20 %)

(a) Using autogenous shrinkage obtained at 20 °C as ε_A

(b) Using estimated value as ε_A

Fig. 4 Measured value of ($\varepsilon_f - \varepsilon_A$)

modified by shrinkage - maturity curve to get an estimated
value of autogenous shrinkage under varying temperature.

4.3 Applicability of superposition principle to thermal strain and autogenous shrinkage strain

If superposition principle is applicable,

$$\varepsilon_f = \varepsilon_T + \varepsilon_A \tag{1}$$

where, ε_f : free strain of specimen

ε_T : temperature strain

ε_A : autogenous shrinkage strain

Strain corresponding to $\varepsilon_f - \varepsilon_A$ obtained from observation
was plotted against temperature (Fig. 4-(a), (b)).   In
Fig. 4-(a), autogenous shrinkage obtained at constant
temperature 20 ^{O}C was used as ε_A, while in Fig. 4-(b)
estimated value (4.2) was used as ε_A.   Ascending curve and
descending curve in Fig. 4-(b) reduced to a single line.
Hysteresis of temperature deformation, which was pointed
out in old literature, was not observed when we consider
autogenous shrinkage.   Therefore, the following relation is
ideally applicable to thermal deformation.

$$\varepsilon_T = \alpha \cdot \Delta T \tag{2}$$

α : thermal coefficient of expansion

ΔT : temperature difference

For free deformation, principle of superposition might be
applicable to thermal strain and autogenous strain.   The
value of α is obtained from regression analysis of Fig. 4-
(b), then thermal deformation ε_T $(= \alpha \cdot \Delta T)$ can be shown as
in Fig. 5.

Naturally, overall strain shifted toward negative direction when variation of the autogenous shrinkage strain was
larger than that of thermal strain even if temperature of
specimen was still increasing.

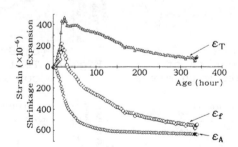

Fig. 5   Strain change of free specimen
(water cement ratio = 20 %)

4.4 Results of cracking tests
Case (a) where effects of autogenous shrinkage is very small

When low heat cement mainly consisting of belite was
used, strain due to autogenous shrinkage of concrete was
very small compared to the one due to temperature change.

Temperature, strain and stress are shown in Fig. 6 versus time from casting. In this case, the maximum compressive stress is generated at the maximum temperature and stress is linearly varied with temperature (Fig. 7).

Case (b) where effect of autogenous shrinkage was very large

When water cement ratio was reduced to 20 % using 10 % silica fume with portland cement, about 500×10^{-6} of autoge-

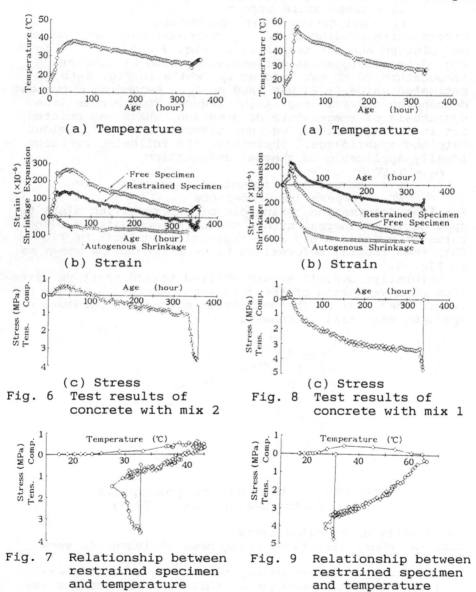

(a) Temperature   (a) Temperature

(b) Strain   (b) Strain

(c) Stress   (c) Stress

Fig. 6 Test results of   Fig. 8 Test results of
        concrete with mix 2   concrete with mix 1

Fig. 7 Relationship between   Fig. 9 Relationship between
        restrained specimen   restrained specimen
        and temperature   and temperature
        (mix 2)   (mix 1)

nous shrinkage was generated at about two days from casting
(Fig. 8). Owing to large amount of autogenous shrinkage,
strain of free and restrained specimen started to decrease
and tensile stress was generated while the temperature in
the specimen was still rising. Relation between temperature and stress turned out curvilinear (Fig. 9). These
data suggest that thermal stress generation is greatly
affected by autogenous shrinkage.

4.5 Stress relaxation due to creep
For the above two cases, creep strain is compared in
Fig. 10. In case (a), overall creep strain was compressive
while in case (b) the creep strain was tensile. Creep
coefficient ($\phi = \varepsilon_c / \varepsilon_e$) obtained for tensile stress developing region was approximately 1.5 for case (a) and 4.2
for case (b). This means that restrained strain due to
autogenous shrinkage is more subjected to creep than the
one due to temperature change.

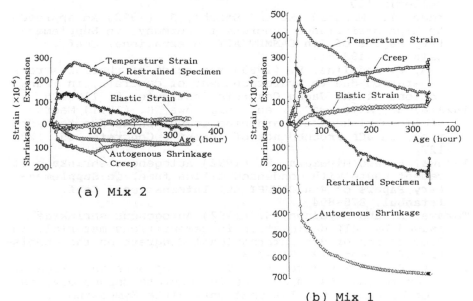

(a) Mix 2

(b) Mix 1

Fig. 10 Strain variations of restrained specimens

5 Conclusions

(1) Autogenous shrinkage strain could be superimposed to
thermal strain to get overall strain of concrete.
(2) Estimation based on maturity is applicable to predict
the amount of autogenous shrinkage under varying temperature.
(3) Autogenous shrinkage that occurs simultaneously with
thermal strain influences self stress generation in early

age concrete.
(4) Autogenous shrinkage strain is more subjected to creep
than thermal strain when it is restrained.

6 Acknowledgement

The authors are grateful to Dr. K. Kawai, Messrs. S. Hashi-
moto and K. Sujino for their efforts spent in experiments
and preparation of the paper.

7 References

Davis, H.E. (1940) Autogenous volume change of concrete, in
    Proceedings of ASTM, 40, 1103-1110
Helmuch, R.A. (1961) Dimension changes of hardened port-
    land cement pastes caused by temperature changes, in **PCA
    Bulletin 129**
Schrage, I., Mangold, M. and Sticha, J. (1992) An approach
    to high-performance concrete in Germany, in **Supplemen-
    tary Papers of 4th CANMET/ACI International Conf.,
    Istanbul**, 493-511
Tazawa, E. and Iida, K. (1983) Mechanism of thermal stress
    generation due to hydration heat of concrete, in **Trans-
    actions of the Japan Concrete Institute**, 5, 119-126
Tazawa, E., Miyazawa, S. and Shigekawa, K. (1991) Macro-
    scopic shrinkage of hardening cement paste due to hydra-
    tion, in **JCA Proceedings of Cement & Concrete**, 45, 122-
    127
Tazawa, E. and Miyazawa, S. (1992) Autogenous shrinkage of
    cement paste with condensed silica fume, in **Supplemen-
    tary Papers of 4th CANMET/ACI International Conf.,
    Istanbul**, 875-894
Tazawa, E. and Miyazawa, S. (1992) Autogenous shrinkage
    caused by self desiccation in cementitious material, in
    **Proceedings of 9th International Congress on the Chemis-
    try of Cement**, IV, 712-718
Tazawa, E. and Miyazawa, S. (1993) Autogenous shrinkage of
    concrete and its importance in concrete technology, in
    **Proceedings of 5th International RILEM Symposium on
    Creep and Shrinkage of Concrete, Barcelona**, 159-168

28 HIGH PERFORMANCE CONCRETE: EARLY VOLUME CHANGE AND CRACKING TENDENCY

E. SELLEVOLD and Ø. BJØNTEGAARD
Norwegian Institute of Technology, Division of Concrete, Trondheim, Norway
H. JUSTNES and P.A. DAHL
SINTEF Structures and Concrete, Trondheim, Norway

Abstract
A variety of experimental techniques concerning early volume change and crack sensivity have been applied to HPC and equivalent binder phases.

It appears from the results that the low w/b-ratios lead to more pronounced volume reductions at early ages, earlier build-up of internal tensile stresses and greater sensitivity to early cracking under conditions of strong evaporation from the concrete surface. Field experience has shown that cracking may occur even when great efforts have been made to avoid evaporation. We believe that chemical shrinkage and the consequent self-desiccation play a major role in the crack sensivity of HPC.
Keywords: Chemical Shrinkage, Cracking, High Performance Concrete, Pore Water Pressure, Shrinkage Rig.

1 Introduction

High Performance Concrete (HPC) here denotes concrete with water-to-binder ratio of 0.40 or less. Such concrete is used more and more frequently in Norway today, in order to obtain increased durability of concrete structures in aggressive environments. However, a new problem has appeared with widespread use of HPC, particularly on horisontal surfaces such as bridge decks, which have shown great sensitivity to cracking at early ages, most often less than about 12 hours after casting. Field observations and experiences with the problem are described in detail by Kompen (1994) at this conference. The present paper describes a number of experiments using different techniques carried out over the last two years, in order to obtain a better understanding of the phenomenon of early cracking in HPC. The work was initiated and partly sponsored by the Norwegian Road Research Lab., Oslo.

Plastic shrinkage cracking is a well known phenomenon, associated with evaporation from the surface of newly placed concrete. A laboratory equipment to measure the sensitivity of concrete to this type of cracking has been developed and used a number of years at NTH/SINTEF, Johansen et al (1993). The technique was employed extensively in this work, in spite of the fact that in field situations cracking may well occur even though every possible precaution is taken to prevent evaporation from the surface. The technique was used for lack of a more relevant "macromethod", and with the hope that a meaningful rating of the factors influencing crack sensitivity could be obtained, also for situations when surface evaporation is not the main driving force.

Thermal Cracking in Concrete at Early Ages. Edited by R. Springenschmid. Published 1994 by E & FN Spon, 2–6 Boundary Row, London SE1 8HN, UK. ISBN: 0 419 18710 3.

The other methods used here were of a "microtype", focusing on measures of early volume change; "total chemical shrinkage" (le Chatlier shrinkage) and "external chemical shrinkage" (bulk volume shrinkage of cement paste isolated from the environment by a thin rubber membrane). Measurements of early relative humidity (RH) development in the pore system, and the development of pore water pressure both before and after final set of cement paste were also carried out.

The choice of these methods was based on the consideration that chemical shrinkage and self-desiccation becomes more important as the w/c-ratio is reduced, as is the case for HPC binders.

The purpose of these microtype experiments was to gain a better understanding of the early age behavior of the binder phase in HPC, and, if possible, to identify important driving forces to cracking. Several researchers and students have been engaged in this work; this article will only give brief summaries of the experimental work done so far, and should be seen as a report on work-in-progress on a very complex problem.

Note that cracking induced by thermal effects have not been mentioned so far, because these are not considered to be major effects in this particular problem, and that intensive work on the thermal problem is currently going on at several research centers.

2 Micro methods

2.1 Relative Humidity Measurements
One way to detect the presence of capillary tension in the pore water due to selfdesiccation is to measure the RH-development during isolated and isothermal hydration. Fig. 1 shows an example of such a measurement on a cement paste with a water-binder ratio of 0.30, and containing 8% silica fume. This low ratio was chosen since such a mixture is known to experience significant self-desiccation at very short times, and therefore tell us if it is possible to detect RH-effects within the first 12 hours. The Rotronic measuring system with Hygrolyt sensor was used. Fig. 1 demonstrates, as expected, very significant RH-decreases during the first two weeks, but no significant change during the first 12 hours.

It is useful to consider the relationship (Kelvin-Laplace equation) beetween RH and capillary tension (ΔP):

$$\Delta P = -133 \bullet \ln(RH) \text{ , where } \Delta P(MPa) \text{ is positive for tension.}$$

A very small lowering of RH results in a very large tension in the pore water. F.inst., RH=0.99 gives ΔP=1.3 MPa, while RH=0.90 results in ΔP=14 MPa tension. RH=0.9 is achived in the pore system after about 200 hrs hydration (Fig. 1), thus clearly chemical shrinkage effects must be quite important in shrinkage of HPC. However, RH-measurements used to detect the much smaller effects present during the first 12 hours is not practical with the present accuracy of RH-measurements, which is expected to be no better than ±2% RH in this high range. In addition it must be taken into account that RH of the pore water is influenced by dissolved ions as well as by capillary effects. The effect may be several %RH, Hedenblad (1993). Thus RH-measurements may be useful for longer time measurements, but at short times a more sensitive technique appears to be to measure pore water pressure directly.

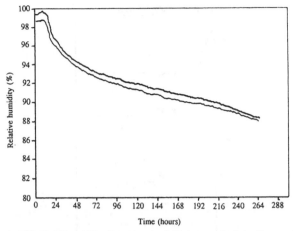

Fig. 1 Relative humidity development during sealed, isothermal hydration of cement paste with w/b=0.30 and 8% silica fume (two parallel samples).

2.2 Pore Water Pressure Measurements

A very sensitive method to measure pore water pressure directly has recently been developed and applied extensively by Radocea (1990). The system consists of a water filled syringe connected to a very sensitive pressure transducer. Measuring on cement paste, the syringe is inserted to a certain depth in the paste, fixed in position, and the pressure recorded over time. The paste surface may be sealed or allowed to dry according to choice. Fig. 2 shows typical results for 3 norwegian cements at w/c=0.40, sealed condition, carried out by Radocea in Trondheim. The pressure is in units of mm water, and 0 indicates pure hydrostatic water pressure, i.e. the initial value of 30 mm in fig. 2 shows that for this paste with a fresh density of about 2.0, the syringe tip is inserted 30 mm vertically into the paste. The initial pressure decrease for the 3 pastes indicates the transition from a fluid paste to the establishment of a solid self-supporting skeleton. This transition involves both settlement and growth of hydration products on the cement surface. In fig. 2 the process is complete after 40-80 minutes, depending on the cement type, i.e. both on fineness and chemical composition. The length of the dormant period also varies with cement type (2-4 hrs), succeeded by further reduction in pore water pressure, which is taken to be a result of cement hydration and the consequent chemical shrinkage. It is unclear to what extent the reduced pressure is a pure "vacuum effect", created by the formation of contraction pores, or a capillary meniscus effect.

The pressure reduction has been reported by many researchers, however, values below absolute vacuum have not been reported, which we believe reflects a limitation in the measuring system. In any case we find the importance of the lower portion of fig. 2 to be that less than 0 mm water pressure indicates a situation where the solid skeleton in the paste is sufficiently strong to allow partly empty contraction pores to form. Before this point the total chemical shrinkage simply leads to "collapse" of the paste volume, i.e. the external chemical shrinkage is equal to the total chemical shrinkage.

The experimental system of Radocea is very simple, and to our mind yields very

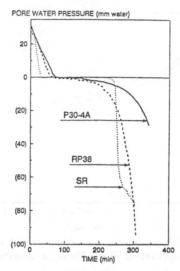

Fig. 2 Changes in pore water pressure due to
sedimentation and cement hydration.

important information on early age behaviour of cement paste quite relevant
regarding crack sensivity; it is probable that the transition from a semiliquid state to a
solid skeleton represents a very sensitive period in that the skeleton initially is very
weak with a low strain capacity and thus vulnerable to any external influence, that
will add to the already existing internal stresses caused by vacuum/capillary effects.

2.3 Total- and External Chemical Shrinkage
The transition of the paste from a semiliquid to a solid with a coherent skeleton can
also be recorded using the two techniques discussed here. Total chemical shrinkage
is the easiest to measure accurately. In the present case a 10 mm layer of cement
paste was placed in a small bottle, with water on top extending up in a capillary tube.
The total chemical shrinkage is simply recorded as the level change in the tube.
Knudsen and Geiker in RILEM (1982) have pioneered a variety of applications for
the method since the early 1980's. For example the total chemical shrinkage is a
direct measure of the degree of hydration for a given cement. To measure the
external chemical shrinkage is, however, much more difficult and the litterature
contains differing results; see a number of papers in the proceedings from RILEM
"Concrete of early ages" (1982), and Setter and Roy (1978). Very careful linear
measurements have recently been reported in combination with RH-measurements
by Jensen (1993), however, only from the time of final set.

In priciple, the total and the external chemical shrinkage are identical as long as
the paste is fluid and continually "collapse" as the hydration reduces the volume.
When a self-supporting skeleton is formed, the external chemical shrinkage rate is of
course much reduced, and the two curves will deviate. Fig. 3 illustrates this
behaviour. Total chemical shrinkage measurements are usually given in terms of ml
pr. 100 gram cement, and have been found to vary little with w/c-ratio the first day.

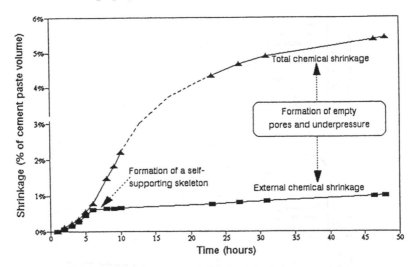

Fig. 3 Total- and External Chemical Shrinkage for cement paste with w/c=0.40.

When expressed as volume % of the binder (fig. 3), however, decreasing w/c-ratio means increasing volume change, probably contributing to the greater cracking sensivity of HPC.

The measurements of external chemical shrinkage are made with the cement paste contained in a rubber membrane (condom) which is immersed in a water bath at constant temperature. The volume change is recorded as a change in the buoyancy. The major problem with the method is an extreme sensitivity to bleeding. With bleeding, the water collects on the top, and when a skeleton is formed this excess water is gradually sucked into the pores and therefore erroneously recorded as an external volume change. To avoid the bleeding problem a system was made where the paste is rotated under water until final set. The extreme influence of bleeding is shown in fig. 4, Verboven and van Gemert (1994). In fig. 4a the w/c-ratio appears to have a very large influence on the external chemical shrinkage, however, fig. 4b demonstrates clearly that the main effect of the w/c-ratio is on the bleeding characteristics, and not, somewhat surprisingly, on the time and the neccessary degree of hydration to form a self-supporting skeleton.

The external chemical shrinkage behavior of course depends on cement type, admixture type and dosage and additions such as pozzolans.

Presently, there exist a fairly large number of experimental results on these factors. which are in the process of being analyzed. We believe that the early volume change in the binder phase is related to the early cracking sensivity, but no model explaining the relationship to a "macro"-method which directly rates the cracking sensivity of a concrete exists now. One such a macro method is discussed in the following, the "Plastic Shrinkage Rig".

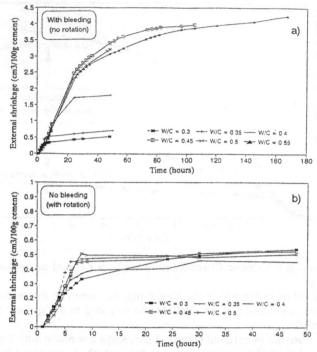

Fig. 4   External chemical shrinkage of cement paste with different
w/c-ratios. Principle illustration, a) and b) are not from
the same cement. Verboven and van Gemert (1994).

3 Macro methods: The Plastic Shrinkage Rig

The Rig is shown in fig. 5. Fresh concrete is contained between two rigid steel rings.
The concrete surface is exposed to a laminar air stream with speed of 4.5 m/sec.
The resulting evaporation activates plastic shrinkage and possibly cracking, because
of the restraint provided by the stiff steel rings. Crack sensivity is characterized by
the Crack index (C_i) measured as the average accumulated crack width around two
concentric circles on the concrete surface (fig. 5).

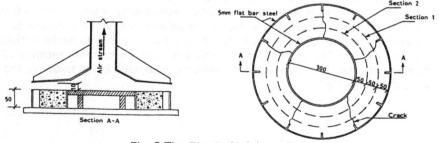

Fig. 5 The Plastic Shrinkage Rig.

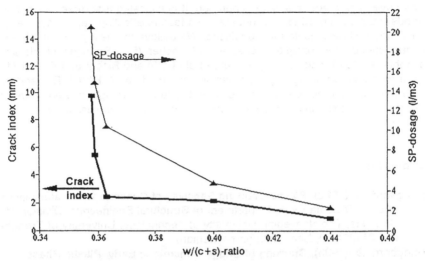

Fig. 6 Crack sensitivity (C_i) in the Plastic Shrinkage Rig as a function of w/b-ratio. SP (superplastisizer) dosage required for 18 cm slump is also given.

The apparatus has been utilized and improved continuously over the past decade, Johansen et al (1993), and is able to give at least a rough relative rating of crack sensivity. However, the C_i has a relatively high standard deviation, and is extremely sensitive to the initial slump of the concrete. Thus, careful control is necessary to obtain systematic results, and, in order to rate f.ex. a certain cement/admixture combination tests have to be carried out at several w/b-ratios, Bjøntegaard (1993).

Fig. 6 shows clearly the very large effect a reduction in water-binder ratio below a critical value has on the crack sensivity, Bjøntegaard (1992). Other cement types give different critical values, opening the possibility to rate binder compositions in terms of critical w/b-ratios; the lower value indicating smallest crack sensivity. Note however, in fig. 6 that the critical value could be assigned to the dosage of SP-admixture as well as to the w/b-ratio. Considering also the great influence that the initial slump of the concrete has on C_i, it is obvious that the situation is very complex, and modesty is called for when drawing conclusions. However, fig. 6 very clearly indicates that the problem of cracking would be expected to increase strongly at low w/b-ratios; excactly what is found in practice with HPC.

4 Concluding remarks

A variety of experimental techniques have been applied to HPC and to equivalent binder phases.

It appears from the results that the low w/b-ratios lead to more pronounced volume reductions at early ages, earlier build-up of internal tensile stresses and greater sensitivity to early cracking under conditions of strong evaporation from the concrete surface. Field experience has shown that cracking may occur even when great efforts have been made to avoid evaporation. It is indicated that more realistic test methods must be applied to map out the main factors affecting early crack sensivity of HPC under close to isotermal conditions. We believe the best possibility is to continue work on the principles pioneered in München (f.ex. Springenschmid and Breitenbücher 1990) and used by Paillere et al (1989) and Bloom and Bentur (1993). Such a restrained shrinkage rig is presently being made at NTH/SINTEF, and the testing will focus on the relationship between the tensile strength development and the corresponding self-induced tensile stress in HPC as a measure of crack sensivity.

5 References

Bjøntegaard, Ø. (1992), **Plastic Cracking Sensivity of Concrete with water-binder ratio of 0.40**, Diploma Work, Department of Structural Engineering, Division of Concrete, The Norwegian University of Technology, University of Trondheim, No. 071264.00, Dec. 1992 (in norwegian).

Bjøntegaard, Ø. (1993), **Cracking of Bridge Concrete in Early Plastic Phase**, Department of Structural Engineering, Division of Concrete, The Norwegian University of Technology, University of Trondheim, Report 14, Oct. 1993 (in norwegian).

Bloom, R. and Bentur, A. (1993), **Restraint Shrinkage Of High Strength Concrete**, National Building Research Institute, Faculty of Civil Engineering Technion, Israel Intitute of Technology, From the Conference: High Strength Concrete 1993, Lillehammer, Norway, 20-24 June 1993.

Hedenblad, G. (1993), **Water Pressure in Fresh and Young Cement Paste**, Nordic Concrete Reasurch, No. 9, Dec. 1990, Oslo.

Jensen, O.M. (1993, **Autogen Deformation og RF-ændring**, Building Mat. Lab., TR284/93, Technical Univ. of Denmark, Lyngby (in danish).

Johansen, R. Dahl, P.A. and Skjøldsvold, O. (1993), **Control of Plastic Shrinkage in Concrete at Early Ages,** Quality in Concrete and Structures, Singapore, Aug. 1993, Vol XIII.

Kompen, R. (1994), **High Performance Concrete: Field Observations of Cracking Tendency at Early Age**, This Conference.

Paillere, A.M. Buil, M. and Serrano, J.J. (1989), **Effect of Fiber Addition on the Autogenious Shrinkage og Silica Fume Concrete**, ACI Materials J., V. 86, 1989, pp. 139-144.

Radocea, A. (1990), **Water Pressure in Fresh and Young Cement Paste**, Nordic Concrete Research, No. 9, Dec. 1990, Oslo.

RILEM (1982), **International Conference on Concrete at Early Ages**, Ecole Nationale des Ponts et Chaussées, Paris, 6-8 April 1982.

Setter, N. and Roy, D.M. (1978), **Mechanical Features of Chemical Shrinkage of Cement Paste**, Cement and Concrete Research, Vol. 8, No. 5, 1978.

Springenschmid, R and Breitenbücher, R (1990), **Cement with Low-Crack-Susceptibility**, Advances in Cementitious Materials, American Ceramic Society Conference, Gaithersburg, Md, USA, 1990.

Verboven, F. and Van Gemert, A. (1994), Diploma Work, Katholieke Universiteit Te Leuven, Belgium, Carried out at NTH, Trondheim, Fall 1993.

29 FACTORS INFLUENCING EARLY CRACKING OF HIGH STRENGTH CONCRETE

I. SCHRAGE
Institute of Building Materials, Technical University of Munich, Germany
Th. SUMMER
Allianz AG, Munich, Germany (formerly at Institute of Building Materials)

Abstract
Concretes with compressive strengths between 75 and 105 MPa (HSC) were investigated. The mixes with a 16 mm maximum grain size contained 450 kg/m³ of cementitious material including up to 8% silica fume (SF). The w/c-ratio was 0.30.

The tests were performed in a cracking frame. During early hardening the concrete could neither expand nor contract in length. Evaporation was prevented. The different origin of restraint stresses was studied under semi-adiabatic and isothermal conditions. Stress development was monitored until crack formation.

HSC was found more liable to cracks than traditional concrete. This adverse feature was caused by the high cement content, the low w/c-ratio and the addition of superfines. Cooling the mix and partially substituting cement with fly ash or filler reduced the cracking tendency.

Thermal cracking at early ages was not the predominant problem of HSC. The impact on stress formation resulting from chemical shrinkage and the consequent selfdesiccation was at least equivalent to that of the thermal origin.
Keywords: Blended Cement, High-Performance Concrete, Restraint, Selfdesiccation, Silica Fume.

1 Introduction

HSC is rather prone to cracking at early ages. Under restraint there are two origins of stress development during hardening. The first one is the release and dissipation of hydration heat. The second is chemical shrinkage and the consequent selfdesiccation.

Volume loss due to hydration is customarily an internal effect in traditional concrete. The outer dimensions of the element remain nearly constant. The empty space is in the bulk of the hardening cement paste. These voids either dessicate or attract water from the surroundings dependent on the curing conditions, Czernin (1960). But in mixes with

Thermal Cracking in Concrete at Early Ages. Edited by R. Springenschmid. Published 1994 by E & FN Spon, 2–6 Boundary Row, London SE1 8HN, UK. ISBN: 0 419 18710 3.

w/c-ratios of about 0.4 or below, chemical shrinkage de-
forms the whole element, Jensen (1990), Bloom and Bentur
(1993). This process is connected with a decrease of the
relative humidity in the pore system and the formation of
capillary tension. There are indications that even water
curing fails to compensate for this selfdesiccation,
Radocea (1990). Though there is lack of knowledge about the
proper mechanism it must be assumed that "external" chemical shrinkage is an inherent feature of any HSC. The coincidence of shrinkage and temperature makes such concrete
more vulnerable to cracks than traditional concrete.

One cause of cracking in HSC is the high content of
cementitious material which augments the temperature,
Breitenbücher (1989), Springenschmid (1993). Another important cause is the low w/c-ratio. Paillère et al. (1989),
Schrage et al. (1992), Summer (1993). There is evidence
that the addition of silica fume plays a special role,
Springenschmid (1994).

The present study examines the main parameters of stress
formation. The tests include the partial replacement of
cement with fly ash and filler. This strategy to reduce
cracking has been proposed by de Larrard (1989). The study
is part of a program sponsored by the Funds of Industrial
Research (AIF) under grant no. 8798.

2 Experimental

2.1 Equipment and test procedure
Testing under restraint conditions was done in a cracking
frame as described in detail by Springenschmid et al.
(1994) at this conference. Fresh concrete was placed and
compacted in the test set-up, where it could neither expand
nor contract in length during early hardening. Any loss of
moisture was prevented. The temperature, in the center of
the concrete beams (15 x 15 cm² cross-section), and the
longitudinal restraint stresses were recorded. A sudden
drop of the longitudinal stress indicated that the concrete
had cracked. If no early cracking occurred artificial
cooling was started after 4 days. The lower the temperature
which caused the crack the better the concrete will perform
in practice.

In the standard test the concrete temperature increased
under semiadiabatic conditions. The testing conditions
represent the thermal situation in elements 40 to 50 cm in
width. It is done without any artificial heating or cooling
during the first 4 days using only appropriate thermal
insulation.

To separate stresses of chemical origin from temperature
stresses, isothermal tests were executed. The testing
conditions represent the situation in rather thin elements
about 10 cm in width. By external cooling the hardening

temperature of the concrete is maintained at a very low and mostly constant level.

2.2 Materials and mix characteristics

The calcitic gravel had a 16 mm maximum grain size. The sand passing 4 mm with only 10% fines passing 0.25 mm was purely quartz. The sand content was 36% of total aggregates.

The rapid hardening Portland cement PZ 45 F (CEM I 42.5 R) with Blaine-value of about 3000 cm²/g had a high C_3S content. In two tests a part of this cement was replaced by limestone filler (d'=50 μm) or low calcium fly ash. The replacement rate was 20% or 30% thus taking into account the composite cements of EN 197 E. The blast furnace cement HOZ 35 L with Blaine-value of about 4500 cm²/g contained 50% slag (CEM IIIA 32.5). It originated from the same plant as the Portland cement.

The silica fume (SF) was a 50:50 slurry. The superplasticiser was based on melamine naphthalene. A max. addition of 4% with respect to the total content of cementitious material resulted in high slump.

In the mix design the entire plasticiser was regarded as part of the liquid phase (w) and all mineral additives as part of the cementitious material (c).

The high strength mixes contained 450 kg/m³ of cementitious material including up to 8% silica fume. The w/c-ratio was 0.30. They reached compressive strengths of 75 to 105 MPa using 20 cm cubes after 28 days in thermal insulated moulds. This curing corresponds to the standard test conditions in the cracking frame.

The reference mixes of traditional concrete contained 340 or 450 kg/m³ of cementitious material at a w/c-ratio of 0.48. They reached a compressive strength of 55 MPa using 20 cm cubes after 28 days normal curing.

Mix characteristics are recorded in table 1. Some of the experiments were executed by Dingethal (1993).

Table 1. Mix Properties

Mix No.	Cement Type	w/c	sf/c	Cube Strength (MPa) Curing 20°C			Curing 30/50°C*		
				1 d	7 d	28 d	1 d	7 d	28 d
R1	PC**	0.48	0	–	–	54	–	–	–
R2	PC	0.48	0.08	21	42	55	–	–	–
H1	PC	0.30	0	30	68	85	40	69	75
H2	PC	0.30	0.04	41	71	101	47	80	95
H3	PC	0.30	0.08	48	90	112	–	–	105
H4	BFSC 50% slag	0.30	0.08	12	46	74	20	77	103
H5	PC+20% filler	0.30	0.08	–	–	–	37	85	100
H6	PC+30% f.a.	0.30	0.08	–	–	–	25	80	95

* insulated moulds
** cement content 340 kg/m³

2.3 Standard tests at different strength levels (fig. 1)

For the reference mixes with 55 MPa compressive strength and a w/c-ratio of 0.48 the maximum temperature was 50% higher when the binder content was switched from 340 kg/m³ to 450 kg/m³. The latter mix included 8% silica fume. Despite the higher hydration heat, it developed only 50% of the compressive prestress. The first cracks occurred after only 2 days and 28°C compared to the corresponding mix which cracked at 9°C. This mix had to be cooled artificially in order to induce cracking.

Cracking at this early stage was defined by an abrupt release of restraint stresses surpassing, by a single drop or the sum of several consecutive ones, the arbitrary limit of 0.25 MPa. That is equivalent to a crack opening of about 5 μm.

The high-strength concrete with 105 MPa compressive strength and w/c=0.30 had a heat release and compressive prestress similar to the 55 MPa mix with the higher cement content. But tensile stresses developed quite faster and early cracking occurred at 27°C.

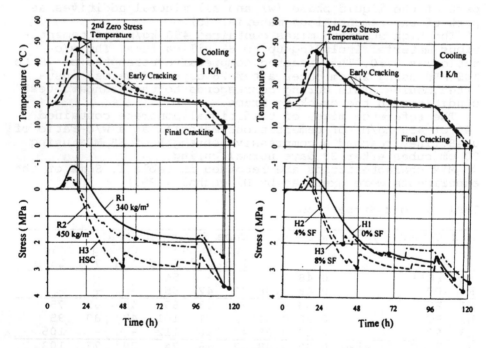

Fig. 1.
Temperatures and stresses in the cracking frame: Normal and high-strength concrete in standard test.

Fig. 2.
High-strength concrete with different additions of silca fume in standard test.

2.4 Standard tests with and without silica fume (fig. 2)
The mixes with w/c=0.30 reached a compressive strength of
74 to 105 MPa in insulated moulds (85 to 110 MPa at normal
cure). In the cracking frame tests with 4% and 8% silica
fume showed about the same development of hydration heat.
Their peak temperature was 30% higher than without silica
fume. Compressive prestress was the same for all three
mixes. First cracks occurred at 28°C and 33°C. The mix
without silica fume cracked at 22°C.

2.5 Isothermal tests with and without silica fume (fig. 3)
Companion specimens of the above mixes were compared under
isothermal conditions. Since the cooling system could not
completely prevent any temperature rise within the first
day there was a certain deviation in the temperatures of
the fresh concretes.
   Both concretes with silica fume exhibited tensile
stresses much earlier. These stresses were much larger than
those in the concrete without silica fume. The concretes
with silica fume cracked at 7°C and 12°C during artificial

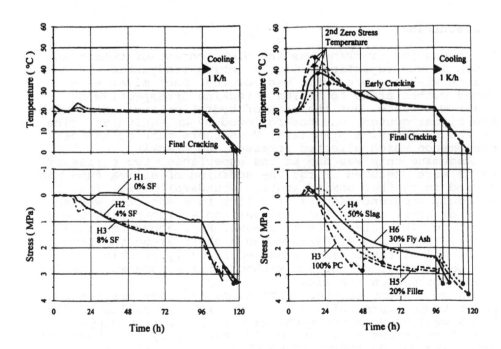

Fig.3.
High-strength concrete with
different additions of silica fume under isothermal
conditions.

Fig.4.
High-strength concrete with
different binder mixes in
the standard test.

cooling. The mix without silica fume had not cracked when
the experiment was stopped at the freezing point of water.

Chemical shrinkage and consequent selfdesiccation seems
to be the main reason for this type of failure. Temporary
swelling reduces the external shrinkage of the mix without
silica fume. The cause could be a reaction of gypsum.

2.6 Standard tests with blended cements (fig. 4)
Mixes with blended cements and w/c=0.30 including 8% silica
fume reached a compressive strength of 95 to 105 MPa. That
was about the same level as for the concrete with pure
Portland cement (strength refers to curing in isolated
moulds). In the cracking frame the slag cement (50% slag)
produced only half of the hydration heat and slightly less
compressive prestress. Tensile stresses were delayed but
early cracking was reached at 24°C. Thus there was no
substantial difference to the mix with pure Portland cement.

A partial replacement of Portland cement with 30% fly
ash or 20% limestone filler postponed cracking until the
concretes had cooled down to 13°C and 16°C respectively.

3 Discussion

One cause for cracking was the high content of cement and
cementitious superfines, fig. 1, which readjusted the paste
volume when the w/c-ratio was lowered. Such mixes were
found rather prone to thermal cracking irrespective of
their w/c-ratio. But the heat of hydration increasing with
the binder content was not followed by an equivalent rise
of the compressive prestress. Thus there must be another
phenomenon which affected stresses and cracking.

Assuming this was due to the superfines, two series of
tests were executed. First the addition of silica fume was
varied in further semiadiabatic (standard) tests, fig. 2.
The only apparent results of silica fume addition was an
earlier start of the heat release and a higher peak temperature. The stress curves however kept about the same
shape. Their only difference was a shift along the time
axis. All of these tests resulted in early cracking. Even
the performance of the mix without silica fume was insufficient for use on the site.

In subsequent isothermal tests on the same mixes any
major temperature influence on cracking was excluded, fig.
3, as well as moisture exchange with the surroundings.
Therefore stress formation could result only from chemical
processes during hydration which must be attributed to
selfdesiccation. These non-thermal induced stresses proved
to be about 50% of the total stresses in the comparison
standard test. The addition of silica fume led to higher
stresses in both mixes for it suppressed the temporary
swelling which reduced the selfdesiccation effect. These

isothermal tests confirmed on the other hand that cooling during hardening is a powerful tool against cracking.

Another way of reducing this adverse tendency was found in partial replacement of Portland cement with fly ash or limestone filler, fig. 4. Slag cement was not equally effective. This difference suggested an activation of slag but not of limestone or fly ash during early hydration. Thus it may be reasonable to regard those latter additives as filler. Referring to the rather high w/PC-ratio of 0.41 and 0.47 in this case the strength of about 100 MPa was considerable.

4 Conclusion

Cracking of high-strength concrete is a combined effect of thermal and chemical stresses. Both early hydration heat and external chemical shrinkage are adversely influenced by the addition of silica fume. Cracking can be reduced by limestone or fly ash cements without major influences on high strength. Slag blended cement was not found equally effective.

5 References

Bloom, R. and Bentur, A. (1993) Restrained Shrinkage of High Strength Concrete. Conference: High-Strength Concrete, Lillehamer, Norway, June 1990, Proceedings, Vol. 1, pp. 1007-1014.

Breitenbücher, R. (1989) Zwangsspannungen und Rißbildung infolge Hydratationswärme. Doctoral Thesis. Technical University of Munich (in German)

Czernin, W. (1960) Zementchemie für Bauingenieure. 3rd. ed. 1977, Wiesbaden, Germany: Bauvlg., pp. 194 (in German)

Dingethal, C. (1993) Diploma Work, Technical University Munich

Jensen, O.M (1990) Schwinden infolge Eigenaustrocknung in Zement-Silika Mörtel bei niedriger Porosität. VMPA-Conference, Munich, Germany. Oct. 1990, Proceedings, Vol. C, pp. 124-135 (in German)

de Larrard, F. (1989) Ultrafine Particles for the Making of Very High Strength Concretes. Cement Concrete Research. Vol. 19, 1989, pp. 161-172.

Paillère, A, M., Buil, M. and Serrano, J.J. (1989) Effect of Fiber Addition on the Autogenous Shrinkage of Silica Fume Concrete. ACI Materials J., Vol. 86, 1989, pp. 139-144

Radocea, A. (1990) Water Pressure in Fresh and Young Cement Paste. Nordic Concrete Research. No. 9, Dec. 1990

Schrage, I. Mangold, M. and Sticha, J. (1992) An Approach to High-performance Concrete in Germany. Conference: Fly Ash, Silica Fume, Slag, and Natural Pozzalans in

Concrete, Istanbul, Turkey, May 1992, Suppl. paper pp. 493-511

Springenschmid, R., Schrage, I.(1993) Hochfester Beton: Zeitabhängige Verformungen, Rißempfindlichkeit und Widerstand gegen chemische Einwirkungen. DBV-Conference: Forschung, Wiesbaden, Germany, July 1993 (under publication, in German)

Springenschmid, R. (1994) The Influence of Cement, Pozzolanes and Silica Fume on Cracking Tendency of High Strength Concrete. RILEM-Workshop: High-Strength Concrete, Vienna, Austria, Feb. 1994 (under publication)

Springenschmid, R., Breitenbücher, R., and Mangold, M. (1994) Development of the Cracking Frame and the Temperature - Stress Testing Machine. This Conference.

Summer, Th., Schrage, I. (1993) Hochfester Beton-Schwinden, Kriechen und Reißneigung. DAfStb-Conference: Forschung, Munich, Germany, Oct. 1993, Proceedings, pp. 147-150 (in German)

PART SEVEN

COMPUTATIONAL ASSESSMENT OF STRESSES AND CRACKING

(Simulation numérique des contraintes
et de la fissuration)

30 EXPERIENCE IN CONTROLLED CONCRETE BEHAVIOUR

E. MAATJES
HBW, Hollandsche Beton- en Waterbouw bv, Gouda,
The Netherlands
J.J.M. SCHILLINGS
STRUCOM, Structures and Computers Ltd, Utrecht, The Netherlands
R. De JONG
Ingenieursbureau Nederlandse Spoorwegen, Utrecht, The Netherlands

Abstract
This paper describes the main characteristics and the use of the
concrete curing control program FeC₃S. The theoretical background is
explained and an example of a test is given and discussed. Daily
application of the program is illustrated by an example of strength
prediction using mix optimisation, insulation or external heating.
Some QA-aspects are discussed.
Keywords: Control of Curing, Numerical Simulation, Strength
Prediction, Stress Prediction, Temperature Prediction.

1 Introduction and background

In the last one or two decades the working methods in concrete
construction work gradually become more industrial.
This industrial approach is characterised by design for repetition,
detailed planning, frequent (daily) concrete pours and shorter cycle
times. This change is mainly caused by the need for:

(a) reduction of direct costs
(b) reduction of the overall construction time
     (indirect costs)

Since the reduction of cycle times generally leads to an equal
reduction in the time available for curing, techniques such as the
maturity concept and temperature controlled curing of concrete cubes
or cylinders were introduced in order to proof that sufficient
strength is gained. Also strength prediction based on recorded temperatures and maturity was introduced. The accuracy of this method was
limited due to the inability to predict the temperature-development in
early-age concrete. This lack of accuracy as well as the need for
crack control lead to computerprograms for the prediction of the
development of temperatures, strength and stresses. Within HBG a 1-D
program, known as VBS, was developed. The characteristics and
applications of this program are described by Horden, Maatjes en de
Sitter (lit.).

The five-year experience gained with this program showed that
temperature and strength predictions were sufficiently accurate. In
order to improve the prediction of stresses, the temperature and
strength part of the VBS program is now combined with the finite
element program ANSYS. The new program FeC₃S, Finite element Concrete
Curing Control System, is a 2-D temperature and a 3-D stress program.

Thermal Cracking in Concrete at Early Ages. Edited by R. Springenschmid. Published 1994
by E & FN Spon, 2–6 Boundary Row, London SE1 8HN, UK. ISBN: 0 419 18710 3.

2 Theoretical background and material laws of FeC₃S

2.1 General
The computerprogram FeC_3S uses the finite element program ANSYS as a basis. A 2-D temperature model and a 3-D structural model can be generated using standard ANSYS-commands. The heat of hydration is modelled as a time dependant internal heat source. The development of compressive strength is based on isothermal laboratory tests and the degree of hydration. The development of tensile strength and Youngs Modulus can be derived from compressive strength, or alternatively be based on laboratory tests as well. The calculation of stresses is a standard feature of ANSYS.

2.2 Heat development
The heat development of hydrating concrete is determined in an adiabatic test which lasts up to 7 days. The adiabatic circumstances are simulated by keeping the temperature in a water basin equal to the temperature of an insulated concrete cube, which is cured in this waterbasin. The purpose of the insulation is to limit heat exchange due to unavoidable temperature differences betwee the concrete and the waterbasin. The heat development used in the calculations is based on the Mandry-formulation (lit.) A schematised set-up of the adiabatic test-equipment is given in figure 1.

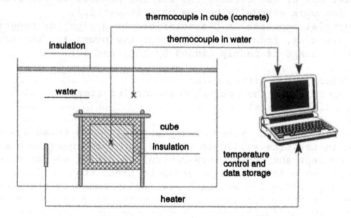

Fig.1. Setup of adiabatic test.

The testresults are fitted to formula (1), which is used as input for the calculations.

$$W = W_{max} * (1 - e^{\frac{(t-t_*)}{t_0}}) \qquad (1)$$

with:

W	=	amount of liberated heat	[kJ/kg]
W_{max}	=	maximum heat of hydration in a 7-day test	[kJ/kg]
t_*	=	dormant stage	[h]
t_0	=	constant, depending on the mix	[h]

An example of an adiabatic test is given in figure 2. The deviations between the concrete and the watertemperature which are in the order of magnitude of 0.1 °C do not show in this graph. The concrete temperature is fitted to formula (1) and the result is also plotted in figure 2.

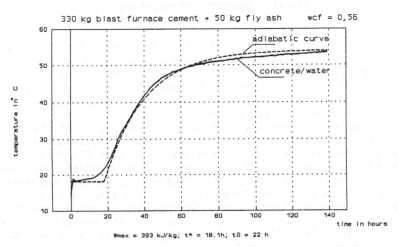

Fig.2. Results of an adiabatic test.

2.3 Thermal properties and boundary conditions

The thermal properties, coefficient of heat transfer, specific heat and the coefficient of thermal expansion, which may be changed at discrete time points are normally kept constant throughout the calculation. The default values are:

(a)	thermal conductivity:	2.6	[kJ/m°C]
(b)	specific heat:	1.1	[kJ/kg°C]
(c)	coefficient of thermal expansion:	$12*10^{-6}$	[-]

The thermal boundary conditions may be changed at discrete time points in order to simulate the removal of formwork or the application of insulation blankets. The effect of the heat capacity of formwork is neglected. The chilling effect of wind is accounted for. Daily variations in air temperature are not taken into account.

2.4 Semi-adiabatic heat development

Due to heat exchange with the environment the heat development in a concrete structure deviates from the adiabatic heat development. The difference between the actual and the adiabatic temperature is used to calculate a fictitious time-step. The adiabatic heat liberation during this fictitious timestep is used as heat generation in the actual timestep. The length of the fictitious timestep is calculated with a variation on Rastrups' law:

$$\Delta t_{fictitious} = \Delta t_{actual} * 2^{\frac{(T_{actual} - T_{fictitious})}{10}} \tag{2}$$

with:

$\Delta t_{fictitious}$ = fictitious time   [h]
Δt_{actual} = real time   [h]
T_{actual} = real concrete temperature   [°C]
$T_{fictitious}$ = adiabatic concrete temperature   [°C]

2.5 Artificial cooling

Artificial cooling of concrete with cast-in cooling pipes can be simulated by prescribing fixed temperatures in the nodes of the element model. The heat dissipation into the cooling pipes can be summarised in order to calculate the heat flow in the cooling pipes.

2.6 Development of strength and stiffness

Using the adiabatic heat development as a starting point the iso-thermal heat liberation at 20 °C can be calculated. Assuming that the degree of hydration is equal to the degree of heat liberation, the degree of hydration can easily be expressed as a function of time at 20 °C. The compressive strength development can also be expressed as a function of time at 20 °C. Combination of both functions leads to a compressive strength development as a function of the degree of hydration. This relation is used to calculate the strength on a semi-adiabatic basis. The effect of the curing temperature on the 28-day strength is not accounted for.

The development of the tensile strength and the modulus of elasticity are calculated from the compressive strength according to the 1984 version of the Dutch Concrete Code.

$$E = (1800 - 4f_{cc})\sqrt{10f_{cc}} \tag{3}$$
$$f_{ct} = 0.78(1 + \frac{f_{cc}}{20}) \tag{4}$$

with:

E   = modulus of elasticity   [MPa]
f_{cc} = cube compressive strength   [MPa]
f_{ct} = tensile strength   [MPa]

2.7 Calculation of stresses and relaxation

The calculation of stresses is a standard ANSYS application. Stresses are calculated on the basis of:

$$\Delta\sigma_i = E_i (\Delta\epsilon_i - \Delta\epsilon_i^{th}) = E_i (\epsilon_i^{el} - \epsilon_{i-1}^{el} + \epsilon_i^{cr} - \epsilon_{i-1}^{cr}) \tag{5}$$

with,

$\Delta\sigma_i$ = relaxed stress increment
E_i = modulus of elasticity
$\Delta\epsilon_i$ = total strain-increment at t_i
$\Delta\epsilon_i^{th}$ = thermal strain-increment at t_i
ϵ_i^{el} = elastic strain at t_i
ϵ_{i-1}^{el} = elastic strain at t_{i-1}
ϵ_i^{cr} = creep strain at t_i
ϵ_{i-1}^{cr} = creep strain at t_{i-1}

The creep strain is calculated on the basis of the actual stress level
and on an assumed average age at loading. The average age at loading
is recalculated at every time step. Since the method is still
experimental, no further details are given.

3 Example of temperature and stress calculations

In 1988/1989 a 230.000 m3 storage facility for chemical waste was
build near Rotterdam. In order to prevent leakage of pollutants
through unforeseen cracks in the concrete walls or concrete floors the
inside of the storage facility had to be lined with HDPE. Due to the
presence of the lining limited cracking (maximum crack-width 0.15 mm)
was allowed. From handmade calculations it appeared that artificial
cooling was required. It was agreed that temperatures and cracking had
to be monitored in order to be able to improve the cooling procedure.
The temperature development was monitored for the duration of the
cooling proces (72 hours) only. The monitored values as well as the
calculated values are plotted in figure 5. Apparently the maximum
temperature and the time at which the maximum temperature is reached
coincide well for the thermocouples 1 and 3. Due to the steep
temperature gradients near the cooling pipes the temperature in
thermocouple 2 may be influenced by a deviation in the exact position
in the thermocouple.

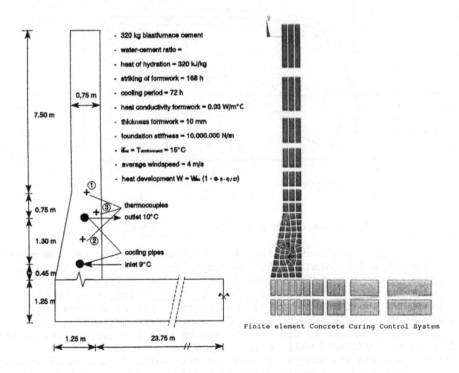

Fig.4. Cross-section with mixdata and element distribution.

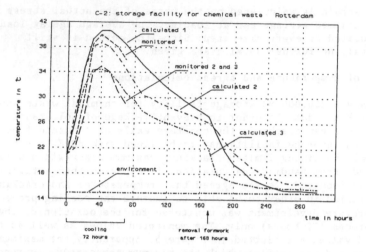

Fig.5. Calculated and monitored temperatures

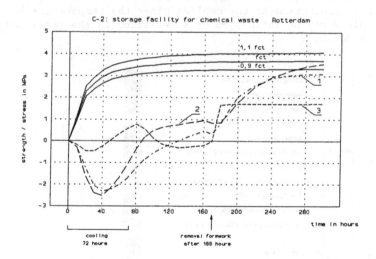

Fig.6. Predicted strength and stress-development

The strength development in figure 6 is calculated from the compressive strength using formula (4). The compressive strength is based on average values for the whole cross-section. The tensile strength development will therefore not be exact. This is indicated by the lines 1.1 fct and 0.9 fct. From the figure it can be derived that cracking may possibly occur after 270 hours. From monitoring it appeared that untill 10 days after casting no cracks were found. After 24 days all monitored walls were cracked.

Remark: In case the program is used to determine the amount of cooling required to prevent cracking a value of 50 % of the tensile strength is used.

4 Use of FeC₃S

FeC$_3$S is used in design, cost estimation, preparation of work procedures and execution. Dependant on the situation FeC$_3$S may be used:

(a)   to determine the strength development of a given mix (striking of formwork, prestressing)
(b)   to optimise a mix with respect to strength or stress development
(c)   to determine the amount of artificial cooling
(d)   to optimise work schedules

As an example the optimisation of strength development for a tunnel deck is chosen. The time schedule for the tunnel deck allows for a 64 hour curing period. The required average compressive strength for the removal of formwork is 14 MPa. Under winter conditions and with standard mix designs the required strength may not be achieved in time. As a precaution the strength development under winter conditions was determined. Also a number of methods to improve the strength development was investigated. These methods were:

(a)   using a mix with 280 kg blastfurnace cement and 50 kg Portland C in stead of 330 kg blastfurnace
(b)   insulation of the topside of the deck
(c)   external heating from the underside of the deck

The effect of the methods on the temperature development can be seen in figure 7. The effect on the strength development is given in figure 8. Assuming that a margin of 25 % = 3.5 MPa is required in case of unforeseen circumstances both the heating and the adjusted mix design are succesfull.

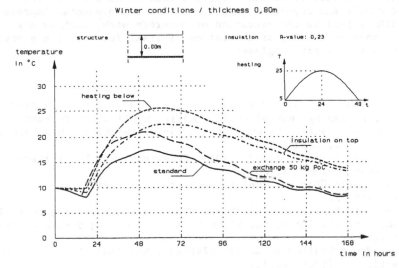

Fig.7. Effect of curing control on temperature development

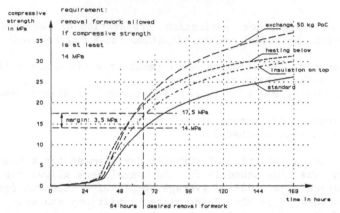

Fig.8. Effect of curing control on strength development

5 QA-aspects

5.1 QA of the FeC₃S-program

The formulas used for the development of temperature, strength and stiffness are based on 5 years experience with VBS. The calculation of stresses takes place within the already existing ANSYS-program, which is continuously improved under QA. The correct transfer of data is checked by recalculating monitored situations. All formulas as well as the program are extensively documented.

5.2 QA of the application

The engineer responsible for a set of temperature-stress and strength calculations is also responsible for the monitoring program, the required checks and reporting suggestions for improvements. Improvements with respect to the execution of concrete work, such as mix changes, enhanced cooling or insulation of steel formwork are always discussed with the site engineer.

6 Literature

Horden, W.C. Maatjes, E. and Berlage, A.C.J. (1989) **A computerised concrete hardening control system and its application in tunnel construction**, Conference on immersed tunnel techniques Manchester, Thomas Telford, London.

Maatjes, E. and Berlage, A.C.J. (1989) **Beheersing van het verhardingsproces**, Cement, 3, 56-62.

Mandry, W. (1961) **Über das Kühlen von Beton**, Springer Verlag, Berlin.

De Sitter, W.R. Ramler, J.P.G. (1989) **The concrete hardening control system**, Contribution to the RILEM Workshop "Testing during concrete construction".

Anonymous, **Voorschriften Beton VB 1974/1984, NEN 3880**, Nederlands Normalisatie Instituut, Delft.

31 NUMERICAL SIMULATION OF CRACK-AVOIDING MEASURES

J. HUCKFELDT, H. DUDDECK and H. AHRENS
Technical University of Braunschweig, Germany

Abstract
In order to avoid or at least to control early thermal cracking of massive concrete, structural, technological and constructive measures can be applied by the designing engineer. The optimal aftertreatment varies depending on the structure at present. If numerical models are calibrated by experiments, they can predict the performance and the efficiency of the measures according to the real structure. This paper demonstrates that numerical investigations applied during the planning stages enable the designing and the construction of crack-free massive concrete structures.
Keywords: Cooling Systems, Damage due to Cracks, Damage due to Thermal Stresses, Effects of Cracks on Hydration, Numerical Simulation.

1 Introduction

During the hydration process the concrete develops its mechanical characteristics and is exposed to thermal actions. That may result in stresses, if inner and/or outer restraints exist. If these stresses reach the tensile strength, cracks may occur. Depending on the kind and depth of the cracks, durability will be impaired. In order to control thermal cracking during the hydration, there are structural, technological and constructive methods. Numerical simulations of such aftertreatments become increasingly important. The efficiency of crack-avoiding methods may be checked already during the planning stage.

2 Process Kinetics

Processes on microscale are transformed to macrolevel by applying the concept of evolutionary equations. The definition of the heat development rate $\dot{q}$ is the starting point of the phenomenological description of the process kinetics. Assuming that only exothermic reactions take place during the hydration process, the heat development rate $\dot{q}$ and the progress of the degree of hydration $\dot{\alpha}$ are proportional to each other.

$$\dot{q} = H_C^{ad} \, C \, \dot{\alpha} \qquad \text{with:} \qquad \begin{array}{ll} H_C^{ad} & : \text{heat of hydration of cement} \\ C & : \text{cement content of concrete} \end{array} \qquad (1)$$

For the functional description of the degree of hydration, the following approach

Thermal Cracking in Concrete at Early Ages. Edited by R. Springenschmid. Published 1994 by E & FN Spon, 2–6 Boundary Row, London SE1 8HN, UK. ISBN: 0 419 18710 3.

has been verified: The temperature dependency on the process is modelled by an
ARRHENIUS-function. The temperature history of the process is related to an
isothermal reference process at $20\ ^{\circ}C$ and is integrated arriving at the equivalent
age τ_e:

$$\tau_e \;=\; \int_t K\, dt \quad \text{where:} \quad K \;=\; e^{\frac{A}{R}\left[\frac{1}{293} - \frac{1}{273 + \vartheta}\right]} \qquad \text{(s. Fig. 1a).} \quad (2)$$

The activation energy A may be a function of the temperature as suggested by
FREIESLEBEN ET AL. or a complex function of the temperature, the degree of
hydration and the C_3S-content as proposed by VAN BREUGEL [2]. R represents
the gas constant.

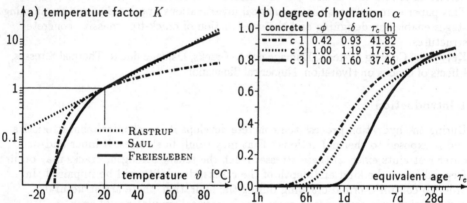

Figure 1: a) Comparison of different functionals for the temperature factor K,
see Eq. (2).   b) Degree of hydration for different concretes, see
Table 1 and Eq. (3). Based on adiabatic temperature measurements
performed by ROSTÁSY ET AL. [4], [6], [7].

Table 1: Description of three concrete compositions for the numerical investigations.

concrete	cement	[kg/m³]	additive	mixing proportion	heat of hydration
c1: B25	240 PZ	35 F	25 % fly ash	1 : 5.60 : 0.58	524.05 J/g
c2: B35 WU	390 HOZ	35 LNW		1 : 4.53 : 0.47	348.09 J/g
c3: B35	240 HOZ	35 LNW	20 % fly ash	1 : 6.47 : 0.53	353.13 J/g

It is useful to express the temperature development of adiabatic calorimetry by the
transformed time scale of the equivalent age τ_e . Then the following function given

by JONASSON describes the temperature increase and, due to Eq. (1), the degree of hydration α, see Fig. 1b.

$$\alpha = e^{\phi \left\{ \ln \left[1 + \frac{\tau_e}{\tau_c} \right] \right\}^{\delta}} \qquad \text{where:} \quad \tau_c , \ \phi , \ \delta \quad \text{are constants.} \qquad (3)$$

3 Material Model

3.1 Hypo-Elasticity

As an example, the temperature development shown in Fig. 2a is applied to a completely restraint beam. The resulting developments of stress and strength are shown in Fig. 2b.

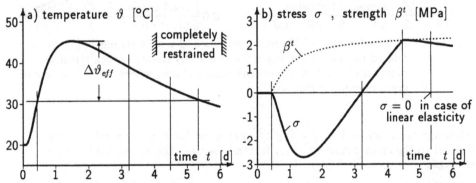

Figure 2: a) Temperature development and   b)   resulting stress and tensile strength β^t over actual time t (in days) in case of a completely restrained beam.

Of special significance is the development of the Young's modulus with the degree of hydration. The temperature rise from 20 to 30 °C within the first hours of young concrete is without any effects. The following temperature increase is causing compressive stresses, because of the in time growing stiffness (Young's modulus). Then decline of temperature sets in yielding tensile stresses (here after 3.2 days, s. Fig. 2b). After 4.5 days the current tensile strength is reached and cracks may occur. Experiments, carried out by LAUBE [4] and ROSTÁSY/ONKEN [7] confirm the progressive development of the Young's modulus as shown in Fig 3a. In case of linear elastic behavior the stress zero will reach after the effective temperature increase is convected to the surroundings - here after 5.3 days, s. Fig. 2a.

3.2 Cracks

Cracks are modelled by a local crack band approach. Cracks occur if the maximum principal stress reaches the current tensile strength. For the numerical continuum model, the discrete crack openings w are "smeared" over the length of the fracture process zone l_{PZ} , [1], [5]. The post failure behavior is described by a declining exponential function, see Fig. 3b.

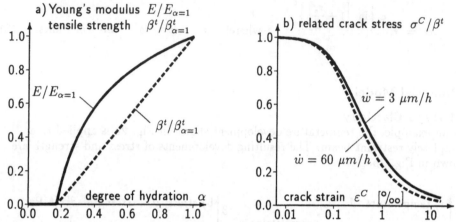

Figure 3: a) Development of Young's modulus and tensile strength, [7], [4], related to final full strength values.   b) Related crack stress as function of crack strain in post failure states, where $\dot{w}$ is the rate of crack opening width w.

Although the crack width $w = l_{PZ}\,\varepsilon^C$ may grow further, the resulting stress normal to the crack-plane σ^C may increase because the hydration is still in progress.

3.3 Viscosity

Experiments, performed by LAUBE [4] and ROSTÁSY ET AL. [8], verify that the rheological behavior of hydrating concrete can be modelled by the following creep function κ :

$$\kappa = \frac{\varepsilon_{viscous}}{\varepsilon_{elastic}} = a\left[\frac{t - t_0}{t_c}\right]^b \qquad \text{where:} \qquad t - t_0 \;:\; \text{"loading" period} \qquad (4)$$

a and b are functional parameters of the degree of hydration α and the constant t_c is set at 1 h. In [3] it is shown that Eq. (4) can be interpreted as a nonlinear MAXWELL-model with elasticity E^* and viscosity η^*. The reversible and irreversible part of the viscous strain require one another. Both coefficients of the rheological model E^* and η^* are functionals of the degree of hydration and the "loading" period $t - t_0$:

$$\dot{\varepsilon}_{viscous} = \frac{1}{E^*}\,\dot{\sigma} + \frac{1}{\eta^*}\,\sigma \qquad \text{where:} \qquad E^* = E/\kappa \qquad \text{and} \qquad \eta^* = [E/\kappa]^{\textstyle\cdot} \qquad (5)$$

Fig. 4 compares the formulation by Eqs. (4) and (5) to a BURGERS-model. In case of the BURGERS-model, viscosities and elasticity depend on the degree of hydration and the state of stress. Both models can be applied equivalently. Bench tests investigating the behavior for unloading are needed in order to decide which model is more suitable. Basically, the BURGERS-model has a better capacity to reproduce the observed material behavior for unloading.

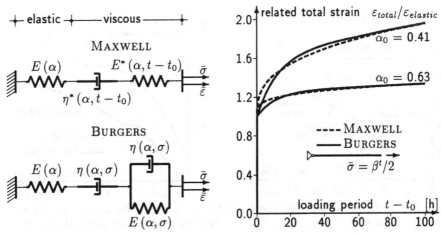

Figure 4:   Comparison of two different models for the description of viscous strain. The two distinct degrees of hydration α_0 are kept constant during the numerical test.

3.4 Longer Lasting Stresses

Laboratory tests by ROSTÁSY and ONKEN [7] confirm that longer lasting stresses influence the mechanical properties of the concrete at early ages - especially the tensile strength. For a consistent description of that damage effect the *equivalent tensile age* τ_σ is defined as an integral over the related tensile stress history, see [3]. Thus the damage of a stress history depends on the stress level as well as on the stress rate.

$$\tau_\sigma = \int_t \left\langle r_\sigma \frac{\sigma_1}{\beta^t} \right\rangle dt \quad \text{where:} \quad r_\sigma = \frac{1}{2} \frac{\sum\limits_{i=1}^{3} \left[|\sigma_i| + \sigma_i \right]}{\sum\limits_{i=1}^{3} |\sigma_i|} = \begin{cases} 0 : \text{compression} \\ 1 : \text{tension} \end{cases} \quad (6)$$

σ_i are the principal stresses and σ_1 is the maximum principal stress. By introducing the equivalent tensile age, the tensile strength is modified in order to consider the damage due to longer lasting stresses:

$$\beta^t_{mod} = \beta^t \, \eta^t \quad \text{where:} \quad \eta^t = 1 - (1 - \bar{\eta}) \left[3\hat{\tau}^2 - 2\hat{\tau}^3 \right] \quad \text{and} \quad 0 \le \hat{\tau} = \frac{\tau_\sigma}{\bar{\tau} \tau_e} \le 1 \quad (7)$$

The parameters $\bar{\eta}$ and $\bar{\tau}$ have to be calibrated to experimental results. Fig. 5b demonstrates the reduction of the tensile strength for a given stress history. The damaging effects of stress histories depend on the degree of hydration. Therefore, the equivalent tensile age τ_σ is related to the equivalent age τ_e . Thus, the same stress history leads to a greater reduction of the tensile strength during isothermal hydration than during adiabatic hydration, see Fig. 5b. The damage caused by longer lasting stresses is naturally coupled to the progress of hydration.

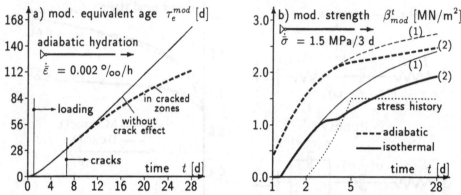

Figure 5:   a) Reduced process-kinetics in frature process zones, see Eq. (8) with $\omega = \bar{\gamma} = 0$ and $\bar{\lambda} = 1\ \%_0$.   b) Reduced tensile strength due to longer lasting stresses, see Eq. (6), (7) with $\bar{\tau} = 0.2$ and $\bar{\eta} = 0.8$, curves (1) without, curves (2) with modification.

3.5 Effects of Cracks on Hydration Progress

Furthermore, the influence of cracks on the progress of hydration in the fracture process zones is another aspect of a consistent mechanical model for concrete at early ages, see [3]. In order to consider the instantaneous ω and the developing γ effect of cracks, two reduction functions have to be defined. By assuming that all mechanical properties can be expressed as functions of the degree of hydration α which itself depends on the equivalent age τ_e, the equivalent age has to be modified in fracture process zones, if the effect of cracks on hydration is taken into consideration, see Eq. (2) and Fig. 3:

$$\tau_e^{mod} = \int_t K\,\gamma\,dt\ -\ \omega\,K \qquad \text{where:} \qquad \gamma = \bar{\gamma} + \left\{1 - \bar{\gamma}\right\}\,e^{-\varepsilon^C/\bar{\lambda}} \qquad (8)$$

Fig. 5a illustrates the delayed process kinetics in fracture process zones with the fitting parameters ω , $\bar{\gamma}$ and $\bar{\lambda}$. Applying Eq. (8) leads to a generalized equivalent age τ_e^{mod} which also considers the damage history - in addition to the temperature history.

4 Examples

Within the numerical investigations the material model as described in section 3 is considered. The rheological behavior is modelled by the modified MAXWELL-modell as given in Eqs. (4) and (5).

4.1 Cooling Pipes

A very efficient measure to reduce the risk of thermal cracking during hydration can be achieved by cooling pipes. If cooling coils are placed in equal distances over the height of a wall the double symmetry of the problem can be taken advantage of, see Fig. 6.

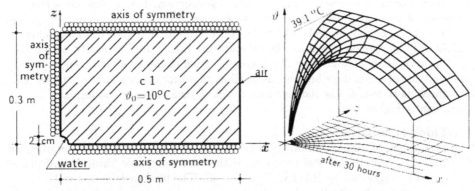

Figure 6: Equidistantly placed cooling pipes (60 cm) over the height of a 1 m thick wall and the corresponding temperature distribution after 30 h.

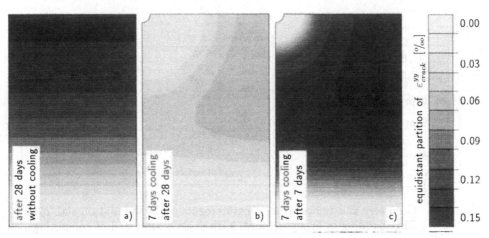

Figure 7: Crack openings normal to the plane ε_{crack}^{yy} a) after 28 days without cooling, b) after 28 days for 7 days cooling, c) after 7 days for 7 days cooling. For geometry and boundary conditions see Fig. 6.

For the example the equal distance of cooling pipes is set to 0.6 m, the considered concrete is of type c1, see Table 1, and the initial temperature is set at

$\vartheta_0 = 10\ ^\circ\text{C}$. The convection towards the air and towards the water inside the cooling pipe is modelled by:

$$\text{air}:\quad \begin{array}{l} \alpha_{air} = 15.2\ \text{W}/(\text{Km}^2) \\ \vartheta_{air} = 15\quad ^\circ\text{C} \end{array} \qquad \text{water}:\quad \begin{array}{l} \alpha_{water} = 450\ \text{W}/(\text{Km}^2) \\ \vartheta_{water} = 10\ ^\circ\text{C} \end{array} \qquad (9)$$

It is assumed that the wall is restrained completely in longitudinal direction. Therefore the main stresses do also occur in y-direction. Fig. 7 illustrates the influence of cooling by pipes on the crack openings normal to the plane ε^{yy}_{crack}. The positive effects of cooling coils is obvious. Furthermore, it is shown that cooling should only be applied to reduce the temperature maximum but not to accelerate the temperature decrease. Even if the greater crack openings after seven days for a cooling time of seven days became as small as those for cooling times between two and three days, see [3], the damage as a result of seven days of cooling is significantly higher.

4.2 Thick Foundation

The foundation of a bank building (BFG) located in Frankfurt a.M. is investigated. The massive concrete slab was concreted with concrete c3, see Table 1, and placed and cured as suggested by ROSTÁSY ET AL. [6]. The final surface was watered after pouring for three to four days. In the plan view the foundation is hexagonal, its diameter varies between 60 and 70 meters and its thickness differs from 3 to 6 meters. For the numerical analysis a rotational symmetric geometry is assumed, see Fig. 8.

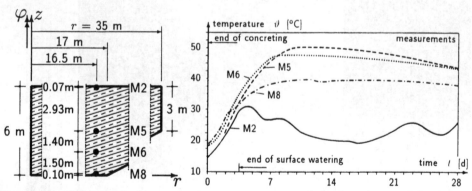

Figure 8:   Geometry and temperature measurements of the foundation slab of the BFG building in Frankfurt a.M. with d: days after end of concreting.

The slab can be considered as thick in comparison to its length. Therefore, the heat convects mainly along the thickness of the foundation. Measurements monitored on site are shown in Fig. 8. For the numerical simulation of the concreting sequence, the duration of pouring of 90 hours was simulated by a constant concreting rate of $v = 6\ \text{m}/90\ \text{h} = 1.6\ \text{m/d}$. The numerical results are shown in Fig. 9. They are in good aggreement with the measurements of Fig. 8. If the bedding stiffness of the ground is assumed to be rigid in normal direction to the slab foundation and zero in tangential direction, the three-dimensional problem can be degenerated to a one-dimensional one. Deformations are shown in Fig. 10 and the residual stresses in Fig. 11.

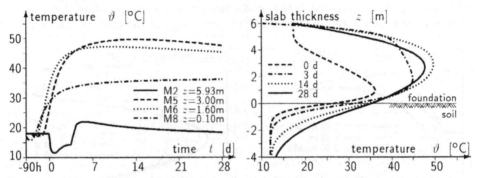

Figure 9:   Results of analysis for temperature distribution with time and across the slab thickness with d: days after end of concreting.

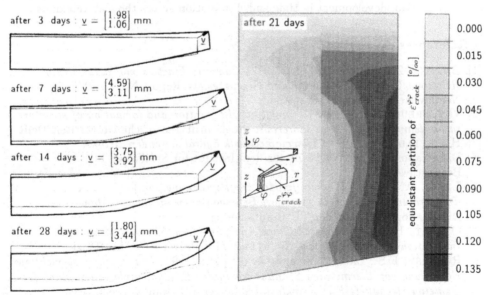

Figure 10:   Deformed shapes of the thick foundation slab (left) and cracks in ring direction at the outer region (from $r = 33$ m to $r = 35$ m) of the slab (right).

Cracks occur in the vicinity of the watered surface of the thick foundation slab. The influence of the assumed bedding effects of the ground are investigated by an axially symmetric FEM discretization. The deformed shapes after 3, 7, 14 and 28 days after placing the final concrete show the rotation of the cross sections when a flexible ground is assumed, see Fig. 10. Caused by the two-dimensional cooling at the outer thinner areas, cracks in ring direction may occur during the cooling, see Fig. 10. Within a simplified one-dimensional analysis only one stress component yields, see Fig. 11. Compared to the rotational symmetric computation this component may be interpreted either as σ_{rr} or as $\sigma_{\varphi\varphi}$ at the centre of the slab.

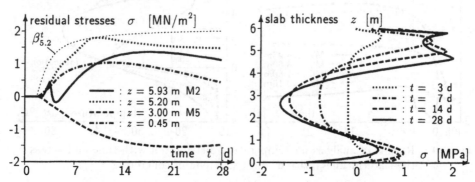

Figure 11:   Results of one-dimensional analysis for residual stresses of the thick slab, development in time and distribution across the slab thickness.

5 References

[1] BAŽANT, Z.P. and OH, B.H.: (1977) *Concrete Fracture via Stress-strain Relations*, Center for Concrete and Geomaterials, Report-No. 81-10/665c, Northwestern University Evanston.

[2] VAN BREUGEL, K.: (1991) *Simulation of hydration and formation of structure in hardening cement-based materials*, Proefschrift Technische Universiteit Delft.

[3] HUCKFELDT, J.: (1993) *Thermomechanik hydratisierenden Betons*, Dissertation, Bericht-Nr. 93-77 aus dem Institut für Statik der TU Braunschweig.

[4] LAUBE, M.: (1990) *Werkstoffmodell zur Berechnung von Temperaturspannungen in massigen Betonbauteilen im jungen Alter*, Dissertation am iBMB der TU Braunschweig.

[5] OŽBOLT, J.: (1992) *Smeared Crack Analysis - New Nonlocal Microcrack Interaction Approach*, Report-No. 4/14-92/19, University of Stuttgart.

[6] ROSTÁSY, F.S.; LAUBE, M. and ONKEN, P.: (1990) *Untersuchungsbericht zur Ermittlung von thermischen Spannungen in der Fundamentplatte der Bank für Gemeinwirtschaft (BFG) Frankfurt*, iBMB der TU Braunschweig.

[7] ROSTÁSY, F.S. and ONKEN, P.: (1992) *Wirksame Betonzugfestigkeit im Bauwerk bei früh einsetzendem Temperaturzwang*, Forschungsbericht DBV-Nr. 126 bzw. AIF-Nr. 7855 aus dem iBMB der TU Braunschweig.

[8] ROSTÁSY, F.S.; GUTSCH, A. and LAUBE, M.: (1993) *Creep and Relaxation of Concrete at Early Ages - Experiments and Mathematical Modeling*, in *Creep and Skrinkage of Concrete* (Editors: Z.P. BAŽANT and I. CAROL), Proceedings of the 5th International RILEM Symposium in Barcelona.

32 THERMAL PRESTRESS OF CONCRETE BY SURFACE COOLING

M. MANGOLD
Institute of Building Materials, Technical University of Munich, Germany

Abstract
Laboratory investigations using the cracking frame proved that is very important whether the temperature gradient occurs during the first hours or at a higher concrete age. When cooling the surface from a very early age, the temperature gradient is very steep, but later on the thermal tensile stress will be small. Thus, by surface cooling, even a thermal prestress of the concrete surface can be achieved, which is still valid when the heat of hydration has dissipated. Surface cooling also improves the long term behaviour of the member, e.g. with regard to tensile stresses (and cracks) caused by drying shrinkage. Investigations concerning the influence of the cooling start time and of the cooling duration as well as the risk of cracking caused by cooling which begins too late are presented in detail.
Keywords: Thermal Prestress, Surface Cooling, Intrinsic Stresses, Cracking Frame.

1 Introduction

During hydration, the hardening temperatures over the cross-section of a concrete member are not uniform. This is due to the different boundary conditions (e.g. bottom placed on soil, bedrock or on a preceding concrete layer, surface exposed to solar radiation) and because of the flow rate of the hydration heat. If the deformations of the concrete are restrained during hydration, stresses arise.

The temperature at which a restrained concrete element is stress free under the given degree of restraint in the direction of restraint is called the zero-stress temperature T_z, see Springenschmid and Nischer (1973). T_z depends essentially on the temperature development within the considered concrete volume during the first day of hydration, see Mangold and Springenschmid (1994).

If the zero-stress temperature T_z of the specified element within the concrete member is known, the stress which occurs at the temperature T is given by the equation $\sigma = E \, \alpha_T \, (T_z - T)$. If the actual concrete temperature T is higher than the zero-stress temperature T_z, compressive stresses occur. If T is lower than T_z, tensile stresses arise. Thus, the stress state of a restrained concrete element depends on the

Thermal Cracking in Concrete at Early Ages. Edited by R. Springenschmid. Published 1994 by E & FN Spon, 2–6 Boundary Row, London SE1 8HN, UK. ISBN: 0 419 18710 3.

size of the temperature difference $\Delta T = (T_z-T)$, which is as a rule not constant over the cross-section of a concrete member. This also means that the distribution of thermal stresses over the cross-section of a member not only depends on the actual temperature distribution, but also on the zero-stress temperature distribution over the cross-section ("Zero-stress temperature gradient").

Surface cooling of thick members at an early age affects the distribution of T_z, particulary the curvature of the zero-stress temperature gradient. If the zero-stress temperature is considerably lower at the surface than in the centre of the member, the concrete hardens with a convex curvature of the zero-stress temperature gradient. If the curvature of the zero-stress temperature gradient is more convex than the actual temperature gradient, compressive intrinsic stresses are present at the surface even though the concrete temperature is lower at the surface than in the centre (fig. 1a). When the heat of hydration has dissipated, the resulting intrinsic stresses produce a favourable prestressing of the considered concrete section (comparable to the state of stress within prestressed glass), i.e. compressive stresses at the surface and tensile stresses in the centre (fig. 1b).

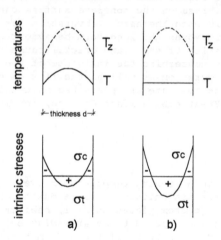

Fig. 1. Inrinsic stresses in the case of a convex curvature of the
        zero-stress temperature gradient
        a) convex gradient of concrete temperature
        b) constant concrete temperature over the member cross-section

2 Determination of Thermal Stresses and Zero-Stress Temperatures of Concrete Members by Laboratory Tests

Thermal stresses and zero-stress temperatures which occur in concrete members during hydration can be determined by cracking frame tests, see Mangold (1994). For this, the temperature development which occurs at characteristic points of the cross-section (e.g. on the surface or in the centre) during hydration is calculated using the finite element method. The calculated temperature changes are then forced upon the concrete specimens in the cracking frame. Thus, each beam specimen simulates the hardening conditions of one point of the cross-section. In the cracking frame, the concrete hardens under a high degree of longitudinal restraint. Therefore longitudinal stress arises which is measured from the very beginning, see RILEM Technical Recommendation (1993).

The corresponding change of the zero-stress temperature can be calculated from the cracking frame test data (development of temperature and stress) using the model developed by Mangold (1994). An example for the development of concrete zero-stress temperature and thermal stress at the surface and in the centre of a wall is shown in fig. 2. Considering the whole cross-section, a symmetrical distribution of thermal stresses can be subdivided into the following types.
- The longitudinal restraint stresses, which arise when the constant deformation of the concrete member, is restrained, see fig. 2. The longitudinal restraint stress has a constant value over the cross-section of the member.
- The intrinsic stresses ("eigenstresses") which occur when the temperature difference ΔT is non-linear over the cross-section, see fig. 1. These stresses are self equilibrating and do not result in external forces. The curvature of the intrinsic stress distribution (compressive stresses at the surface or in the centre) depends on the curvature of ΔT and therefore on the curvature of T and T_z.

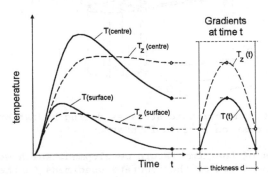

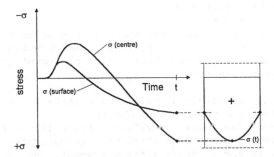

Fig. 2. Development of concrete temperatures (T), zero-stress temperatures (T_z) and stress (σ) at the surface and in the centre of a restrained wall (thickness d). Gradients of T, T_z and σ over the cross-section at time t (schematically)

3 Investigations on Surface Cooling

3.1 Concrete

For the investigations a concrete was used with a cement content of 340 kg/m³ (CEM I 32,5 R) and moraine gravel from the Munich area with a maximum aggregate size of 32 mm (grading AB_{32} according to DIN 1045). The water/cement ratio was 0.50. The hydration heat development of the concrete was determined by adiabatic calorimetry on a concrete volume of 50 l. From this the development of the degree of hydration and of the tensile strength was calculated according to the model proposed by Laube (1990). The development of the Young's modulus (E-modulus) was determined experimentally with the Temperature-Stress Testing Machine, see Schöppel et al. (1994). The coefficient of thermal expansion was calculated according to the method proposed by Dettling (1964).

3.2 Scope of Investigations

The surface cooling of concrete members was studied using the example of a concrete wall of 1 m in thickness, placed in a steel formwork, see fig. 3.

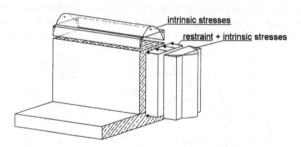

Fig. 3. Restraint and intrinsic stresses during hydration in a wall which is cast on a rigid foundation

The calculations of the temperature development which were used to control the temperature of the concrete specimens in the cracking frame tests were carried out for the following boundary conditions:
- Fresh concrete temperature 20°C
- Ambient air temperature 20°C
- Concrete placement in a formwork, which is precooled to 10°C and kept at this temperature (series 1), or in a formwork which is cooled down to and maintained at 10°C some hours after placement (series 2)
- No solar radiation present

To find out the efficiency as well as the risks of surface cooling, two test series were carried out to check:
a) the influence of the cooling duration (series 1). In this series the following cases of cooling duration were investigated: surface cooling for 0, 8, 16 and 24 hours after placement. The beginning of the cooling was assumed to be immediately at concrete placement.
b) the starting time at which the cooling begins (series 2). For the case of a cooling duration of 24 hours, the time at which cooling started was varied between 0, 4, 8 and 12 hours after concrete placement.

4 Results

4.1 Influence of the Cooling Duration (Series 1)

The temperature calculations demonstrated that surface cooling of up
to 24 hours had a marked effect on the maximum surface temperature,
but only a small influence on the temperature maximum in the centre of
the 1 m thick member. Depending on the cooling duration, the temperature maximum was reduced by 11 K at the surface and up to 4 K in the
centre. Owing to the lower temperature, the degree of hydration and
consequently the mechanical properties of the concrete (e.g. the
Young's modulus) increased slower at the surface than in the centre.
At the end of the cooling period, the degree of hydration at the surface was 12% after 8 hours cooling, 23% after 16 hours cooling and 34%
after 24 hours cooling. A high degree of hydration at the end of the
cooling period (i.e. the longer the cooling period had been) resulted
in conversion of more of the subsequent temperature increase into compressive stresses. After 8 hours cooling the subsequent temperature
increase from 10°C to 33.5°C resulted in a compressive stress of
-0.50 MPa. This corresponds to an average conversion factor of
-0.021 MPa per K. After a cooling period of 16 and 24 hours this
factor amounted to -0,042 MPa/K and -0.067 MPa/K respectively.

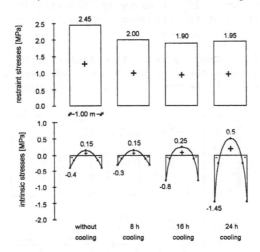

Fig. 4.Restraint and intrinsic stresses remaining within the 1.0 m
thick wall after the heat of hydration has dissipated (at the
age of 7 days) in the case without or with surface cooling
(immediate begin of cooling).

The larger the portion of the temperature increase which was converted into compressive stresses, the smaller was the thermal tensile
stress after dissipation of the heat of hydration. Fig. 4 shows the
distribution of longitudinal restraint stresses and of intrinsic
stresses over the cross-section of the wall at the age of 7 days. As
can be seen from fig. 4, the compressive intrinsic stress superposed
on the tensile longitudinal restraint stress at the surface of the
member. The total tensile stress (sum of intrinsic and longitudinal

restraint stress) at the surface was diminished significantly by surface cooling. Without cooling the tensile stress at the surface was
2.05 MPa, i.e. 85% of the actual concrete tensile strength. In the
case of 8 hours cooling the tensile stress at the age of 7 days was
70% of the tensile strength. For 16 hours cooling it was 50% and in
the case of 24 hours cooling less than 30%.

The cooling duration of 8 hours was too short in order to produce a
durable strongly convex curvature of the zero-stress temperature gradient. Therefore the distribution of intrinsic stresses is the same as
without cooling. Obviously, at the end of the cooling period the concrete was too soft (i.e. relaxation was high and Young's modulus still
low), so that the initially strongly convex curvature of the Tz-gra-
dient could not be maintained. Only if the concrete was cooled long
enough (i.e. longer than 8 hours) a favourable zero-stress temperature
gradient could be "frozen" in the structure.

Due to the lower tensile stress, surface cooling also resulted in a
lower cracking sensitivity (expressed by low cracking temperatures,
see Springenschmid et al. (1994)) of the members. Compared to the member without cooling, the cracking temperatures of the cooled members
were up to 17°C lower at the surface. This means a considerable reduction of the cracking sensitivity, particularly at the surfaces of the
structures. Thus, not only the early age cracking risk, but also the
long-term properties of the structure (durability, resistance against
shrinkage cracks) are improved.

4.2 Influence of the Cooling Start Time (Series 2)
If cooling began late, the concrete had already heated up and thus the
degree of hydration at the beginning of cooling was higher.

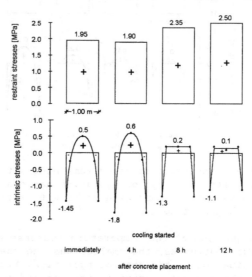

Fig. 5. Restraint and intrinsic stresses remaining within the 1.0 m
thick wall after the heat of hydration has dissipated (at the
age of 7 days) when the cooling started immediately or 4, 8 or
12 hours after placement.

Fig. 5 shows the distribution of restraint and intrinsic stresses over the cross-section of the wall at an age of 7 days. Cooling which began within the first 4 hours after concrete placement had the best efficiency, see fig. 5. When cooling started after 4 hours (i.e. after 8 or 12 hours), the degree of hydration was higher when cooling started and therefore the tensile stress due to surface cooling was higher. When cooling started 12 hours after concrete placement, the degree of hydration was already about 40% at the surface. When the concrete surface was cooled down from 35°C to 10°C, intrinsic stresses were produced, with a maximal tensile stress of 2.25 MPa at the concrete surface. At an age of 12 hours, this would result in severe cracking of the structure, see fig. 6.

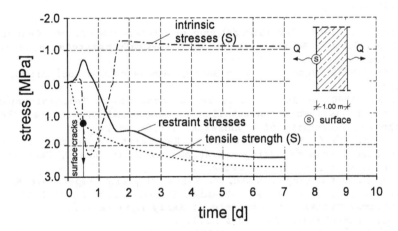

Fig. 6.Cracks were provoked if surface cooling started too late, i.e. later than 4 hours.

5 Conclusion

The investigations reveal the following.
(1) If the surface of a concrete structure is cooled during the first day of hydration, considerable compressive intrinsic stresses ("thermal prestress") will remain at the member surface when the heat of hydration has dissipated. This is comparable to the state of stress which is also present in prestressed glass.
(2) However, in order to achieve a significant thermal prestress it is necessary to cool for a sufficiently long time. For the concrete used in these investigations and for cooling of the surface from 20 to 10°C, a cooling duration of more than 8 hours was required. If surface cooling was stopped too early, the thermal prestress was largely reduced again by relaxation.
(3) Surface cooling must start at a very early age, i.e. within the first few hours after placement (within 4 hours in these investigations). Then it is possible to produce a strongly -favourable- convex curvature of the zero-stress temperature gradient by largely plastic deformations of the young concrete.

(4) If surface cooling starts too late (i.e. at a time when the Young's modulus increases rapidly and relaxation decreases), then surface cooling causes considerable damage. The concrete is too stiff to be sufficiently deformed plastically. Low concrete temperatures are converted into tensile stresses then and not into low zero-stress temperature.

(5) If surface cooling of the young concrete is camed out properly, the cracking sensitivity of the hardened member is reduced. This also improves the long-term properties of the member, such as the resistance against shrinkage cracks.

6 References

Dettling, H. (1964) Die Wärmedehnung des Zementsteins, der Gesteine und des Betons, **Schriftenreihe des Otto-Graf-Instituts,** Heft 3, Stuttgart.

DIN 1045 (1988), Beton und Stahlbeton, Bemessung und Ausführung, (in German).

Laube, M. (1990) Werkstoffmodell zur Berechnung von Temperaturspannungen in Massigen Betonbauteilen im jungen Alter, Doctoral Thesis, Technical University of Brunnswick, (in German).

Mangold, M. (1994) The Development of Restraint and Intrinsic Stresses in Concrete Members During Hydration, Doctoral Thesis, Technical University of Munich, (in German).

Mangold, M. and Springenschmid, R. (1994) Why are Temperature-Related Criteria so Undependable to Predict Thermal Cracking at Early Ages?, in **Thermal Cracking in Concrete at Early Ages** (editor R. Springenschmid), Chapman & Hall.

RILEM Technical Recommendation (1993) Testing of the Cracking Tendency of Concrete at Early Ages, 2nd Draft, December 1993.

Schöppel, K.; Plannerer, M.; Springenschmid, R. (1994) Determination of Restraint Stresses and of Material Properties during Hydration of Concrete with the Temperature-Stress Testing Machine, in **Thermal Cracking in Concrete at Early Ages** (editor R. Springenschmid), Chapman & Hall.

Springenschmid, R. and Nischer, P. (1973) Untersuchungen über die Ursache von Querrissen im jungen Beton, **Beton- und Stahlbetonbau,** Heft 9, pp. 221-226 (in German).

Springenschmid, R.; Breitenbücher, R.; Mangold, M. (1994) Practical Experience with Concrete Technological Measures to Avoid Cracking, in **Thermal Cracking in Concrete at Early Ages** (editor R. Springenschmid), Chapman & Hall.

33 DEFINING AND APPLICATION OF STRESS-ANALYSIS-BASED TEMPERATURE DIFFERENCE LIMITS TO PREVENT EARLY-AGE CRACKING IN CONCRETE STRUCTURES

P.E. ROELFSTRA
Intron SME, Yverdon, Switzerland
T.A.M. SALET and J.E. KUIKS
Intron SME, Houten, The Netherlands

Abstract
In construction specifications and building codes temperature difference limits between structural members (old- and fresh concrete) or areas (inner- and outside) are given to prevent early-age cracking. It is shown that these limits are often not appropriate and can lead to serious damage or exaggerated cooling requirements. The chief cause is that these limits are based generally on too simplified assumptions of the structural restraining. This means that only those temperature difference limits, which are based on reliable stress analysis of the structure, make sense. In this contribution it is demonstrated how such limits can be defined and can be integrated in workprocedures for practice.
Keywords: Cracking, Temperature Difference Limits.

1. Introduction

The phenomenon of thermally induced cracking at early-ages has gained in significance over the past 2 decades. It has become more and more apparent that thermal effects are not limited to massive structures only, but also contribute to crack formation in structures of medium and even small thickness. Cracking in concrete structures can cause, among others, a lack of watertightness (disastrous for tunnel constructions), acceleration of the corrosion of the reinforcement, and reduction of structural rigidity /8/. That is why, for matters of quality insurance, methods are applied to avoid this early-age cracking, especially in Northern Europe.

To prevent early-age cracking the evolution of the temperature in the structure must be controlled. The nature of temperature control has evolved over the years. In the beginning the attention was focused on the limitation of the heat of hydration, resulting in well known low heat cements as a result of the research effort. Nowadays, the application of artificial cooling has become more usual. In practice temperature controlling systems must be designed to satisfy temperature limits, given in building specifications and codes. Commonly used temperature limits are :

D_{nn} = difference between the average temperature of a fresh cast, compared with the average temperature of a previously casted and already hardened adjacent component; in specifications D_{nn} is generally about 12-15°C;

D_{mm} = difference between the surface temperature and the maximum temperature in the interior of a cross section of a concrete body; this limit is sometimes also defined in terms of the difference between the average temperature of a cross-section compared with the maximum temperature in the same cross-section; dependent on the definition, D_{mm} usually is 20°C or 15°C, respectively;

Thermal Cracking in Concrete at Early Ages. Edited by R. Springenschmid. Published 1994 by E & FN Spon, 2–6 Boundary Row, London SE1 8HN, UK. ISBN: 0 419 18710 3.

These temperature limits, however, are non-appropriate and can lead to serious damage or exaggerated temperature controlling systems, if they are not based on reliable analysis of the evolution of stresses and strength in the structure. This important point has already been recognized by research workers in the early 1960's. A historical overview is described by Emborg /1/. Perhaps one of the most important reasons that simply defined temperature limits are still being used, is the lack of validated computer programs for reliable stress analyses of practical problems.

For this reason, a special purpose program for PC's has been developed, called HEAT /2,3,4,5/. This program is based on the ideas developed at the Laboratory for Building Materials of the Swiss Federal Institute of Technology (FEMMASSE) /10/. The program has been applied succesfully to several important construction projects like i.e. the Storebealt bridges in Denmark, Ras Laffan Harbour in Qatar, Storm barrier, Wijkertunnel and Piet Hein tunnel, all in the Netherlands and to many other projects. These different types of construction projects enabled to make extensive validations of the implemented models.

This contribution starts to explain, by means of a practicle example, that simply defined temperature limits are non-appropriate criteria to avoid early-age crack formation. Next a framework is described, along the lines of the models implemented in program HEAT, by which reliable stress-analysis-based temperature difference limits can be defined. Finally, it is outlined how these reliable limits can be integrated in temperature controlling systems at construction sites.

2.   Practical consequences of applying simply defined temperature difference limits

It is known that early-age cracking takes mainly place in the cooling down phase of the hydration process. In this phase concrete has become a certain rigidity and, consequently, any restraining of thermal shrinkage will be translated into stresses. The evolution of stresses depends therefore on the evolutions of:

a.   the temperature;
b.   the rigidity (relaxation included);
c.   the structural restraining.

Temperature difference limits in construction procedures and building codes are often defined with too simplified assumptions concerning points b. and c. The practical consequences of these simplified assumptions is demonstrated by the following example, computed with the help of program HEAT.

Example
The example concerns the construction of precasted tunnel segments. The segments consist of a base and deck slab, three inner walls, and (obviously) two outer walls. The length of the tunnel segment is about 25 m. In this example two different casting sequences (A and B) are considered. Also an alternative construction method (C) of the tunnel without segmentation is analyzed. For these three cases it is assumed that the soil constant is small (0.015 N/mm^3) and that the base slab is casted long before the walls and deck slab. The difference between sequence A and B is:

Sequence A: Inner-, outer walls, and deck slab casted in one pour;

Sequence B: Outer walls and deck slab casted in one pour. Inner walls have been casted earlier.

In the alternative construction method (C) it is of less importance whether the inner walls are casted earlier or not, because of the length of the construction.

In addition, for each construction the following season condition and demoulding times are examined.

1. Walls and deck casted in summertime and demoulded after 3 days;
2. Walls and deck casted in wintertime and demoulded after 3 days;
3. Walls and deck casted in summertime and demoulded after 14 days;
4. Walls and deck casted in wintertime and demoulded after 14 days.

Physical boundary and initial conditions have been imposed such that the maximum temperature difference D_{nn} between the hardening concrete and the base slab is about 22°C. The stress distributions in the centre of the outer wall after the cooling down phase is plotted in Fig.1.

It can be seen in Fig.1 that the magnitude as well as the stress distribution differ largely, despite similar temperature difference D_{nn}. This example, taken from practice, makes clear that no general temperature difference limits can be defined. Temperature difference limits must always be linked to the stress evolution, which is structural dependent.

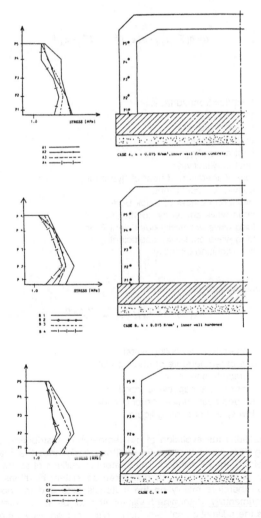

Figure 1 Computed stress distribution over the height of the outer wall for each tunnel alternative, different seasons and demoulding times.

3.   Framework to define reliable stress-analysis-based temperature difference limits

The consequence of the conclusion, drawn in the preceding chapter, is that e.g. measurements to avoid early-age cracking must be designed on the basis of reliable stress analysis. The reliability of stress analysis depends on the consistency of the employed constitutive relations as well as on the reality of assumed physical and kinematical boundary conditions. These important items are discussed in detail here. The analysis consists of solving a boundary value problem for the temperature as well as for stresses. The mathematical formulation of the boundary value problem for the temperature is given by Eqs.1-3.

In problem domain:

$$C\dot{T} = \nabla\cdot(\lambda\nabla T) + \dot{S} \tag{1}$$

On boundary:

$$q = h(T - T_a) + \sigma\left(\epsilon_b T^4 - \epsilon_a T_a^4\right) - \epsilon_s J \tag{2}$$

where:

q	=	heat flux from surface;
$\dot{T}$	=	temperature;
T	=	temperature rate;
T_a	=	air temperature;
C	=	heat capacity;
λ	=	heat conductivity;
$\dot{S}$	=	rate of liberation of heat of hydration;
h	=	convection coefficient;
σ	=	radiation number of black body;
ϵ_s	=	short wave emissivity coefficient;
ϵ_a	=	long wave emissivity coefficient (from);
ϵ_b	=	long wave emissivity coefficient (to);
J	=	sun radiation intensity.

In addition for artificial cooling (convective flow):

$$C_c\left(\dot{T}_w + v\nabla T_w\right) + h_c\left(T_w - T_s\right) = 0 \tag{3}$$

where:

$\dot{T}_w$	=	temperature of cooling water;
T_w	=	rate of temperature of cooling water;
T_s	=	surface temperature;
C_c	=	lumped heat capacity of cooling;
h_c	=	lumped heat convection of cooling;
v	=	flow speed of cooling liquid.

With this system of equations the evolution of the temperature distribution can be computed by time integration, starting from an initial state. For arbitrary geometries the integration is performed numerically by means of e.g. the FE-method. The computer facilities of to-day (even PC's) allow to use a sufficient number of finite elements to guarantee consistent solutions. Whether the results are meaningful depends therefore mainly on the materials data and the boundary conditions. It can be said that experimentally determined values should be taken preferentially. The most essential data concerns the adiabatic heat evolution. This evolution allows e.g. to fit parameters in mathematical expressions, describing the liberation of heat of hydration. An expression, which has been integrated in code HEAT, and which has given excellent results in many practical applications, is given by:

$$\dot{S} = S\frac{M}{a+M}$$ (4)

where:

$\dot{S}$	=	rate of liberation of heat of hydration;
S	=	total heat of hydration;
a	=	time constant;
M	=	maturity.

Constant a is the maturity at which half of the total heat of hydration is liberated. With maturity M the influence of the temperature on the rate of hydration is taken into account. A common expression (Arhenius) for maturity M is:

$$M = \int_{t_d}^{t} e^{\frac{Q}{R}\left(\frac{1}{T_r} - \frac{1}{T}\right)} dt$$ (5)

where:

Q	=	activation energy;
R	=	universal gas constant;
T_r	=	reference temperature in K (generally 293 K);
t_d	=	dormant period.

This means that in total 4 parameters must be fitted with a measured adiabatic heat evolution: S, a, t_d and ratio Q/R. The convective and emissivity coefficients need, however, not to be measured. Reliable values can be found in literature /11/. It is important to note that analyses with different possible climatological scenario's should be performed in order to find the worst case situation.

After the computation of the temperature distributions, stress analysis can be carried out. The mathematical formulation of the boundary value problem is given by:
In the domain:

$$\dot{\sigma}_{ij} = S_{ijkl}^{ve}\left(\dot{\epsilon}_{kl}^{total} - \dot{\epsilon}_{kl}^{initial}\right)$$ (6)

At boundaries:

$$u_i - \overline{u}_i = 0$$ (7)

$$\sigma_{ij} - \overline{\sigma}_{ij} = 0$$ (8)

where:

$\dot{\sigma}_{ij}$	=	stress rate tensor;
S_{ijkl}^{ve}	=	visco-elasticity tensor;
$\dot{\epsilon}_{kl}^{total}$	=	total strain rate tensor;
$\dot{\epsilon}_{kl}^{initial}$	=	initial strain rate tensor;
$\overline{u}_i$	=	prescribed displacements;
$\overline{\sigma}_{ij}$	=	prescribed stresses.

This system of equations can be solved again by performing a numerical time integration. The link with the previous analysis is made by the initial strain rate (thermal dilatancy and plastic shrinkage), the influence of the temperature on the visco-elastic behaviour, and the influence of the maturity on the mechanical properties. In program HEAT the visco-elastic behaviour has been modelled by means of an ageing (maturity dependent) Maxwell chain model. Also here it should be remarked that the model parameters (maturity dependent distributions of the elasticity over the Maxwell units) must be fitted from experimental results. Not before long, parameter estimations could only be made from test results of the evolutions of the strength and elasticity. Presently, more useful experimental data can de achieved by means of the testing rig of Springenschmid /12/.

The kinematical boundary conditions determine the structural restraining. Assumptions on these conditions are at least even important as assumptions on materials data. In full 3D analysis kinematical boundary conditions can be imposed realistically. In 2D analysis, however, special care must be taken for the kinematical boundary conditions in the direction perpendicular to section. In the 2D version of program HEAT the following plane strain condition has been implemented:

$$\epsilon_{zz} = a + bx + cy \tag{9a}$$

$$\gamma_{xz} = \gamma_{yz} = 0 \tag{9b}$$

(Note: Z is out of plane direction)

Parameter a is the dilatancy of the structure in the Z-direction, while b and c are the "curvatures" around the Y and X axis respectively. The corresponding "forces" are the total force in the Z-direction, and the moments around the Y and X axis. In addition normal and moment spring elements can be attributed to these so-called degrees of freedom (dofs). From experience, it can be said that the reliability of stress analysis depends strongly on the choice of the kinematical conditions to be imposed on these dofs. In practical cases, where they are somewhat difficult to define, such as in the case of short constructions and or weak foundations, additional so-called side-view analysis is needed, from which reliable spring stiffnesses can be computed. The computed evaluation of the stresses must be compared with a tensile stress-limit in order to determine the risk of early-age cracking. A good experience has been achieved by the CEB /7/ like equation:

$$f_t = c + 0.3 * f_c [M]^{\frac{2}{3}} \tag{10}$$

where:

f_t	=	tensile stress limit;
f_c	=	characteristic compressive strength;
M	=	maturity;
c	=	reduction coefficient.

Reduction factor c varies between 0.5 and 0.65, depending on the tensile stress gradient. Obviously, the model is not sufficient if a certain amount of cracking is allowed, since the influence of the reduction of restraining, due to crack formation, on the stress distribution is not taken into account. In case that the tensile stress limit is exceeded preventive measurements must be designed. One of the possibilities is the application of an artificial cooling. This is, of course, an iterative process, as can be seen from the following flowchart.

After a convenient solution has been found, the corresponding computed temperature distributions can be exploited to establish stress-analysis-based temperature difference limits. These limits must next be integrated in work procedures concerning the control of the hardening process on construction sites.

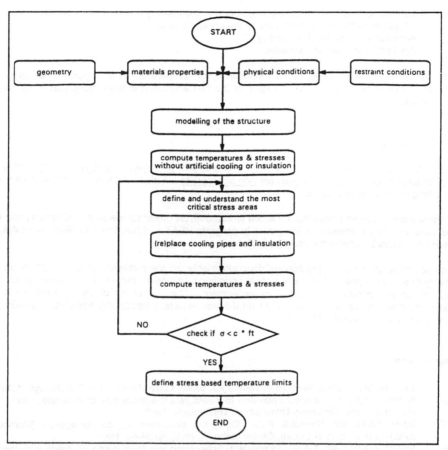

Figure 2   Flowchart for an artificial cooling design.

4.   Control of the hardening process during construction

In order to succeed in preventing early-age cracking, the stress-analysis-based temperature difference limits must be translated into a clear workprocedure. This workprocedure should consist of the following parts.

1.   Verification of the assumptions made for the computer simulations
For the simulations various assumptions have been made with regard to geometry, materials properties, physical- and kinematical boundary conditions. These assumptions must be described in the workprocedure and verified before casting. Amongst all parameters, in particular the initial mix temperature and the inlet temperature of a cooling system must be verified. Obviously, these assumptions will have been discussed ahead between the computer simulation expert and the construction engineer.

2.   Control during construction
The workprocedure should describe how the preventive measurements, which have been designed for the worst possible case, must be adapted for other weather and season conditions. These changes may consist of:

* Increasing or slowing down the speed flow of the cooling liquid;
* Changing of the inlet temperature of the cooling liquid;
* Accelerated or delayed shuttering;
* Application of thermal insulation.

The controlling must be performed by measuring the temperature in the hydrating concrete, as well as the temperature of the cooling liquid at the inlet and outlet of the cooling circuit (by means of thermocouples).

5.   Conclusions

It has been demonstrated, by means of an example, that temperature difference limits, which are not based on reliable stress analysis, are non-appropriate criteria to determine the risk of early-age cracking in concrete structures.

A framework has been presented, by which stress-analysis-based temperature difference limits can be obtained. This framework is integrated in program HEAT and has been applied succesfully to many important European construction projects.

The reliability of computer predictions depends mainly on the materials data as well as on the reality of assumed physical and kinematical conditions. Values for materials parameters of models should be determined preferentially by experiments. The testing rig of Springenschmid is an excellent device to obtain essential data on the visco-elastic-(plastic) and fracturing behaviour of young hardening concrete /12/.

References

/1/   Emborg, M., Thermal stresses at early ages, thesis Lulea University of Technology, 1989.

/2/   Roelfstra, P.E., A numerical approach to investigate the properties of concrete - numerical concrete, thesis Lausanne University of Technology, 1989.

/3/   Salet, T.A.M. and Roelfstra, P.E., Computer simulaties op de bouwplaats (Computer simulations on site; in Dutch), Civiele techniek, no.2, pp.36-40, 1993.

/4/   Visser, J. and Salet T.A.M., Temperatuurbeheersing van jong beton op basis van computermodellen (Temperature control based on computermodellen; in Dutch), Cement, no.2, pp.16-22, 1992.

/5/   Roelfstra, P.E. and Salet, T.A.M., Prevention of early-age cracking in construction practice, poster session RILEM symposium 119 TCE Avoidance of thermal cracking at early-ages, 1994; available on paper;

/6/   Thermische coëfficiënten en relaxatie van beton in het temperatuurgebied van -5°C tot 80°C; literatuur studie (Thermal parameters and relaxation of concrete in the temperature range of -5°C till 80°C); survey, CUR/VB IroMaTS, no.84-3, 1984.

/7/   CEB Model Code 1990

/8/   Metha, P.K., Durability of concrete exposed to marine environments, a fresh look, 1988

/9/   Manual of the pc-version of FEMMASSE, 1994 ·

/10/   Roelfstra, P.E. and Wittmann, F.H., Computer codes for material science and structural engineering. Proc.IABSE Coll., Delft, The Netherlands, pp. 293-305, 1987.

/11/   Bulletin d'information N°167, Contribution à la 24ᵉ Session Plénière du C.E.B., "Thermal effects in Concrete Structures", Rotterdam, The Netherlands, 1985.

/12/   Springenschmid, R., Breitenbücher, R. and Ballardini, P., "Vergleich zwischen Berechnungen und Messungen von Zwangspannungen in jungem Beton", Beton- und Stalbetonbau, 83, pp. 93-97, 1988.

34 NUMERICAL SIMULATION OF TEMPERATURES AND STRESSES IN CONCRETE AT EARLY AGES: THE FRENCH EXPERIENCE

J.M. TORRENTI
CEA, Saclay, France (formerly at LCPC)
F. de LARRARD, F. GUERRIER and P. ACKER
LCPC, Paris, France
G. GRENIER
Campenon Bernard, France

Abstract
In this paper we present the thermal and mechanical behaviour of concrete at early ages. Because of a weak coupling a two staged procedure is adopted: we calculate the temperature and the degree of hydration then the stresses. Examples of application for different part of French bridges are presented.
Keywords: Mechanical Behavior, Numerical Simulation, Thermal Simulation.

1 Introduction

The hydration of cement is a highly exothermal reaction and is thermo-activated. The result, in the hours that follow the preparation of the concrete, is high temperature rises in massive structures. The concrete then sets hot, resulting in thermal strains during cooling, accompanied by temperature gradients.

These strains, if restrained, and these gradients induce stresses that will be called thermal stresses in what follows. They are of an intensity such that they can lead to cracking. This cracking affects the durability of the concrete structures and often requires expensive treatment.

For several years, there has been a major experimental and modelling effort for the purpose of predicting stresses of thermal origin (SPRINGENSCHMID [1986], CHUI [1993], EMBORG [1984], ACKER [1985], BOGERT [1987], HARADA [1990], BREUGEL [1991], etc.). The CESAR-LCPC calculation code (DUBOUCHET [1992]) provides two tools for predicting the effects of the hydration of cement: the TEXO and MEXO modules. The first one allows numerical simulation of the evolution of the temperature fields and of the degree of hydration of the concrete, taking account both of the heat of hydration and of the various boundary conditions. The MEXO module simulates the mechanical effects resulting from the temperature fields and from the evolution of the mechanical characteristics of the concrete with its degree of hydration.

In a first part, we will describe these two tools; we will then give examples of application to actual structures: the ELORN and NORMANDIE bridges.

Thermal Cracking in Concrete at Early Ages. Edited by R. Springenschmid. Published 1994 by E & FN Spon, 2–6 Boundary Row, London SE1 8HN, UK. ISBN: 0 419 18710 3.

2 Simulation of the generation of heat and of the resulting stresses.

2.1 The TEXO module

This module predicts the evolution of the temperatures in concrete structures during the hydration of the cement. After presenting the equations governing our problem, we shall look at the specific way the generation of heat is taken into account.

The equation that governs our problem is the equation of conduction of heat :

$$\rho \, C \, dT/dt = \text{div} \, (\, k \, \text{grad} \, T \,) + \dot{q} \tag{1}$$

where $\bar{q}$ is the vector flow of heat, T the temperature, t the time, C the specific heat, ρ the voluminal mass, and k the tensor of heat conductivity. In our model, ρ and k are constants. The term $\dot{q}$ can cover very different origins such as the dissipation of energy due to irreversible strains in solids. In our application, $\dot{q}$ represents the rate of heat generation.

The boundary conditions are generally described by:

$$\bar{q} \, . \, \bar{n} = \lambda \, (T - \text{Text}) \tag{2}$$

where n is the normal outgoing vector, λ the coefficient of exchange, and Text the external temperature.

Equation 1 undergoes a classic treatment (discretization in space by the finite-element method and discretization in time by a CRANCK-NICHOLSON diagram). The specific part of our program is the introduction of the heat of hydration in our calculation. Several works have shown that the rate of heat generation $\dot{q}$ follows approximately ARRHENIUS's law (REGOURD [1980], BYFORS [1980], UCHIDA [1987]). This law expresses the fact that $\dot{q}$ depends only on the quantity of heat already released and the temperature T :

$$\dot{q} = f(q).\exp(-E / RT) \tag{3}$$

where R is the ideal gas constant and E the energy of activation.

This law has two consequences. The first is that parameter q can not be eliminated between equations 1 and 3; in other words, the thermal state of the concrete can not be described using the temperature only. It is necessary to know the heat production q or the degree of hydration $r = q / q(t=\infty)$. ARRHENIUS's law is a veritable law of evolution of this parameter.

The second consequence is that it is sufficient to perform a calorimetric test to predict the rate of heat generation under different conditions (assuming that the specific heat of the concrete remains constant and that the temperature in the test specimen is homogeneous).

A semi-adiabatic test has been developed for this purpose; in it, the temperature of a concrete specimen placed in a calorimeter is monitored from the time of preparation [ACKER, 1988]. Function f(q) and the rate of heat generation can be calculated from the adiabatic curve whatever the conditions. From a numerical viewpoint ARRHENIUS's law is discretized in the same way as the equation of heat.

2.2 The MEXO module

From the temperature field and the degree of hydration, known at each step of the thermal calculation, we can perform the mechanical calculation. The constitutive law used is a thermo-elastic ageing law :

$$\dot{\varepsilon}_{ij} = (\alpha \dot{T} - \beta \dot{r}) \delta_{ij} + ((1+v)/E(r)) \dot{\sigma}_{ij} - (v/E(r)) \dot{\sigma}_{kk} . \delta_{ij} \tag{4}$$

where ε_{ij} is the strain tensor, σ_{ij} the stress tensor, $E(r)$ the YOUNG's modulus, v the POISSON's ratio, α the coefficient of thermal expansion, and β the final endogenous shrinkage determined by a shrinkage test.

The YOUNG's modulus is a function of the degree of hydration. The function used is that proposed by BYFORS [1980] and depends on the strength of the concrete at 28 days. In our case the coefficient of thermal expansion is constant but could be a function of r [BOULAY, LAPLANTE, 1993].

Since our module does not take into account the non linear behavior due to cracking and creep strains the stresses obtained are only qualitative (cf GUENOT [1994]). It is not simple to estimate the error due to the fact that we use an elastic model because this error depends on the boundary conditions: if we have restrained strains, compressive then tensile stresses are generated. And during the change from compression to tension taking into account creep will conduct to higher tensile stresses than the elastic ones.

3 Example of simulations

Thanks to the TEXO and MEXO modules, many applications can be planned, such as the minimization of a thermal cycle in precasting [ACKER, 1985], the optimization of formwork and its use, the location of zones at risk of cracking, the choice of concrete placement cycle, the choice of a suitable concrete mix design, the impact of a change of cement etc.

Economic necessity (fast rotation of forms and therefore choice of fast cements) and environmental constraints (exceptional structures, massive parts) have led to very sustained demand for calculations of the TEXO and MEXO type in recent years. We present examples here.

3.1 The towers of the ELORN bridge

The ELORN bridge is a bridge having a central span 400 m long suspended by two axial sets of cables. The towers supporting these cables, 120 m high, are made of high-performance concrete (strength greater than 80 MPa). The high cement content (450 kg/m^3) and the thickness of the tower (cf. figure 1) obviously lead to high temperature rises (more than 50°C), accompanied by large gradients and therefore risks of cracking [LE BRIS et al, 1993]. Since the structure is on the edge of the sea, it is particularly necessary to limit this cracking.

The mechanical calculations based on the thermal simulations enabled us to inform the builder concerning the risks of cracking according to the plan adopted for form release. Figure 2 gives two examples of the evolution of the tensile stresses in the skin, compared to that of the tensile strength, for two form release modes. It is then easy to deduce which mode does less harm to the structure.

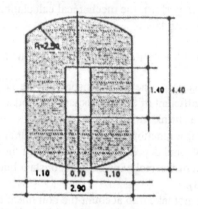

Fig. 1. Cross section of the tower

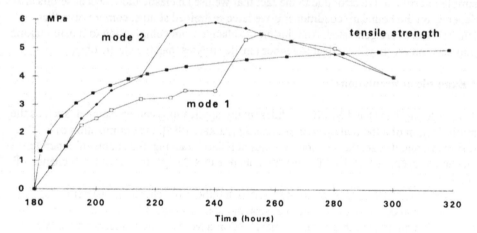

mode 1: form removal is scheduled 72 hours after the pouring of concrete
mode 2: the forms are unstuck at 24 hours and removed at 48 hours.

Fig. 2. Comparison of tensile stresses according to form release mode

3.2 The Normandy's bridge

The Normandy's bridge, on the Seine estuary, constitutes a major technological innovation. With a central independant girder of 800-m length, it will constitute a world record for the cable-stayed bridges. We will study two parts of it: the foundation slabs and the segments.

The foundation slab is a massive structure (3.50-m height, 15-m width and 20-m length), located at the feet of the pylones (figure 3) and for wich thermal stresses were expected. We have recorded the temperature inside one slab since the concrete casting. Figure 4 shows the results obtained along a vertical line in the center of the structure. The important role played by the ground has to be noticed. That is why, in our mesh, we took also the ground into account.

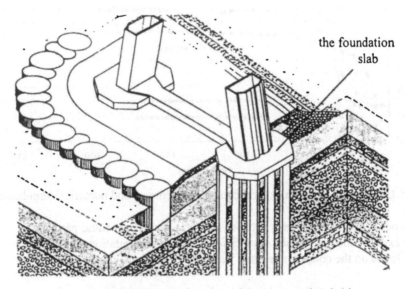

the foundation slab

Fig. 3. The foundation slab of the Normandy's bridge

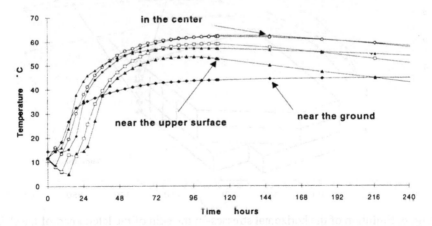

Fig. 4. Experimental measurements of the temperature along a vertical line in the center of the foundation slab.

The comparison between experimental results and numerical simulations shows very limited differences (figure 5).

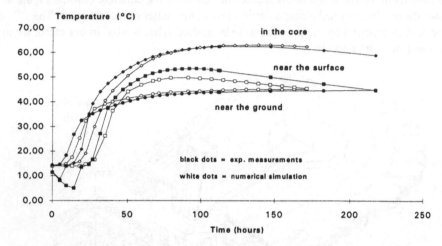

Fig. 5. Comparison between experimental measurements and numerical simulations.

The presence of thermal gradients between the surface and the central parts of the structure explains the high tensile stresses found by the computation (figure 6). The cracking found on the constructive work was quite conform to such a calculation.

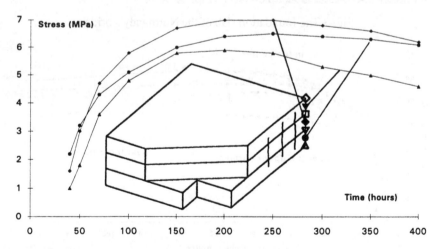

Fig. 6. Evolution of the horizontal stresses on the skin of the lateral part of the slab

The access viaducts and part of the central span are made of segments in high performance concrete. In the more massive parts of the segment (the edge girder in particular) the observed temperature is much higher than that of the other parts of the segment (figure 7).

Such difference induces consequences on a mechanical point of view: the edge girder is submitted to a bending in the horizontal plane, as its expansion is restrained by the other parts of the segment, which are not so hot. The resulting tensile stress can exceed, very locally, the tensile strength of the concrete (figure 8). Some cracks, very slightly open, have been observed on the edge girders of the constructive work.

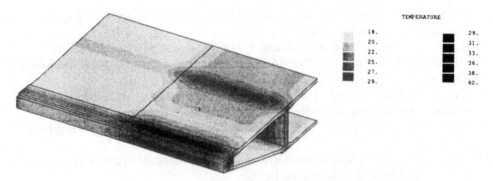

Fig. 7. Numerical simulation of the temperature in the segments

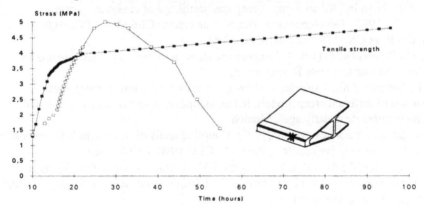

Fig. 8. Evolution of the stress (point * of the edge girder) compared to the strength.

4 Conclusion

It is now possible to predict the temperature evolution of concrete structures at an early age. The heat given off by the hydration of the cement is taken into account in this prediction using ARRHENIUS's law. The rate at which the heat is given off is determined from an adiabatic or semi-adiabatic calorimetric test.

It is also possible to predict the strains and stresses related to these releases of heat. These predictions are useful in locating cracking risks or estimating the impact of a concrete mix design.

References

Acker, P., Foucrier, C., Malier, Y. (1985), Temperature-related mechanical effects in concrete elements and optimisation of the manufacture process, ACI symposium on Properties of Concrete at Early Ages, Chicago, PP. 33-47.

Acker, P. (1988), Comportement mécanique du béton: apports de l'approche physico-chimique, rapport de recherche LPC n°152.

Bogert, P.A.J. van den, Borst, R. de, Nauta, P. (1987) Simulation of the mechanical behaviour of young concrete, Proc. IABSE Colloquium on Computational Mechanics of Reinforced Concrete Structures: Advances and Applications, Delft.

Boulay, C, Laplante, P. (1993), Evolution du coefficient de dilatation thermique du béton en fonction de sa maturité, **Matériaux et Constructions**.

Byfors, J. (1980), Plain concrete at early ages, Swedish Cement and Concrete Institute, Stockholm.

Breugel, K. van (1991), Simulation of hydration and formation of structure in hardening cement-based materials, PhD thesis, Delft.

Chui, J. J., Dilger, W. H. (1993), Temperature stress and cracking due to hydration heat, Proc. of the Fifth Int. Rilem Symp. Creep and shrinkage of concrete, Barcelona.

Dubouchet, A. (1992), Développement d'un pôle de calcul: CESAR-LCPC, **Bull. liaison Labo. des P. et Ch.**, 178, mars-avril, pp.77-84.

Emborg, M., Bernander, S. (1984), Temperature stresses in early age concrete due to hydration, Nordic Concrete Research n°3.

Guenot, I., Torrenti, J.M., Laplante, P. (1994), Stresses in concrete at early ages: comparison of different creep models, Rilem symposium on avoidance of thermal cracking in concrete at early ages, Munich. .

Harada, S., Suzuki, Y., Maekawa, K. (1990), Coupling analysis of cement heat generation and diffusion for massive concrete, Proc. of SCI-C 1990, Zell am See.

Lebris, J., Redoulez, P., Augustin, V., Torrenti, J.M., Larrard, F. de (1993), Verys high performance concretes in the Elorn bridge, ACI fall meeting, session "High performance concrete in severe environnement".

Regourd, M., Gauthier, E. (1980), Comportement des ciments soumis au durcissement accéléré, **Annales de l'ITBTP**, n°387, pp. 65-96.

Springenschmid R, Breitenbücher R (1986), Are low heat cements the most favourable cements for the prevention of cracks due to heat of hydration, **Betonwerk+Fertigteil-Technik**, Heft 11.

Uchida, K., Sakakibara, H. (1987), Formulation of the liberation rate of cement and prediction method of temperature rise based on cumulative heat liberation, Concrete library of Japan Society of Civil Engineers, n°9, pp. 85-95.

35 A PRACTICAL PLANNING TOOL FOR THE SIMULATION OF THERMAL STRESSES AND FOR THE PREDICTION OF EARLY THERMAL CRACKS IN MASSIVE CONCRETE STRUCTURES

P. ONKEN and F.S. ROSTÁSY
Technical University of Braunschweig, Germany

Abstract
The usual methods for the control of early-age thermal cracking of massive concrete structures due to the heat of hydration are mainly based on practical experience. In many cases the only applied measure is the limitation of the temperature rise within the structure or the temperature differences within a pour. To verify the efficiency of specific curing methods already during the design phase, the constructional engineer needs a reliable analytical planning tool which is applied to practice. In the following a method for the calculations of temperature fields, stresses, strength and the probability of early-age thermal cracking in hardening concrete structures is presented. It allows to consider the aspects of concrete technology, construction and execution of the construction in order to enhance the quality and performance of the structure.
Keywords: Crack Prediction, Heat of Hydration, Mass Concrete, Material Laws, Modelling of Thermal Stress, Numerical Simulation, Restraint, Temperature Fields, Thermal Stress Simulation.

1 Introduction

The estimation of thermal stresses requires the knowledge of the thermal and mechanical properties of the early-age concrete such as heat development, elasticity and viscoelastic behaviour. The material laws have to take into account the time and maturity dependency of these properties. Furthermore a realistic mathematical model for the continously changing strength, based on probabilistic evaluations, is needed for the estimation of the cracking risk.

Dealing with complex nonlinear material models, in 2 or 3-dimensional FEM-programs is rather complicated and time-consuming, furthermore an increasing amount of output data has to be analyzed. Although all structures are three-dimensional, a computation model assuming uniaxial stress and restraint conditions can applied for the most practical purposes without any noticeable loss of accuracy. The ther-

Thermal Cracking in Concrete at Early Ages. Edited by R. Springenschmid. Published 1994 by E & FN Spon, 2–6 Boundary Row, London SE1 8HN, UK. ISBN: 0 419 18710 3.

mal stress analysis performed by the laminar element method shows much more efficiency than 2D or 3D FEM analysis. Provided realistic assumptions of the external restraint, the risk of cracking may be computed conveniently for massive concrete structures with a predominant main stress direction.

2 Determination of the temperature field and degree of hydration

2.1 General

Although the hardening process of the concrete is a coupled thermo-mechanical problem, the slowly hydration process of the cement allows to distinguish between the problem of the heat conduction on the one hand and the strength and stress problem on the other hand. That means, if the temperature development of a structure is known, the calculation of the thermal stresses due to external or internal restraint can be solved as a seperate problem.

The calculation of temperatures in hardening concrete can be subdivided into two steps: The determination of the heat of hydration to postulate the internal heat source and the computation of the temperature field.

2.2 Heat of hydration

The essential thermal parameter for the determination of the temperature distribution in hardening concrete is the time-dependent heat of hydration $Q(t)$, which is liberated by the reaction of the cement constituents with water. The potential heat of hydration of Portland cement liberated at complete hydration can be determined by summing up the amounts of heat liberated by the individual constituents, v. Breugel (1991):

$$Q_{ce}^{max} = \sum_{i}^{n} q_i \left(\% C_3S, \% C_2S, \% C_3A, \% C_4AF, \% CaO, \% MgO \right) \qquad (1)$$

The accuracy of the calculated maximum heat depends on the accuracy with which the clinker composition and the potential heat of the constituents have been determined.

The development of the heat of hydration can either be predicted by simulation programs based on microstructure models as introduced e.g. by van Breugel (1991) or even more accurately by measuring the heat release with experimental methods. The experimental approach has the advantage, that all chemical, physical and stereological aspects of the cement, admixtures and additives can be considered. A commonly used method for measuring the heat release of a concrete mix is the adiabatic calorimetry.

The rate of heat evolution can be expressed mathematically by following equation:

$$q(t) = \frac{dQ(t)}{dt} \quad \text{with} \quad Q(t) = Q_{ce}^{max} \cdot C \cdot \alpha(t) \qquad (2)$$

where C is the amount of cement per m^3 and $\alpha(t)$ the time-dependent degree of hydration.

2.3 Degree of hydration

For practical purposes it is convenient to define the degree of hydration as the quotient between the liberated heat Q(t), which can be derived by the adiabatic calorimetry and the maximum heat of hydration Q^{max}:

$$\alpha(t) = \frac{Q(t)}{Q^{max}} \quad \text{with} \quad Q^{max} = Q_{ce}^{max} \cdot C \qquad (3)$$

For numerical reasons $\alpha(t)$ should be described by an analytic expression. The slightly modified exponential function, originally presented by Jonasson (1984), fits the shape of the hydration curve best:

$$\alpha(t_e) = e^{-\left(\ln\left(1+\frac{t_e}{t_1}\right)\right)^c} \qquad (4)$$

with t_1 and c being constants for the specific concrete mix, determined by adiabtic tests. The degree of hydration eq. (4) depends on the equivalent age t_e, which refers to curing temperature T = 293 K. It is commonly known, that the curing temperature does merely affect the hydration process. The temperature sensitivity of the rate of reaction requires the formulation of a maturity function. As stated by Marx (1986), Laube (1990), Rostásy (1994) and others, the maturity function proposed by Arrhenius was found to be best suited.
    The shape of the hydration curve in eq. (4) is defined by the factors t_1 and c, which depend on the clinker composition of the cement, the particle size distribution, the water/cement ratio and the initial reaction temperature. Therefore for designing purposes a database for the heat of hydration of different concrete mixes can be obtained from adiabtic tests defining the material factors as a function of these parameters:

$$t_1, c = f\{C_3S, \text{Blaine}, \text{w/c}, T_0\} \qquad (5)$$

Nevertheless the quality of the prediction of the heat evolution can significantly improved by testing a chosen concrete.

2.4 Calculations of temperature field and degree of hydration

The fundamental basis for prediction of temperature fields in hardening concrete is the differential equation of the

heat conduction, extended by the term of the internal heat
source eq. (2).

The time-dependency of the material parameters requires
a step-wise solution. Commonly the FEM and the Difference
Method are used. For many practical cases the problem can
be simplified to a one- or two-dimensional heat flow.

Attention has to be payed to the formulation of the
boundary conditions, which requires assumptions concerning
the heat transfer properties between the surface of the
early-age concrete layer and the surrounding area as well
as the course of the ambient temperature.

Fig. 1 shows the fields of temperature and degree of
hydration of a wall on foundation due to the evolution of
heat of hydration at the age of 42 hours. The distribution
of the degree of hydration is needed for the generation of
the material properties; it can be calculated simultaneously.

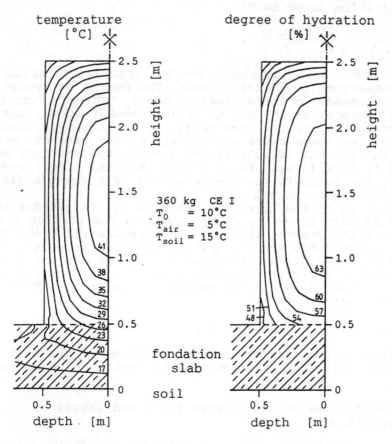

Fig. 1: Distribution of temperature and degree of hydration
in a wall on foundation (numerical example)

3 Material laws for the mechanical behavior at early ages

The mechanical properties of the young concrete distincti-
vely depend on age and curing conditions. Many investigations dealt with this problem. Usually the mechanical properties are described as a function of the equivalent time
t_e applying the maturity concept. In recent research reports the strength growth and other mechanical properties
have been expressed in terms of the degree of hydration.
    Laube (1990) and Rostásy (1994) developed relationships
between mechanical properties and the degree of hydration
based on a series of tests. Fig. 2 shows the normalized
relations between the axial tensile strength f_{ct}, the compressive strength f_{cc} and the tension E-modulus E_{ct} and the
degree of hydration. The validity of these relations is
proven for all kinds of concrete mixes. Differences only
arise with respect to the evaluation of the fictitious
values for complete hydration ($\alpha = 1$, $t_e \rightarrow \infty$) and the degree
of hydration at zero strength, which denotes the end of the
so-called *dormant phase*. These parameters depend on the
composition and constituents of each concrete and have to
be determined by regression.
    As proposed by Laube (1990) and Rostásy (1993) not only
the elastic behaviour of the hardening concrete, but also
the the viscoelastic and viscoplastic behaviour may be
related to the degree of hydration. Similar correlations
may even be applied to thermal properties, like the specific heat and the thermal conductivity.

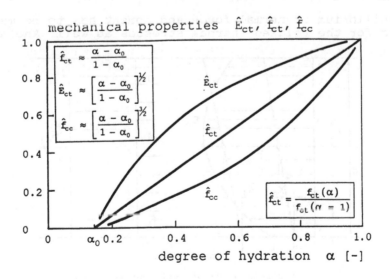

Fig. 2: Material laws for the mechanical properties

4 The computation of thermal stresses

4.1 Calculation Model

The calculation of thermal stress development in the structure is performed by a finite laminae model as shown by Emborg (1989). Each of the discrete laminar element strips with varying width can be attributed with age-depending mechanical properties, Fig. 3. The input temperatures required for the calculation of the free thermal strain can be taken over for each time step from the thermal evaluations as mentioned above. Plain strain conditions are assumed when considering the internal as well as external restraint. With the condition of equilibrium of normal force and moment the resultant deformations and actions can be determined incrementally.

4.2 The superposition principle of thermal stresses

Each strain step $\Delta\epsilon_{el,ki}$ causes a spontaneous stress response σ_{ki} according to the stiffness of each individual layer:

$$\Delta\epsilon_{el,ki} = \Delta\epsilon_{res,i} + \Delta\kappa_{res,i} \cdot z_k - \Delta\epsilon_{0,ki} \qquad (6)$$

The incremental stress response decreases due to relaxation. The age-dependent E-modulus E_{ki} and relaxation function ψ_{kni} are related to the degree of hydration according to section 3.2:

$$\sigma_{ki} = \Delta\epsilon_{el,ki} \cdot E_{ki} \cdot \psi_{kni} \qquad (7)$$

As equilibrium of normal force and moment has to be maintained for the relaxed thermal stresses as well. Another

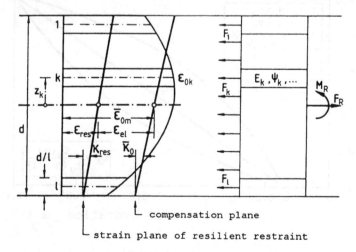

Fig. 3: Finite laminae method

essential condition of the superposition method is the consideration of the total history of stress for any incremental step.

4.3 Restraint actions and crack criterion

Since restraint actions may lead to severe cracks, the realistic assessment of the degree of restraint is important. Therefore the calculation model has to consider different kinds of restraint actions, i.e. the axial and bending restraint, the base restraint, modeled by friction between the structure and soil and the most common type of restraint of a wall on foundation.

All parameters for the determination of thermal stresses are described by deterministic models, nevertheless they are scattering quantities. The definition of a crack criterion has to consider the probalistic aspects of these parameters as well as the loss of the tensile strength within the structure in comparison to the tensile strength tested by specimen.

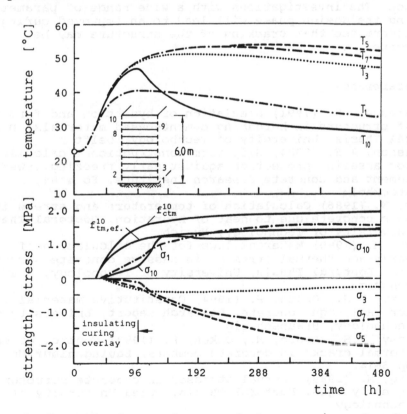

Fig. 4: Concrete temperatures and stresses in a foundation slab dependent on time (numerical example)

5 Applications

The application of the outlined planning tool is depicted in Fig. 4. A massive concrete foundation slab is modeled by a representative linear strip, which is subdivided into ten discrete layers. One dimensional heat flow and total bending restraint κ_{res} = 0 have been assumed, friction between slab and subbase can be neglected.

One-dimensional calculations have been performed with about 30 sets of different parameters, i.e. placing temperature, environmental conditions, thermal properties, curing treatment und curing period, which are all affecting the temperature distribution and thermal stresses decisively. The example in Fig. 4 shows the effect of the curing method, using an insulating curing blanket, which covers the surface of the slab during the first four days after the pour. After removing the curing overlay the restraint stress of the surface layer is growing rapidly and reaches the effective tensile strength within the structure, cracking of the surface layer will be the consequence. The investigations with a wide range of parameters during the design phase will lead to an improved curing treatment, so that cracking of the structure may be prevented.

6 References

v. Breugel, K. (1991) Simulation of hydration and formation of structure in hardening cement-based materials. **Doctoral Thesis**, University of Technology, Delft

Jonasson, J.E. (1984) Slip form construction - calculation for assesing protection against early freezing, Swedish Cement and Concrete Research Institute, Fo 4·84, Stockholm

Marx, W. (1986) Calculation of temperature and stress in mass concrete due to heat of hydration. **Doctoral Thesis**, University of Technology, Munich.

Laube, M. (1990) Material laws for the calculation of early-age thermal stresses in massive concrete structures. **Doctoral Thesis**, University of Technology, Braunschweig

Rostásy, F.S., Onken, P. (1994) Constitutive material model for early-age concrete. **Research Report**, University of Technology, Braunschweig

Rostásy, F.S., Laube, M., Onken, P. (1993) Control of early thermal cracks in concrete members. **Bauingenieur No. 68**, pp 5-14.

Emborg, M. (1989) Thermal stresses in concrete structures at early ages, **Doctoral Thesis**, Lulea University of Technology

36 PREDICTION OF TEMPERATURE AND STRESS DEVELOPMENT IN CONCRETE STRUCTURES

E.S. PEDERSEN
DTI Concrete Centre, Danish Technological Institute, Taastrup,
Denmark

Abstract
When a cast concrete is hardening, energy is released in
the form of heat. As a result of temperature variations,
the concrete will try to expand/contract. If the surface
the concrete is cold compared to the interior of the concrete, the contraction of the surface may result in tensi
stresses exceeding the tensile strength and cracks are
formed. This is a typical result when the temperature dif
ference between the interior of the concrete and the surface is maximum.
When a structure is cast on an existing slab the
stiffness of the latter will influence a possible crack
formation during the hardening period. Crack formation du
to restraints from adjoining structures will typically ta
place in the newly cast structure during the cooling
period.
This paper describes a method for calculating the thre
dimensional stress situation in a hardening concrete stru
ture. The method is implemented in an Finite Element Meth
program (CIMS-2D) and the results from a practical exampl
are presented.
Keywords: Numerical Simulation, Hardening Concrete, Stress
Prediction, Temperature Prediction.

1 Introduction

The hydration process in concrete results in the producti
of heat. Depending on ambient conditions an exchange of
heat between the concrete surface and the surroundings wi
take place leading to a time dependent and non-uniform te
perature distribution and thus to temperature deformation
in the concrete. The temperature deformations may be cons
dered as an eigenstrain creating stresses which will depe
on the boundary conditions for the hardening concrete par
and on the concrete properties changing with time.

A calculation of the stress distribution is possible i
the principle by means of several standard computer pro-

Thermal Cracking in Concrete at Early Ages. Edited by R. Springenschmid. Published 1994
by E & FN Spon, 2–6 Boundary Row, London SE1 8HN, UK. ISBN: 0 419 18710 3.

grams offering 3-dimensional finite-element solutions taking into account material properties, which are depending
on the degree of hydration of the concrete. In practice,
however, two types of difficulties arise.

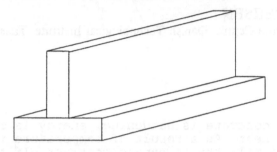

Fig.1. Wall cast on existing deck.

One is the pure volume of the calculations needed, making a
full 3-dimensional calculation a very time consuming job.
The other one is the lack of data on the material properties in the very early age of the concrete, which may make
the calculations efforts less productive. The present paper
deals primarily with the first difficulty suggesting that
in a lot of practical cases a 3-dimensional analysis can be
replaced by a 2½-dimensional analysis without loss of accu-
rancy and with considerably smaller time consumption. The
second problem about data on material properties will be
dealt with very briefly.

The example in fig. 1, showing a wall cast on top of an
existing slab, illustrates the type of cases that can be
treated. It represents a bar-shaped structure with constant
cross section and ambient conditions that may vary within
the section but are constant from section to section. A
full 3-dimensional analysis of temperature stresses in such
a structure will typically lead to results as shown in fig.
2. Except for a small region at the ends of the wall the
stresses will be almost constant from section to section.

2 Stress calculation

Away from the end zones the stress distribution in this
type of structure can be determined by superimposing two
sets of stresses:

a) disc stresses in the plane of the cross section
b) normal stresses perpendicular to the cross section
   assuming that these stresses are constant along the
   axis and that plane sections remain plane.

However, the two sets of stresses are dependent, as the distributions are tied together by the Poisson's ratio of the concrete. If this connection is ignored, the analysis mentioned under item b) corresponds to the Compensation Plane Method according to T.Tanabe.

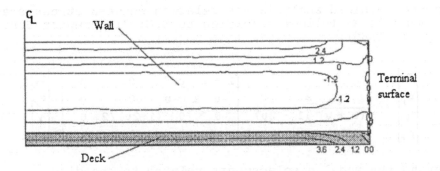

Fig. 2. Isocurves for the longitudinal stress component.

The strain distribution according to Tanabe's theory can be described as (see fig. 3):

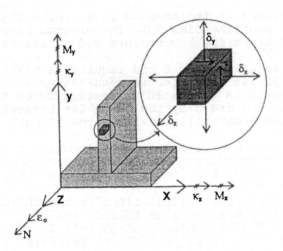

Fig. 3. Stress-components.

$$\epsilon_z = \epsilon_o + \kappa_x\, y - \kappa_y\, x \tag{1}$$

where ϵ_0 is the strain at the origin
        κ_x, κ_y are the bendings around the x- and y-direc tions, respectively.

In the combined analysis the relation between stresses and strains is as follows according to O.C. Zienkiewicz & R.L. Taylor (1989):

$$
\begin{bmatrix} \sigma_x \\ \sigma_y \\ \tau \\ \sigma_z \end{bmatrix}
=
\frac{E}{(1 + \upsilon)\,(1 - 2\upsilon)}
\begin{bmatrix}
1 - \upsilon & \upsilon & 0 & \upsilon \\
\upsilon & 1 - \upsilon & 0 & \upsilon \\
0 & 0 & (1-2\upsilon)/2 & 0 \\
\upsilon & \upsilon & 0 & 1 - \upsilon
\end{bmatrix}
\begin{bmatrix} \epsilon_x \\ \epsilon_y \\ \gamma \\ \epsilon_z \end{bmatrix}
\tag{2}
$$

Beyond the nodal displacements related to the disc mentioned in item a) three unknown quantities have been added : ϵ_0, κ_x and κ_y. For determining the 3-dimensional stress-situation the total amount of unknowns is now reduced to the nodal displacements in one cross section plus 3 values describing the axial movements an the curva ture of the structure.

3   FEM-computer program

The described method is implemented in a Finite-Element- Method computer program (CIMS-2D). The program is capable of carrying out transient temperature and stress calcula tions.
    The program is written in C and requires a WINDOWS 3.1 environment. A PC-386 or better is required.
    The element mesh is generated automatically on the basis of the cross section drawn by the user. The triangular elements describe a parabolic distribution of both tempera ture and strain.
    The temperature calculation is two dimensional and pro vides for thermal boundary conditions (including cooling pipes) as functions of time, casting time and casting tem perature.
    The load in the stress calculations is the thermal strains obtained from a previous thermal analysis.
    By organizing the hardening process into a number of small time intervals, it is possible to determine the stress increase within each time interval on a normal lin ear-elastic basis.
    The calculations require assumptions on several material properties and their development with time. Among these the the most important seems to be:

Adiabatic heat of hydration
Thermal expansion coeffient
E-modulus
Creep

The risk of cracking due to hardening temperatures is often maximum within the first 3-4 days after casting. Thus the development of material properties in that period must be reflected in the input data. This statement is also valid, by the way, for other kinds of calculations including full 3-dimensional FEM calculations. The genarel lack of such data will at the moment often necessitate testing of the specific concrete to be used. The usual testing methods may have to be somewhat modified in order to facilitate measurements within the first few days. In CIMS2D the properties are assumed to be functions of the maturity function described in ASTM C-1074-87 (1987).

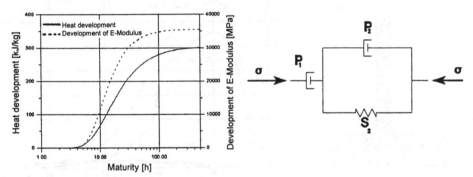

Fig. 4. Development of heat   Fig. 5. Creep-model.
         of hydration and E-
         modulus.

Fig. 4 is showing examples of development of adiabatic heat and E-modulus. Creep is taken into account by using a rheological model as shown in fig. 5. When a test concrete cylinder is loaded and unloaded at different ages, deformations as shown in fig. 6 are observed. On the basis of these observations the development of the component-properties in the rheological model can be determined. The calculated results are also shown in fig. 6.

4 Example

The wall on a slab from fig. 1 is treated. The wall is 3.9 m high and and its thickness is 0.5 m. The slab is 5 m wide and its thickness is 1.1 m. Fig. 7 shows the automatic meshing of the cross section. It is assumed that bottom of the slab is restrained so that vertical deflections are prevented ($\kappa_x = 0$).

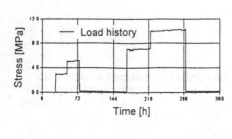

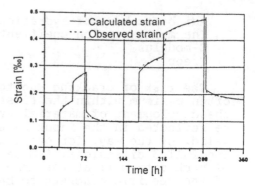

Fig. 6. Load history, observed and calculated deformations.

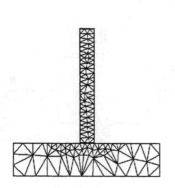

Fig. 7. Geometry –
Finite element mesh.

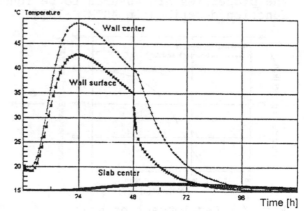

Fig. 8. Temperatures during
the hardening period.

The initial temperature of the slab is 15 °C and the casting temperature of the wall is 20 °C. A 19 mm plywood formwork is removed from the wall 48 hours after casting. The ambient temperature is 15 °C and the wind velocity i 5 m/sec.

The thermal expansion coefficient is assumed to be $1.1 \cdot 10^{-5}$ (independent of time), the E-modulus and the adiabatic heat development correspons to fig. 4. Creep is not taken into account in this example.

Some results are shown in figure 8-12. The temperature in the wall (fig. 8) rises rather rapidly and reaches its maximum 24 hours after casting. Temperatures in the slab remains almost unchanged. Removal of the form increases the cooling of the wall so that a larger temperature difference occurs between the center and the surface of the wall.

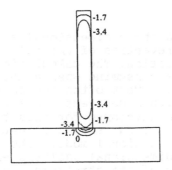

Fig. 9. Normal stress distribution acting perpendicular to the cross section 24 h after casting.

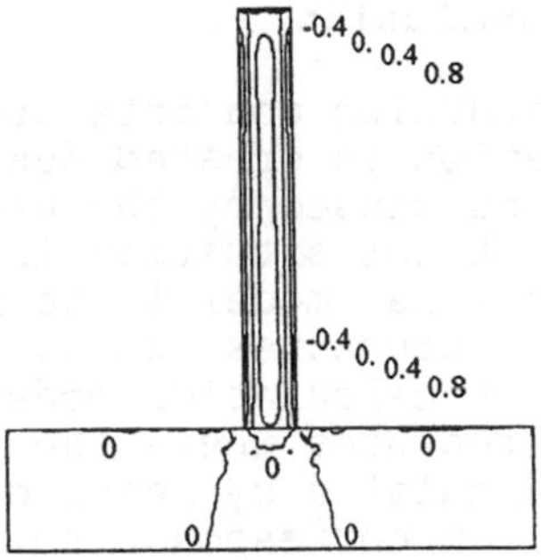

Fig. 10. Distribution of vertical normal stresses 24 h after casting.

The normal stresses acting perpendicular to the cross section (fig. 9) at first are compression stresses because elongation of the wall due to the temperature rise in the wall is prevented by the slab. As the wall temperature decreases the compression stresses also diminish. At form removal the temperature difference between center and surface of the wall creates tensile stresses at the wall surface, while stresses in the center more gradually change from compression to tension stresses. As the temperatures approach the ambient temperature, the wall is left with increasing tensile stresses because the E-modulus is much higher during the cooling process than during the initial heating. Such stresses, however, will gradually decrease because of creep.

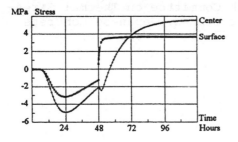

Fig. 11. Normal stresses acting perpendicular to the cross section during the hardening period.

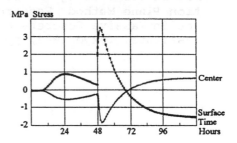

Fig. 12. Vertical normal stresses during the hardening period.

5   Conclusion

In hardening concrete structures a three-dimensional stress situation is created due to the prevention of thermal strains caused by the heat of hydration. The calculation of this stress situation is very time consuming when a three-dimensional model is to be analyzed. Very often the concrete structures are bar-shaped with a constant cross section. A calculation model has been developed for this type of structures where the three-dimensional stress situation is calculated by means of a two-dimensional model. The calculation corresponds to a full 3-dimensional analysis except in a small region at the ends of the structure. In many cases it is not the stress situation in these areas that are decisive for crack-formation. In these areas the stress level is reduced as a consequence of the fact that the terminal surfaces may deform.

The size of the two-dimensional element-method-problem is only a fraction of the corresponding three-dimensional problem. This means that the calculation times are considerably reduced and that it will be possible to carry out analyses along with practical tasks and not only scientific investigations.

6   References

ASTM-Standard C1074-87 (1987) **"Standard Practice for Esti-mating Concrete Strength by the Maturity Method"**.

Zienkiewcz, O.C. and Taylor, R.L. (1989) **The Finite Element Method**, McGraw-Hill, London.

Tanabe, T. (1985) **Report on Review of Studies on Thermal Stresses and Proposal for a Calculation Method of Thermal Stresses using the Compensation Line and Compensation Plane Method**, Research Committee on Thermal Stresses in Massive Concrete Structures, Japanese Concrete Institute, Tokyo 1985, pp. 47-61 (in Japanese)

37 SENSITIVITY ANALYSIS AND RELIABILITY EVALUATION OF THERMAL CRACKING IN MASS CONCRETE

K. MATSUI
Tokyo Denki University, Saitama, Japan
N. NISHIDA, Y. DOBASHI and K. USHIODA
Nishimatsu Construction Co. Ltd, Kanagawa, Japan

Abstract
Reliability thermal analysis of thick concrete sections subjected to heat of hydration and surface heat transfer is presented. The temperature rise due to heat of hydration is calculated based on transient heat-flow analysis and thermal stress is computed according to the method of compensation line. Thermal and mechanical properties used in the analysis contain uncertainty in quantities. Considering variability of the properties, the probability of crack occurrence is analytically computed and compared with the results of Monte Carlo simulation.
Keywords: Heat of Hydration, Heat Transfer, Mass Concrete, Reliability of Cracking Analysis, Sensitivity Analysis of Cracking.

1 Introduction

The reaction of cement with water liberates considerable quantities of heat, which is called heat of hydration, during a curing period of approximately 30 days. Nonlinear temperature distributions due to the heat released by hydration leads to tensile stresses that may exceed the strength of early age concrete and cause cracking to occur. Factors affecting temperature rise resulting from the release of hydration heat have been well investigated. ACI committee 209 produced guideline for predicting average thermal response of mass concrete. JCI task committee on thermal stresses in mass concrete has proposed a practical procedure to compute thermal stresses.

However it is well known that thermal and mechanical properties depend on cement constituents, concrete mixes, water-cement ratio, curing temperature among others, and contain uncertainty in their quantities. Thermal stress evaluation based on the standard values selected from past data for reference or the average values of experiments conducted for the purpose does not constitute a rational approach to predict whether thermal cracking occurs or not.

In order to investigate the effect of variabilities of the properties on thermal cracking, their expected values are chosen from references. However, since their standard deviations have not been readily found in references, four different coefficients of variation are considered in this

Thermal Cracking in Concrete at Early Ages. Edited by R. Springenschmid. Published 1994 by E & FN Spon, 2–6 Boundary Row, London SE1 8HN, UK. ISBN: 0 419 18710 3.

paper. Transient thermal analysis is conducted with finite element discretization in space and modal analysis in time, and thermal stresses are computed by the method of compensation line. By employing the first order approximation, the corresponding standard deviations are evaluated. The results are compared with those from Monte Carlo simulation.

2 Thermal analysis

2.1 Transient heat flow analysis

Mass concrete extended to infinity in horizontal directions is lying on a half space rock base as illustrated in Fig.1. Heat of hydration released in a mass concrete body is transmitted to its ambient environment. The heat conduction equation and its initial and boundary conditions can be written as,

< Concrete >

Heat Conduction Equation:

$$K_c \frac{\partial^2 T_C}{\partial x^2} + \dot{Q}(t) = \rho_c C_c \frac{\partial T_C}{\partial t} \qquad (1)$$

Initial Condition:

$$T_C(x,0) = T_{c0} \qquad (2)$$

Boundary Condition:

$$K_c \frac{\partial T_C}{\partial x} + \alpha_A(T_C - T_A) = 0 \qquad (3)$$

Heat Generation Model:

$$\dot{Q}(t) = \rho_c C_c Q_\infty \gamma e^{-\gamma t} \qquad (4)$$

< Rock >

$$K_R \frac{\partial^2 T_R}{\partial x^2} = \rho_R C_R \frac{\partial T_R}{\partial t} \qquad (5)$$

$$T_R(x,0) = T_{R0} \qquad (6)$$

$$T_R(x_0, t) = T_B \qquad (7)$$

T_c: temperature distribution in mass concrete, K_c: heat conductivity of concrete, ρ_c: mass density of concrete, C_c: specific heat of concrete, α_A: convection heat transfer coefficient, T_A: air temperature, Q_∞, γ: hydration heat model parameters, T_R: temperature distribution in rock, K_R: heat conductivity of rock, C_R: specific heat of rock, ρ_R: mass density of rock, T_{R0}: initial rock temperature, T_B: rock temperature at a fixed temperature boundary.

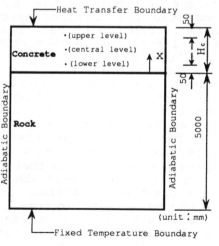

Fig.1. Model

2.2 Thermal stresses

Thermal stresses are computed based on the method of compensation line. Let the stresses at two successive stages of age t_i and t_{i+1} be denoted by σ_i and σ_{i+1}. Then,

$$\sigma_{i+1}(x) = \sigma_i(x) + \Delta\sigma(x) \tag{8}$$

where $\Delta\sigma(x)$ refers to the thermal stress increment between the ages t_i and t_{i+1}. The stress increment will be computed from,

$$\Delta\sigma(x) = E_e(t)\{\alpha\Delta T(x) - \Delta\bar\epsilon - \Delta\phi(x - H_c/2)\} \tag{9}$$

$$+ R_N E_e(t)\Delta\bar\epsilon + R_M E_e(t)\Delta\phi(x - H_c/2)$$

$$\Delta\bar\epsilon = \frac{1}{H_c}\int_0^{H_c}\alpha\Delta T(x)\,dx \tag{10} \quad \Delta\phi = \frac{12}{H_c^3}\int_0^{H_c}(\alpha\Delta T(x) - \Delta\bar\epsilon)(x - \frac{H_c}{2})\,dx \tag{11}$$

in which α is the thermal coefficient of expansion for concrete, R_N is the coefficient of axial restraint and R_M is the coefficient of bending restraint. $E_e(t)$ refers to an effective Young's modulus which is defined as,

$$E_e(t) = \psi(t) \times 1.5 \times 10^4 \sqrt{f_c'(t)} \tag{12}$$

$$\begin{cases} t < 3 \text{ days} & : \psi(t) = 0.73 \\ 3 \text{ days} < t < 5 \text{ days} & : \psi(t) = 0.135t + 0.325 \\ t > 5 \text{ days} & : \psi(t) = 1.0 \end{cases}$$

$f_c'(t)$ is the compressive strength of concrete at tth day and can be written as,

$$f_c'(t) = \frac{t}{4.5 + 0.95t} f_c'(91) \tag{13}$$

by using its compressive strength at the **91**st day.

From Eqs. (8) and (9) thermal stress at time t can be given as,

$$\sigma(t) = \sigma_I(t) + R_N\sigma_N(t) + R_M\sigma_M(t) \tag{14}$$

where $\sigma_I(t)$ is stress due to internal constraint, $\sigma_N(t)$ is stress due to the axial restraint and $\sigma_M(t)$ is stress due to the bending restraint.

According to JSCE design specification for concrete, tensile strength of concrete is related to its compressive strength as follows.

$$f_t(t) = 1.4 \times \sqrt{f_c'(t)} \tag{15}$$

3 First order approximation

Temperature and thermal stress in concrete body are considered as functions of random variables affecting them as well as time and space. Let describe random variables by $b = \{b_1, b_2, \ldots, b_n\}^T$, space by x and time by t. Then temperature and stress are expressed in the following form.

$$Y = g(b,x,t) \tag{16}$$

Expanding Eq. (16) into Taylor series around the mean $\bar{b}$ and taking the first order, one will obtain,

$$Y = g(\bar{b},x,t) + \sum_{i=1}^{n}\left(\frac{\partial g}{\partial b_i}\right)_{\bar{b}}\left(b - \bar{b}_i\right) + \cdots \tag{17}$$

where $\left(\partial g/\partial b_i\right)_{\bar{b}}$ implies the partial differentiation evaluated at $\bar{b}$. Expectation of Y can be written as,

$$E(Y) \cong g(\bar{b},x,t) \tag{18}$$

If there is no correlation among $b_i s$ ($i=1,\ldots,N$), the variation of Y becomes,

$$\text{Var}(Y) = \sum_{i=1}^{n}\left(\frac{\partial g}{\partial b_i}\right)_{\bar{b}}^{2}\cdot\text{Var}(b_i) \tag{19}$$

Hence the expectation and variation of temperature and thermal stress can be computed from the mean values of variables involved and their corresponding variations.

Concrete body will be considered to have cracked if its tensile strength is less than the tensile stress induced in it. The probability p_{cr} of cracking can be stated in the following way.

$$p_{cr} = P(Z \leq 0) \tag{20}$$

where $Z = f_t - \sigma$ is called a safety margin. If Z is normal random variable, Eq. (20) then becomes,

$$p_{cr} = \Phi(-\beta) \tag{21}$$

where Φ is the standard normal distribution and β is defined as the safety index

$$\beta = \mu_z / \sigma_z \tag{22}$$

μ_z and σ_z are given as

$$\mu_z = E(f_t) - E(\sigma) \tag{23}$$

$$\sigma_z^2 = \text{Var}(f_t) + \text{Var}(\sigma) \tag{24}$$

4 Analysis conditions

Four different thickness of mass concrete are considered, which are 0.75m, 1.5m, 2.5m and 5.0m. Values needed for the analysis are summarized in Table 1, in which the parameters numbered in the right-hand column are considered as random variables. The values for the random variables refer to their mean values. Since the effective Young's modulus given by Eq. (12) and the tensile strength by

Table 1. List of Values Used for Thermal Stress Analysis

Parameters		AVE.	Variable Parameters
THERMAL ANALYSIS CONDITIONS			
K_c	Heat Conductivity of Concrete (W/m℃)	2.67	①
C_c	Specific Heat of Concrete (J/kg℃)	1172.1	②
Q_∞	Adiabatic Temperature Rise	46.0	③
γ	Empirical Constant	1.10	④
α_A	Convection Heat Transfer Coefficient (W/m²℃)	12.79	⑤
K_R	Heat Conductivity of Rock (W/m℃)	2.3	⑥
C_R	Specific Heat of Rock (J/kg℃)	795.3	⑦
T_A	Temperature of Air (℃)	15.0	⑧
T_B	Rock Temperature at Fixed Temperature Boundary (℃)	15.0	⑨
T_{CO}	Initial Concrete Temperature (℃)	20.0	⑩
T_{RO}	Initial Rock Temperature (℃)	15.0	⑪
ρ_c	Mass Density of Concrete (kg/m ³)	2300	×
ρ_R	Mass Density of Rock (kg/m³)	2600	×
THERMAL STRESS ANALYSIS CONDITIONS			
α	Thermal Coefficient of Expansion of Concrete (1/℃)	10.0×10^{-6}	⑫
$f'_c(91)$	91st Day Compressive Strength of Concrete (MPa)	35.3	⑬
ε_1	ε_1 of Equation(25)	0.0	⑭
ε_2	ε_2 of Equation(26)	0.0	⑮

Eq.(15) are obtained from experimental data and it contains some ambiguity, they will be expressed as,

$$E_e^* = E_e(t) \ (1.0 + \varepsilon_1) \quad (25)$$

$$f_t^* = f_t(t) \ (1.0 + \varepsilon_2) \quad (26)$$

in which ε_1 and ε_2 are random variables. Their means ε_1 and ε_2 are also given in Table 1. R_N and R_M are, however, considered as constant because they are by nature factitious and not suitable for being treated as random variables. Considering that although an infinite horizontal extension is assumed for the thermal analysis, actual mass concrete has a finite length and that R_N and R_M depend on the length to hight ratio of mass concrete, the length of L=10 m is assumed and the coresponding values of R_N and R_M are summarized in Table 2 when the hight H_c varies.

Table 2. Coefficient of Restraints

L=10.0m, Ec/Er =10 (Er = 2548MPa)				
Hc(m)	L/Hc	R_N	R_M (BP)*1)	R_M (AP)*2)
0.75	13.33	0.42	1.125	1.07
1.5	6.67	0.22	1.185	1.60
2.5	4.0	0.13	0.95	1.65
5.0	2.0	0.075	0.70	1.00

*1) R_M(BP) : Bending Restraint before peak temperature
*2) R_M(AP) : Bending Restraint after peak temperature

5   Computed results

5.1 Sensitivity analysis

Mean stresses at upper and middle levels in Fig.1 are
computed by using the values given in Table 1. It has been
known that crack at upper level will appear at early age
and that at middle level occurs after many days. Hence
sensitivities of stresses are evaluated by the COV of each
parameter in Table 1 at the day when f_t/σ becomes minimum
and at 2000 days. Their results are presented in Figs.2a
and 2b.

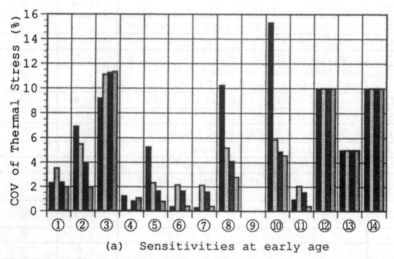

(a)   Sensitivities at early age

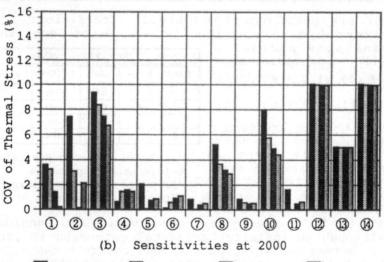

(b)   Sensitivities at 2000

■ H=0.75m   ▨ H=1.5m   ▨ H=2.5m   ▨ H=5.0m

Fig.2.   Sensitivities of Thermal Tensile Stress

5.2 Probability of cracking

The probability of cracking is computed by employing the first order approximation theory. Mean μ_z and standard deviation σ_z are computed from Eqs.(23) and (24), using the values in Table 1 and Table 2. For the random variables, they are presumed to have the same coefficient of variation (COV) but four different magnitudes 5%, 10%, 20% and 40% are considered.

In order to confirm the accuracy of the results, Monte Carlo simulation is also utilized. 200 sets of random variables that have the mean values in Table 1 with the

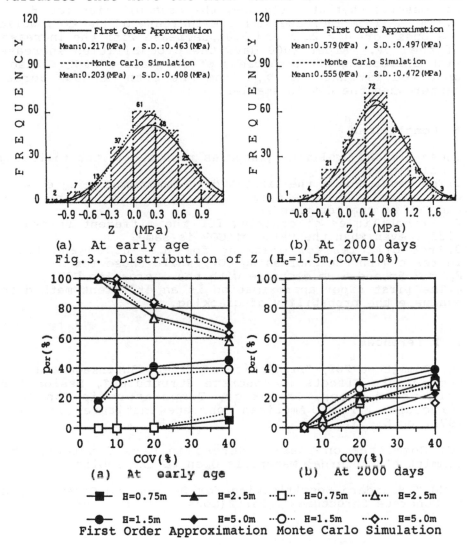

(a) At early age   (b) At 2000 days
Fig.3. Distribution of Z (H_c=1.5m, COV=10%)

(a) At early age   (b) At 2000 days

First Order Approximation   Monte Carlo Simulation

Fig.4. Probability of Cracking

four different COVs are generated. Then the thermal
stresses are computed, from which their means and standard
deviations are obtained.

Figs. 3a and 3b illustrates the normal distribution
curve of Z when the COV is 10% and their histograms from
the simulation.   It may be stated that the both
distributions show a good agreement. Figs. 4a and 4b show
the relationship between a probability of cracking and the
COV for H_c = 0.75m, 1.5m, 2.5m and 5.0m. The figures state
that the probability changes with a thickness of concrete
even when the COV takes the same value. From Fig.4a, one
can observe that at early age, the probabilities for H_c =
0.75m and 1.5m increase with the increase of the COV, while
the ones for H_c = 2.5m and 5.0m decrease with the increase
of the COV.   After 2000 days the probabilities increase
with increasing COV regardless of the values of H_c. Also
observed is that the p_{cr}s from the both methods agree
better when the COV is smaller.

6 Conclusions

These limited preliminary studies resulted in the following
conclusions:
1) The results from the first order approximation agree well
with those from Monte Carlo simulation, particularly when
the COV is smaller than 20%.
2) Probabilities of cracking for the different thickness
differ even when the value of COV is same.
3) Probabilities for cracking for H_c = 0.75m and 1.5m
increase with the increasing COV, while those for
H_c = 2.5m and 5.0m decrease with the increasing COV.
4) The first order approximation is an efficient method to
evaluate the probability of cracking.

7 References

ACI Committee 209, "Prediction of Creep, Shrinkage and
Temperature Effects in concrete Structures", Design for
Effects of Creep, Shrinkage, Temperature in Concrete
Structures, SP-27, American Concrete Institute, Detroit,
1971, pp.51-93.

JCI Committee on Thermal Stress in Mass Concrete, The
Committee's Internal Materials Open to the Public.

JSCE," Standard Specification for Design and Construction
of Concrete Structures", Part2(Construction), 1986.

38 DEFORMATIONS AND THERMAL STRESSES OF CONCRETE BEAMS CONSTRUCTED IN TWO STAGES

R. SATO
University of Utsunomiya, Japan
W.H. DILGER
The University of Calgary, Canada
I. UJIKE
University of Utsunomiya, Japan

Abstract
Massive structures like base slabs of reinforced concrete tanks for storage of liquefied natural gas are concreted in two or more stages. In such cases the deformation due to temperature change of the younger concrete is restrained by older concrete.

In the present study concrete beams were cast in two stages to investigate thermal stresses in the new concrete and the resulting flexural deformation. Temperature change simulating that observed in actual massive concrete structures was generated in the new concrete section.

Curvatures and strains obtained by the finite element method and the beam theory were compared with those obtained in the experiments. The CEB/MC90, the model by Iwaki et al. and "effective elastic modulus" by JCSE were used to consider creep and a step by step method was adopted in the computations. Cracking moments of the beams subjected to the effect of hydration heat were noticeably smaller than those of the beams not subjected to temperature history of two stage construction. The difference is explained by the stress due to heat of hydration.
Keywords: Effective Elastic Modulus, Heat of Hydration, Numerical Simulation, Structural Members (Two-stage Constructed Beams).

1 Introduction

Massive structures like base slabs of reinforced concrete tanks for storage of liquefied natural gas are concreted in two or more layers. In such cases the deformation due to temperature change is restrained by existing concrete and thermal stress is produced in the new concrete. Moreover, concrete structures continue to deform permanently due to creep of the new concrete even after the temperature in the new concrete reaches ambient temperature. Therefore, additional stresses are produced in the new concrete when the residual deformation of the structures are restrained by foundation as well as by dead load.

In order to accurately predict thermal stresses in massive concrete structures which are partially restrained, a proper estimate of the deformation due to temperature change at early ages is required. To date only a few investigations have been performed on the deformation of concrete members built in two or more stages(Tsukayama, R. 1974).

In the present study four concrete beams were cast in two stages to investigate the effect of cross sectional area of the stage 1 concrete on the stress and deformation. Strains of the new concrete and curvatures obtained by computation were compared with those obtained in the experiments. Two methods were used for computation:the finite element method and the beam theory. The CEB/MC90(1991), the model by Iwaki et al.(1982) and effective elastic modulus by JCSE(1991) were

Thermal Cracking in Concrete at Early Ages. Edited by R. Springenschmid. Published 1994 by E & FN Spon, 2–6 Boundary Row, London SE1 8HN, UK. ISBN: 0 419 18710 3.

Table 1 Outline of specimens and properties of concrete

Speci-mens	Reinforcing bars	A_s (cm^2)	ρ (%)	T_o (°C)	stage 2 concrete			stage 1 concrete		
					$f_{cc.28}$ (MPa)	$f_{ct.28}$ (MPa)	$E_{c.28}$ (GPa)	$f_{cc.28}$ (MPa)	$f_{ct.28}$ (MPa)	$E_{c.28}$ (GPa)
T-20	$2d_s16+1d_s13$	4.33	0.96	21.0	49.4	4.6	30.9	44.8	3.7	31.7
L-20	$2d_s16+1d_s13$	4.36	0.97	18.0	48.5	4.0	31.3	44.8	3.7	31.7
T-50	$2d_s16+1d_s13$	4.35	0.97	24.2	55.9	5.1	40.4	41.0	3.8	32.6
L-50	$2d_s16+1d_s13$	4.35	0.97	11.0	61.8	4.5	37.8	49.6	3.6	32.1

ρ:percentage of cross section of longitudinal bars to that of stage 2 concrete, T_o:Temperature at casting

Table 2 Casting, curing and testing schedule

	T-20	L-20	T-50	L-50
Age of stage 1 at casting of stage 2(days)	45	23	20	22
Curing of stage 1 concrete(days)	16[1]+29*	16[1]+7*	16[1]+4*	16[1]+6*
Curing of stage 2 concrete(days)	11[2]+5*	10[3]+29*	10[2]+9*	10[3]+8*
Age of stage 1 at flexural test(days)	61	62	39	40

1):curing with wet cloths, 2):days during insulation, 3):curing with vinyl sheet, *:curing in air

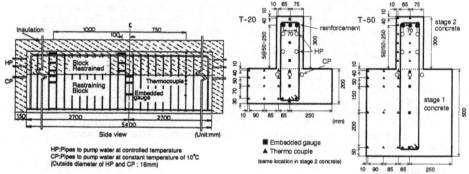

Fig.1 Details of longitudinal cross section   **Fig.2 Details of transverse cross sections**

used to predict creep and a step by step method was adopted in the computations. Theoretical and experimental cracking moments of the beams subjected to the effect of hydration heat and the resulting thermal stresses were compared with those of companion beams not subjected to temperature history of two stage construction.

2 Outline of experiments

To investigate the effect of thermal stresses on cracking of beams built in two stages two pairs of composite beams with the dimensions of Fig.1 and the details of Table 1 were produced.

The main parameters were the dimensions of the cross section of the stage 1 concrete (see Fig.2) and the temperature history. The two beams subjected to a temperature history are designated T, the others L. Beams T-50 and L-50 were 5.40m long and had a 500mm square cross section cast in stage 1 and a 150x300mm section cast in stage 2. Beams T-20 and L-20 had a 500x200mm cross section cast in stage 1 and the same stage 2 section as the others beams. The casting and curing sequence is given in Table 2. The stage 1 concrete was ready mix concrete while the section 2 concrete was mixed in the laboratory. The properties of both concrete at age 28days cured in water at 20±1°C are listed in Table 1. The two sections were interconnected by 10mm ties spaced at 100mm and reinforced longitudinally as shown in Fig.2.

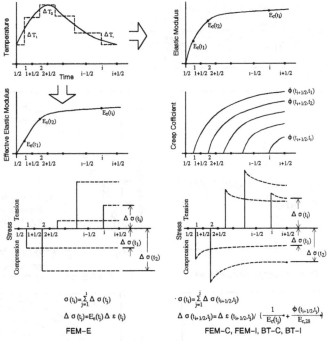

Fig.3 Concept of analysis

Control of the temperature in specimens T was achieved by pipes embedded in the concrete. Water at controlled temperature pumped through the pipes shown in Fig.1 produced the desired temperatures in these specimens which were insulated by styrofoam.

Temperatures and strains in fresh concrete members were measured, respectively, by copper-constantan thermocouples and embedded gauges with elastic modulus of 50MPa at the location indicated in Fig.2. Average curvatures due to hydration heat of cement were obtained from the deflections which were measured by electrical displacement meters located at the center section and at sections distant 10cm from both ends. The cracking moment was established in a flexural test with a constant bending moment zone of 1000mm length. The age of testing is indicated in Table 2.

3 Analysis

3.1 Analysis method

In order to predict deformation of the beams and the concrete stresses the finite element method (FEM) and beam theory (BT) assuming linear strain distribution through full height improving Japan Concrete Institute (JCI) method (1985) were used. Number of elements and nodal points used in FEM are 279 and 171 for T-20, and 414 and 239 for T-50. Three methods are applied to consider the effect of creep. The first is "effective elastic method (EM)" (1991) adopted in Japan Society of Civil Engineering(JSCE), the second CEB/MC90 and the third empirical equation proposed by Iwaki et al.. As is shown in Fig.3, step-by-step method is adopted when

Table 3 Analysis method and creep model

Analysis method	Creep model		
	EM	CEB/MC90	Iwaki
FEM	FEM-E	FEM-C	FEM-I
Beam Theory		BT-C	BT-I

EM:Effective modulus method

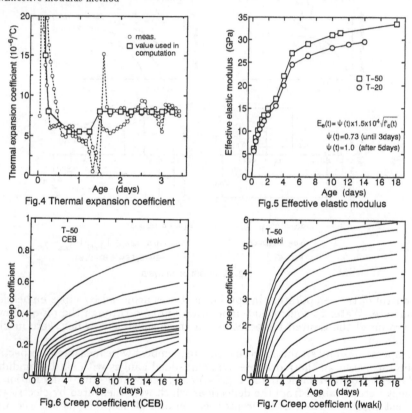

Fig.4 Thermal expansion coefficient

Fig.5 Effective elastic modulus

Fig.6 Creep coefficient (CEB)

Fig.7 Creep coefficient (Iwaki)

models by CEB/MC90 and Iwaki et al. are used. However, step-by-step method simplified in accordance with JSCE is applied when effective elastic modulus considering reduction of elastic modulus due to creep especially at early ages is used (see Fig.5). Five combinations between analysis method and creep model are tabulated in Table 3.

3.2 Materials Properties

1)Elastic Modulus
The variation of the elastic modulus as a function of age and temperature was based on CEB/MC90. The elastic modulus at age of 28days in water with temperature of 20°C was obtained experimentally.
2)Thermal expansion coefficient
Fig.4 shows the thermal expansion coefficient with time obtained from the relationship between free strain and temperature change measured in the specimen with the same temperature history as that of the stage 2 concrete of the beams. This

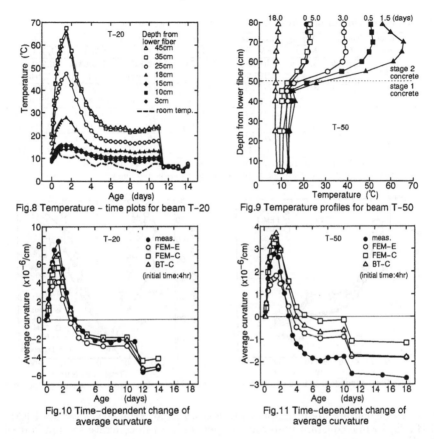

Fig.8 Temperature – time plots for beam T-20

Fig.9 Temperature profiles for beam T-50

Fig.10 Time-dependent change of average curvature

Fig.11 Time-dependent change of average curvature

coefficient is not constant with time and also different temperature rise and drop zones.

3)Creep

As was described previously, creep coefficients were obtained by CEB/MC90 and Iwaki's equation. In using CEB/MC90 the temperature effect is taken into account only for the maturity of concrete. Whitney's rate of creep law is applied in the computation used Iwaki's creep model. Effective elastic modulus and the two creep coefficients adopted in the present study are shown in Figs.5, 6 and 7. $\psi(t)$ in Fig.5 is correction facter for elastic modulus considering the effect of creep at age t.

4 Discussion

Fig.8 shows typical temperature changes with time measured at 7 different levels in T-20 after casting stage 2 concrete. Temperature rises more than 50°C and drops 45°C up to 11days and thereafter drops 20°C to ambient temperature. The sudden drop of temperature is due to removing insulation. Temperature profiles for different ages through the full height are indicated in Fig.9.

Figs.10 and 11 show comparisons between experimental and computed curvature changes with time for T-20 and T-50. All other computed results at peak and stabilized temperatures are listed in Table 4. The computation considers materials

Table 4 Comparison of measured and computed average curvatures

T_i (hours)	Speci- mens	Temp. (°C)	Average curvature ($10^{-6}/cm$)					
			Meas.	FEM-E	FEM-C	BT-C	FEM-I	BT-I
4	T-20	T_p:68.9	8.4	4.0(0.48)	5.6(0.67)	7.0(0.83)	5.3(0.63)	6.9(0.82)
		T_s:7.6	-5.4	-5.1(0.94)	-4.2(0.78)	-5.1(0.94)	-4.4(0.81)	-5.0(0.93)
	T-50	T_p:67.1	3.1	1.8(0.58)	3.4(1.10)	3.7(1.19)	2.8(0.90)	3.0(0.97)
		T_s:8.2	-2.7	-1.8(0.67)	-1.2(0.44)	-1.8(0.67)	-1.7(0.63)	-2.2(0.81)
6	T-20	T_p:68.9	8.4	2.7(0.32)	3.5(0.42)	4.7(0.56)	3.3(0.39)	4.8(0.57)
		T_s:7.6	-5.4	-6.4(1.19)	-6.4(1.17)	-7.3(1.35)	-6.5(1.20)	-6.8(1.26)
	T-50	T_p:67.1	3.1	1.2(0.39)	2.1(0.68)	2.4(0.77)	1.8(0.58)	2.1(0.68)
		T_s:8.2	-2.7	-2.0(0.74)	-2.6(0.96)	-2.9(1.07)	-2.9(1.07)	-2.6(0.96)

T_i:initial time considering temperature and strain changes. T_p:At peak temperature.
T_s:At stabilized temerature, ():Ratio of calculated value to measured one

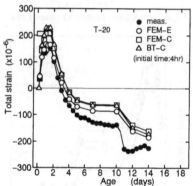

Fig.12 Time–dependent change of total strain

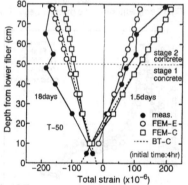

Fig.13 Total strain distributions through the depth

properties depending upon temperature history at an arbitrary level for FEM-C. I and BT-C, I while constant materials properties in fresh concrete section is used for FEM-E.

According to Fig.10, all of FEM-E (JCI 1985), FEM-C and BT-C for T-20 express fairly good agreement with experimental result at stabilized temperature, but underestimate that at peak temperatures. The opposite is true for the resulting of T-50 shown in Fig.11, where computations underestimate the curvatures at stabilized temperature, and computed average curvatures using creep model of CEB/MC90 is larger than experimental one at peak temperature. As tabulated in Table 4, all three model are do not predict experimental values both at peak temperature and at stabilized temperature with satisfactory accuracy. These results indicate that creep depends upon change of stress from increase of compression to decrease of compression (Tazawa, E. et al. 1982) and may be due to the fact that strain does not necessarily distribute linearly through the section as shown in Fig.13.

The average curvature at peak temperature based on CEB/MC90 is larger than those based on Iwaki's model both for FEM and BT because the creep coefficients obtained by CEB/MC90 is smaller than that by Iwaki. As to BT method of analysis predicts average curvatures at peak temperature as well as at stabilized temperature larger than FEM. This can be explained by the fact that in case of BT linearity of strain distribution through the full height is assumed to be valid along full length of the beam, even though FEM shows the nonlinearity of strain in the end zone in which shear stress is transferred from young concrete to old concrete.

As an example the total strain composed of temperature strain and stress related strain with time is shown for T-20 in Fig.12 together with which computed

Table 5 Computed maximum stresses in compression and in tension

T,	Speci-mens	Stress (MPa)	FEM-E	FEM-C	BT-C	FEM-I	BT-I
4	T-20	σ_c	-0.75(1)	-1.25(1)	-0.87(1)	-1.03(1)	-0.74(1)
		σ_t	1.95(12)	1.83(12)	1.68(12)	1.38(12)	1.54(4)
	T-50	σ_c	-1.60(1.5)	-3.35(1.5)	-2.89(1.25)	-2.31(1)	-2.24(0.75)
		σ_t	3.18(11)	2.62(11)	2.56(11)	2.28(11)	3.38(5)
6	T-20	σ_c	-0.51(1)	-0.82(1)	-0.49(1)	-0.69(1)	-0.42(1)
		σ_t	2.18(12)	2.20(12)	2.01(12)	1.50(12)	1.60(4)
	T-50	σ_c	-1.10(1.5)	-2.20(1.5)	-1.85(1.25)	-1.53(1.25)	-1.43(1.25)
		σ_t	3.72(11)	3.63(11)	3.57(11)	2.59(11)	3.51(5)

():days generating maximum compressive stress or maximum tensile stress

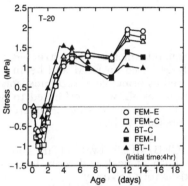

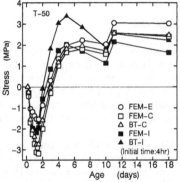

Fig.14 Comparison of thermal stresses with time   Fig.15 Comparison of thermal stresses with time

results. All computed results overestimate the total strain at peak temperature and underestimate those at stabilized temperatures which is different from the case of average curvature(see Table 4). In the case of T-50, however, comparison of experimental and computed strains was similar to that of average curvatures.

A typical example of measured strain distribution over the height of beam T-50 and the linearity of which is a key point for beam theory are shown in Fig.13 compared with those obtained by FEM at midspan of the beam. This figure shows that "plain sections remain plain section after deformation" holds true at central section. This has been confirmed by JCI Committee on Thermal Stress on Massive Concrete. However, the measured strain distribution is not necessarily linear especially at room temperature despite the closely spaced stirrups and the rough deformation made on the surface of the stage 1 concrete.

Figs.14 and 15 indicate comparisons of thermal stresses with time computed by the five methods for T-20 and T-50 putting 4 hours as initial time. Table 5 tabulates maximum stresses in compression and in tension. As can be seen in these figures and in table of the five methods, FEM-C gives the maximum compressive stress because of smaller creep and FEM-E the maximum tensile stress at age of 14days. Compressive and tensile stresses computed by BT do not exceed those by FEM, which differs from the case average curvature. Using Iwaki's model the stresses obtained by both FEM and BT reach maximum tensile stress at 4 or 5days, then decrease remarkably and thereafter fall below those by CEB/MC90 and effective elastic modulus of JSCE at age of 14days, because creep coefficient by Iwaki's model is much higher than that by CEB/MC90 and because the effect of age at stress generation is considered. This also implies that creep is actually affected by change of temperature field from rise to drop.

Table 6 Comparison of measured and computed cracking moment

Speci-mens	Meas. (kN·m)	f_{ct} (MPa)	Calc. (kN·m)					
				FEM-E	FEM-C	BT-C	FEM-I	BT-I
T-20	33.0	3.69		24.7	28.0	30.0	30.8	33.5
L-20	24.0	4.22	34.0					
T-50	35.0	3.30		6.0	30.1	29.5	54.7	26.5
L-50	100.0	4.01	10.9					

Finally, comparison of tested and computed cracking moments are listed in Table 6. The noticeable effect of thermal stress on cracking moment is apparent especially in T-50. The slight difference between tested and computed cracking moments for T-20 is due to compression stress in the upper fiber.

5 Conclusion

Deformation of beams simulating massive concrete members constructed in two stages was investigated experimentally. Analyses based on FEM and beam theory, in which effective elastic modulus method as well as principle of superposition using the creep model proposed CEB/MC90 and Iwaki et al., were performed to predict average curvature of the beams, and stress and strain in the stage 2 concrete. The following conclusions can be drawn from the present study.

1) All three models for creep prediction (CEB/MC90, Iwaki et al. and JSCE) do not predict the average curvatures both at peak and room temperature observed in the present study with satisfactory accuracy.
2) BT predicts the average curvatures at peak temperature and at stabilized temperature larger than FEM.
3) "Plain sections remain plain section after deformation" does not necessarily hold true despite the closely spaced vertical reinforcing bars and the rough deformation made in the surface of fresh concrete of stage 1.
4) Of the five methods FEM-C gives maximum compressive stress before or at peak temperature and FEM-E maximum tensile stress at age of 14days in five methods.
5) The noticeable effect of thermal stress on cracking moment was apparent especially in T-50.

6 References

CEB(1991) **CEB-FIP MODEL CODE** 1990

Iwaki, R., et al.(1982) A study on analysis of thermal stress due to heat of hydration, **Annual report of Kajima Institute of Construction Technology, 28.** pp.45-52.

JCI(1985) The method of calculation of thermal stress in massive concrete structures and its personal computer program

JSCE(1991) **Standard specification for Design and Construction of concrete structures**

Tazawa, E. and Iida, K.(1982) Mechanism of thermal stress generation during hardening of concrete, **Proc. of JCI Colloquium on the Mechanism of Thermal Stress Generation in Massive Concrete Structures** (eds T.Tanabe), Japan Concrete Institute, pp.101-104.

Tsukayama, R.(1974) Fundamental study on Temperature Rise and Thermal Cracks in Massive Reinforced Concrete, Ph.D. Dissertation, Tokyo University.

39 THERMAL STRESSES COMPUTED BY A METHOD FOR MANUAL CALCULATIONS

M. EMBORG and S. BERNANDER
Department of Civil Engineering, Division of Structural Engineering,
Luleå University of Technology

Abstract
By means of a recently refined method for manual calculations it is possible to estimate the
risk of thermal cracking in young concrete. Thus, a model for calculation of thermal stresses
has been formulated and calibrated to laboratory tests and to methods for thermal stress
analysis based on more complex material models. The method is mostly suited for cases
with external restraint and fairly uniform temperatures within the cast structures, i. e. when
there is a risk of through cracks in the cooling phase. Only a few simple formulas are used
and the method is therefore very suitable for more rough evalutations of crack risks e. g. on
the building site and in pilot studies.
Keywords: Mechanical Behavior, Numerical Simulation, Restraint, Risk of Cracking.

1 Thermal stress state

For the analysis of thermal stress states of uniaxial character in early age concrete
Bernander (1973) has proposed a method suited for manual calculations. The method is also
approximately valid for two-dimensional stress fields, where one of the principal stesses is
dominant e. g. walls cast on inflexible supports.

A crucial parameter in the present method of hand calculations is the mean temperature
$T(t_2)$ of the cross section when the concrete is again in a stress-free state after having been
in compression during the heating phase. This condition when the elastic strain, $\varepsilon_{el} = 0$
occurs early in the cooling phase, see Figs 1a and 1b. $T(t_2)$ is often denoted the zero-stress
temperature. Remaining elastic strain ε_t from the continued cooling is then compared with
the ultimate failure strain ε_{tu}. It is also possible to express the crack risk in terms of tensile
stress σ_t during cooling which is then compared with the tensile strength, see Fig 1c.

2 Mathematical formulation

The determination of the zero-stress temperature $T(t_2)$ is based on two alternative
formulations of the maximum viscous (plastic) deformation, ε_v^{max} (see Fig 1). (The analysis

Thermal Cracking in Concrete at Early Ages. Edited by R. Springenschmid. Published 1994
by E & FN Spon, 2–6 Boundary Row, London SE1 8HN, UK. ISBN: 0 419 18710 3.

makes use of the fact that very little viscous deformation takes place from time t_1 to t_2 due to deloading of the hardening concrete).

$$\varepsilon_v^{max} = k_0 \varepsilon_{tot} = k_0 \alpha_e \Delta T_{max} \quad \text{and}$$

$$\varepsilon_v^{max} = \varepsilon_{tot} - \alpha_c \Delta T_2 = \alpha_e \Delta T_{max} - \alpha_c \Delta T_2 \tag{1}$$

where $\varepsilon_{tot}, \varepsilon_{el}$ and ε_v is the total restrained, elastic and viscous strain at T_{max}
   T_{max} is the maximum value of the mean temperature at time t_1
   $T(t_2)$ is the mean temperature at time t_2 when the concrete is stress-free
   T_c is the placing temperature
   T_u is the mean temperature at the end of cooling
   $\Delta T_{max} = T_{max} - T_c$ and $\Delta T_2 = T_{max} - T(t_2)$

Elimination in Eq (1) yields the following expressions for the temperature decrease ΔT_2 from T_{max} to $T(t_2)$ and the zero-stress temperature $T(t_2)$

$$\Delta T_2 = \frac{\alpha_e}{\alpha_c}(1 - k_0)\Delta T_{max} \tag{2}$$

$$T(t_2) = T_{max} - (T_{max} - T_c)(1 - \frac{\alpha_e}{\alpha_c}(1 - k_0)) \tag{3}$$

As may be seen from above, $k_0 = \varepsilon_v / \varepsilon_{tot}$, i e the parameter expresses the viscous part of the total restrained strain. The tensile strain during the cooling phase is expressed as

$$\varepsilon_t = \alpha_c (T(t_2) - T_u) R \tag{4}$$

where R defines the restraint from a support with stiffness S

$$1/R = 1 + E_{eff}(t) A S / L = 1 + (1/J(t,t')) A S / L \tag{5}$$

where E_{eff} is the effective E-modulus of the concrete in the cooling phase
   $J(t,t')$ is the compliance function, see Eq (7)
   A is the crossectional area of the concrete and L is the length of the element

Eq (4) may also be expressed in terms of stresses

$$\sigma_t = \alpha_c (T(t_2) - T_u) E_{eff}(t) R = \alpha_c (T(t_2) - T_u) R / J(t,t') \tag{6}$$

where $J(t,t')$ is the compliance function during cooling, generally defined for a strain $\varepsilon(t,t')$ at time t for a loading with stress $\sigma(t')$ at time t' (here also compared with the commonly used creep coefficient $\varphi(t,t')$)

$$\varepsilon(t,t') = J(t,t')\sigma(t') = \frac{1+\varphi(t,t')}{E(t')}\sigma(t') = \frac{1}{E_{eff}(t')}\sigma(t') \qquad (7)$$

The ultimate strain ε_{tu} at time t_3, with which the computed tensile strain ε_t is compared varies with concrete quality, age of the concrete, loading rate, fracture mechanics etc and is difficult to obtain accurately in tests Little information about the ultimate strain for young concrete is thus reported in literature. Values between 0.08 - 0.16 ‰ may however be found in e. g. Byfors (1980). If we choose the stress formulation according to Eq (6) and look for values of the tensile strength f_t at time t_3 with which the tensile stress $\sigma_t(t_3)$ is to be compared, more abundant test results are found in literature.

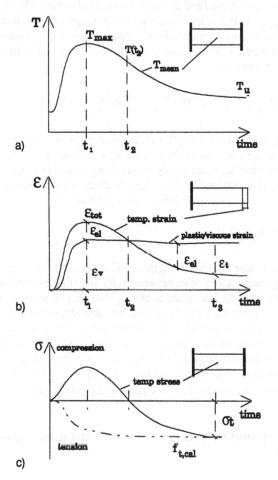

Fig 1   a) Mean temperature in a concrete element.
b) Restrained elastic and viscous (plastic) deformations
c) Induced thermal stresses at 100 % restraint, (notations are explained in the text).

The uniaxial tensile strength $f_t(t)$ at early ages may be related to the compressive strength by e g the following equation (Byfors (1980))

$$f_{ct}(t) = 0.115(f_{cc}^*(t) - 0.022 \quad [MPa] \quad ; f_{cc}^*(t) \le 20\,MPa$$

$$f_{ct}(t) = 0.105(f_{cc}^*(t) - 20.)^{0.839} + 2.28[MPa] \quad ; f_{cc}^*(t) > 20\,MPa \tag{8}$$

where f_{cc}^* is the compressive strength of prisms 100x100x400 mm (= $f_{c,cyl}(t) \approx 0.8 f_{c,cube}$), obtained by standard formulas or standard tendency curves. When estimating the tensile strength $f_t(t_3)$, we should - if the temperature substantially deviates from 20°C - take the influence of temperature on the final strength into consideration. The reduction of the ultimate strength $f_{c,cyl}(28d)$ due to high curing temperatures has been studied in various papers in literature and several mathematical formulations of the phenomenon have been reported, see e. g. Emborg (1994). Here, this reduction is approximated to being dependent on the maximum temperature:

$$f_{c,cyl}(28, T_{max}) = \left[1 - 0.008(T_{max} - 20)\right] f_{c,cyl}(28, 20°C) \tag{9}$$

Further, the computed tensile strength $f_t(t_3)$ should be reduced on account of fracture phenomen related to the very low loading rates. At very low rates of loading, creep effects may lead to failure by tertiary creep implying cracking at loads far below the tensile strengths documented at standard loading rates. Mathematical formulas of this effect may be found in literature. In this context the effect is simply modelled as

$$f_{t,cal}(t_3) = k_3 f_t(t_3) \tag{10}$$

where $f_{t,cal}(t_3)$ is the "true" failure strength including effect of tertiary creep. For loading rates in the cooling phase values of k_3 between 0.65 and 0.70 have been observed in tests.

When tensile stresses occur late in the cooling phase (e. g. thick sections, high ambient temperatures), we may use the 28-day compressive strength value in Eqs (8)- (10). When high tensile stresses occur at an early stage, somewhat lower values should be used. In the same way, different values of the compliance function J(t,t') (or E_{eff}) in Eqs (5) and (6) may be chosen depending on early or late occurrence of tensile stresses, type of cement and concrete quality, see next section. A stress level $\sigma_t(t_3)/f_t(t_3) \le 0.7$ has been found in cracking tests to give a reasonable low risk against thermal cracking.

3 Evaluations of parameters from relaxation tests and from a more sophisticated theoretical model

3.1 General
The formulas above have been calibrated by relaxation tests and by a more sophisticated model of thermal stress analysis. In the relaxation tests small concrete specimens were placed in a testing machine immediately after casting and heated by surrounding water. The

ends of the specimens were fixed in the testing machine. Thermal stresses were recorded in a load cell and time of thermal cracking was documented, see e. g. Emborg (1990) showing examples of evaluated thermal stresses. The relaxation test results serve as a base for the calibration of different models of thermal stress analysis. In addition, laboratory tests have been performed studying thermal and mechanical properties, see e. g. Westman (1994).

The theoretical model is based on a rate-type of viscoelastic law where the thermal properties and the transient mechanical properties of the young concrete are considered. The properties (e. g. heat of hydration, strength, elasticity, creep and fracture mechanics behaviour) are modelled on the basis of the maturity development of the concrete by means of the Arrhenius concept.

3.2 Degree of plasticity in compression expressed by the parameter k_0

By studying results from a study performed by Löfqvist (1946), Bernander (1973) proposed values of the degree of plasticity, k_0, between 0.6 and 0.8 for the Standard Portland cement of Limhamn type. In a recent investigation, the coefficient has been cali-brated to the relaxation tests performed by solving k_0 from Eq (2) and by using values of the coefficients of thermal expansion and contraction obtained in unconstrained thermal volume tests. Thus, for concretes with Standard Portland Cement (type Slite ASTM Type I), grade 25-50, values of k_0 between 0.81 and 0.88 were obtained. For the cement of type Degerhamn (Type II) the values varied somewhat more depending on maximum temperature, cooling rate, concrete strength etc. Values in the range from 0.70 to 0.85 were documented.

Comparisons of crack risks evaluated with the theoretical model described above, have been carried out in a series of about 80 thermal stress computations comprising two types of cements, two concrete strengths, and various combinations of thicknesses of concrete members, placing and ambient temperatures. Values of k_0 were obtained according to Table 1.

In the computations, it was found that the value of k_0 was slightly dependent on the maximum temperature. Hence, approximately, the lowest value of k_0 in Table 1 corresponds to the lowest value of T_{max} and vice versa. However, it should be noted that there is a scatter in the k_0/T_{max} dependence of about ±0.05.

3.3 Viscoelastic response in tension, creep compliance J(t,t')

The creep compliance $J(t,t')$ (or effective modulus) used in the computations of the tensile stress σ_t has been calibrated by means of the tests as well as the more sophisticated theoretical model. It is known that the creep compliance is dependent on the age t_2 and on the time of tensile stress build up $(t_3 - t_2)$. (Correctly, the equivalent age, expressed by the maturity function, should be used instead of t_3 and t_2).

In the comparison with the relaxation tests mentioned above it was observed that lower w/c ratios implied lower compliance (stiffer response). For concretes of Grade K45 - K30 with Type I cement , values in the range $4.8 - 7.0 \ 10^{-11}$ [1/MPa] were obtained. For Type II cement about 20 % higher values were obtained. However, the scatter was considerable.

In the calculations with the theoretical model, lower values of $J(t,t')$ were obtained compared to the relaxation tests. Type I cement and higher concrete grades lead to lower values of $J(t,t')$, see Table 1. The effects of age at zero stress and time duration of tensile stress growth on the compliance seems to counteract each other. This means that an early time of tensile stress initiation (giving a high value of the $J(t,t')$) is often followed by a short time of tensile stress build-up (giving a lower value of the compliance).

Table 1. Calibration of hand calculation method to a more sophisticated rate-type model. Evaluation of k_0 and $J(t,t')$. T_{max} depends on placing temperatures (here: 10 and 20 °C), air temperatures (here: 5-25 °C) and dimensions (1.0-6.0 m), 45 in K45 etc means 28 day cube strength (values within bracket - obtained scatter in computations),

Concrete	T_{max} [°C]	k_0 [-]	$J(t,t')$ [1/Pa]
Grade K45 (w_0/c=0.42), Std Degerhamn (ASTM Type II)	45 - 65	0.71 (0.65 - 0.75)	4.9×10^{-11} $(4.2\text{-}6.0) \times 10^{-11}$
Grade K45 (w_0/c=0.42), Std Slite (ASTM Type I)	50 - 80	0.72 (0.65 - 0.75)	3.2×10^{-11} $(2.2\text{-}4.8) \times 10^{-11}$
Grade K30 (w_0/c=0.55), Std Degerhamn	35 - 50	0.69 (0.65- 0.75)	5.2×10^{-11} $(4.0\text{-}6.1) \times 10^{-11}$
Grade K30 (w_0/c=0.55), Std Slite	40 - 60	0.71 (0.60 - 0.75)	4.1×10^{-11} $(3.2\text{-}4.6) \times 10^{-11}$

3.4 Tensile strength

Besides concrete quality, the tensile strength adjusted for tertiary creep, $f_{t,cal}$, also depends on the equivalent age at time t_3. In most of the evaluations performed, the concrete had reached a considerable degree of maturity when the highest tensile stresses occurred. Hence, we may approximately use 0.9 - 1.0 times the 28-day value of f_{cc} in Eq (8) for the computation of tensile strength. In practical engineering there may of course be cases where low concrete temperatures and fast temperature drop coincide (e. g. winter concreting in Sweden) in which case a lower degree of maturity must be anticipated.

4 Examples of cracking risk estimations

The described method for hand calculation has been applied for evaluations of the risk of thermal cracking in some typical cases. In all of the examples the following is valid: Concrete: Grade K45, Std Portland cement type Degerhamn (Type II ASTM), cement content 425 kg/m³, w/c=0.42, cube strength at 28 day = 55 MPa, form: 25 mm wood (stripped at seven days), varying member thicknesses. Three parameters were studied:

- a) *Influence of temperatures, 100 % restraint* : A strong influence of placing temperature and air temperature is observed, see Fig 3a and 3b.

-b) *Influence of cooling, 100 % restraint*: Cooling with embedded pipes leads to a considerable reduction of tensile stresses and risk of cracking, see Fig 4.

-c) *Influence of restriant*: Even an insignificant reduction of restraint may imply a considerable decrease of maximum stresses, see Fig 5. The stiffness S (Eq (5)) is computed from the length, E-modulus and cross sectional area of the base support (a slab) i. e. $S = L_s/(E_s A_s)$ where index s denotes support. Here the cross sectional area of the support as well as the thickness and height of the wall are varied giving different values of R.

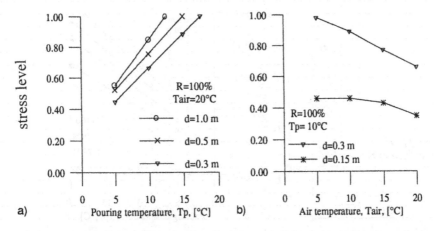

Fig 3. Evaluation of cracking risk (i. e. maximum tensile stress level) by means of manual calculations. Temperature developments evaluated with the HETT program. Influence of pouring temperature (a) and influence of air temperature (b)

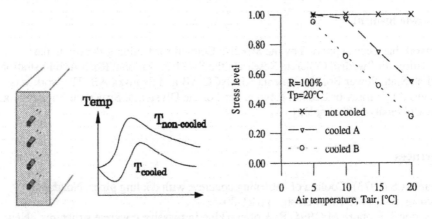

Fig 4. Influence on cracking risk of cooling with embedded cooling pipes.

5 Conclusions

In many situations it is possible and convenient to use the described method for manual computation - even in two-dimensional stress fields where one of the principal stresses is dominant. The method is easy to apply - only a few simple formulas are used. These may, again, be implemented in a minor computer program. By calibration of the coefficients used in the 'hand' method with other methods of thermal stress analysis or with test results, rather accurate predictions of cracking risks may be obtained. However, variations of k_0 and $J(t,t')$ according to the scatter in Table 1 have been found - in a special study - to have some effects on the computed tensile stresses and cracking risks. Therefore, it is recommended to check possible variations of the parameters within reasonable ranges.

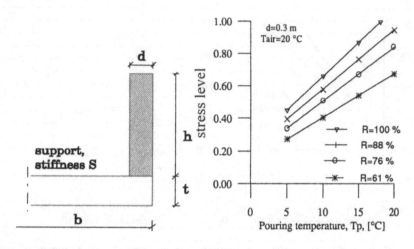

Fig 5. Influence of restraint R from underlying slab on crack risk. (Compare with Fig 3).

Acknowledgement

This work has been supported by the Swedish Council for Building Research, the Foundation for Swedish Concrete Research, the Swedish National Road Administration, the Swedish State Power Board, Cementa AB, NCC AB and Skanska AB. The work was supervised by Professor Lennart Elfgren, head of the Divsion of Structural Engineering at Luleå University of Technology.

References

Bernander S (1973): Cooling of hardening concrete with cooling pipes, **Nordisk Betong**, no 2 1973 (in swedish) , pp 21-30

Bernander S, Emborg M (1994) Risk of cracking in massive concrete structures - New developments and experience. **Proceedings from Rilem International Symposium on "Avoidance of Thermal Cracking in Concrete"**, Munich Oct. 1994, 8 pp.

Byfors J (1980): **Plain concrete at early ages**. Swedish Cement and Concrete Research Institute, Fo 3:80, Stockholm 1980, 350 pp

Emborg M (1990): **Thermal stresses in concrete structures at early ages**, Doctoral Thesis 1989:73 D, Div of Struct Eng, Luleå Univ. of Technology, Luleå 1990, 286 pp

Emborg M, Bernander S (1994) Assessment of the risk of thermal cracking in hardening concrete, **Journ. of Struc. Eng (ASCE)**, (to be published oct 1994),

Löfquist B (1946) **Temperaure effects in hardening concrete** (in swedish), Dissertation, Kungliga Vattenfallsstyrelsen, tekn. Report no 22, Stockholm 1946, 195 pp

Westman G (1994) Basic of creep and relaxation of young concrete, **Proceedings from Rilem Int. Symp. "Avoidance of Thermal Cracking in Concrete"**, Munich 1994.

40 THERMAL EFFECTS, CRACKING AND DAMAGE IN YOUNG MASSIVE CONCRETE

J.P. BOURNAZEL and M. MORANVILLE-REGOURD
Laboratoire de Mécanique et de Technologie, ENS Cachan, Cachan, France

Abstract
This paper gives the framework of a modelling able to describe the behaviour of maturing concrete. Two main state variables are used : - Maturity, which acts on shrinkage, creep and the evolution of elastic properties - Damage, which affects the stiffness of the concrete. Applications on a mass concrete dam and on the testing wall at the nuclear power plant of Civaux show the interest of that approach in analysing the initial state of a construction.
Keywords: Autogeneous Shrinkage, Creep, Damage due to Themal Stresses, Mass Concrete, Maturity/Equivalent Time, Modelling of Hardening Concrete.

1 Introduction

Concrete is always subjected to volumetric variations. Shrinkage or expansion appears for different reasons, linked to internal processes of cement hydration, and water flow due to external conditions of temperature and humidity. When these deformations are not allowed to develop easily, stresses and risk of damage are developed. These phenomena are particularly important in massive structures, due to two reasons : the use of plain concrete which favours the localization of damage and the mass of material concerned, which induces large thermics effects.

We present hereafter a global model developed within the framework of thermodynamics of irreversible processes, which is able to describe the main aspects of the problem induced by volumetric variations for maturing concrete.

2 Maturing concrete

The response of concrete under strain or stress depends on the actual mechanical properties of the solid. These properties are directly linked to the actual state of the microstructure of the medium. We describe the evolution of mechanical properties by maturity which is considered as a state variable called M. Maturing concrete can be

Thermal Cracking in Concrete at Early Ages. Edited by R. Springenschmid. Published 1994 by E & FN Spon, 2–6 Boundary Row, London SE1 8HN, UK. ISBN: 0 419 18710 3.

affected by damage. This damage is characterized by a state variable called D. The model is developed within the framework of thermodynamics of irreversible processes (Lemaitre et al 1990).

2.1 Elastic strain

Deduced from the Helmotz free energy, which defines at any moment the thermomechanical state of the medium, the state laws give the following tensorial equation :

$$\overline{\sigma} = \Lambda(M, D) : \overline{\varepsilon}^e \qquad (1)$$

with $\overline{\varepsilon}^e = \overline{\varepsilon} - (\overline{\varepsilon}^d + \overline{\varepsilon}^c)$

ε is total strain, ε^d is dilatation strain (due to temperature and shrinkage) and ε^c is creep strain, Λ is the elastic tensor, D is damage and M is maturity

Damage is assumed to be isotropic. Then elastic tensor for the damaged material at a given maturity is written as (see Mazars 1986, Mazars et al 1991) :

$$\tilde{\Lambda}(M, D) = (1 - D)\tilde{\Lambda}(M) \qquad (2)$$

Chosen in a simple way, the evolution of the mechanical characteristics is assumed to be linear (Bournazel 1992)

$$E(M) = E_\infty M \qquad (3)$$

E_∞ = long term Young modulus, E is Young modulus at time t

$$v(M) = 0.5 - 0.3M \qquad (4)$$

v is the Poisson's ratio

2.2 State variables

Maturity : We choose for maturity the definition proposed by Regourd and Gautier (1980) and by Carino (1982) :

$$\frac{M}{1 - M} = \int_{t_o}^{t} a e^{\frac{-U}{RT}} d\tau \qquad (5)$$

t_o is the setting time defined on experimental procedure of ultrasonic wave propagation (Bournazel et al 1991), t is time, R is the constant of perfect gaz, T is temperature in K, U is activation energy.

<u>Damage</u> : based on previous works from Mazars (1986)

$$\tilde{\varepsilon} = \sqrt{\sum_i \left\langle \varepsilon_i^e \right\rangle_+^2} \tag{6}$$

where ε_i^e denotes a principal elastic strain; and $\langle X \rangle_+$ corresponds to the positive part of X.

$D = 0$, if $\tilde{\varepsilon} \leq \varepsilon_{do}$ (ε_{do} = initiation damage threshold) $\tag{7}$

$D = \alpha_t D_t + \alpha_c D_c$ if $\tilde{\varepsilon} > \varepsilon_{do}$ $\tag{8}$

which combines the effect of tension Dt, and compression Dc

For the applications presented herafter tension effects prevail. Thus, we write :

$D = D_t(\tilde{\varepsilon})$, then :

$$D = 1 - A_t e^{-B_t(\tilde{\varepsilon} - \varepsilon_{do})} - \frac{(1 - A_t)\varepsilon_{do}}{\tilde{\varepsilon}} \tag{9}$$

At, Bt are constant parameters.

2.3 Non elastic strain

There are two kinds of non elastic strains : those linked to dilatation and shrinkage, and those linked to the viscous response of the medium, creep or relaxation. Some assumptions are necessary to solve the problem. In mass concrete, complete drying of material takes a very long time and will not be considered here. We will take into account and according to Bergstrom (1980) and Bazant (1989), only autogeneous shrinkage and basic creep.

shrinkage

We suppose a decoupling between thermics and mechanics. Based on previous works of Byfors (1980), we consider that the thermophysical parameters are constant in time and temperature. Curves of autogeneous shrinkage versus time have already been plublished (Acker 1988). The shrinkage evolution is directly linked to the growth of crystals and then to the maturity. We chose to use, for roller compacted concrete, the following relation :

$$\tilde{\varepsilon}^{sh} = \varepsilon_\infty^{sh} M(2 - M)\tilde{1} \tag{10}$$

where ε_∞^{sh} is the autogeneous shrinkage for M=1 and $\tilde{1}$ is the unit tensor

creep

In Roller Compacted Concrete massive structure, thermal effects begin just after casting of concrete and reach their critical value (maximum of temperature) after three days. Most of creep and relaxation functions proposed in the calculation codes are available for ages above 5 days. We have developed a serie of creep test under tensile loading at early ages to characterize the viscous behaviour of concrete at early ages. For roller compacted concrete, we have identified a creep function as :

$$\varepsilon^c(t, t_1) = \frac{\Delta\sigma}{E_1} ae^{-bM^c}(1-M)\dot{M}$$ (11)

where t_1 is age at loading, a, b and c are material parameters, $\Delta\sigma$ increasing of stress, E_1 young modulus at time t_1

This relation is similar to those proposed by Laube (1990) and Bazant (1993)

3 Applications

3.1 Case of the Riou RCC dam

The Riou dam is a gravity dam built with the RCC technique without any joint. Its dimensions are 200 meters long, 30 meters high and 18 meters large (center section). Within the national french program "BaCaRa", this construction was used to measure, to control and to observe the thermomechanical effects. Figure 1-a gives the cracks observed on the dam after six months of age. To simulate the phenomena a first three dimensional thermal computation was performed to obtain the history of temperature in the dam. In situ measurements has been done to adjust these imput values and to define the thermics boundary conditions. From the three dimensional mesh used for thermal computations, it is possible to extract a two dimensional mesh which has been used for mechanical computations. Computations use the Cesar-LCPC (1987) finite elements code in which the model, described above, has been introduced. Because the softening behaviour of concrete induces a localization effect, the non local damage theory is used. The damage D is calculated over a representative volume Vr. The size of Vr depends on the characteristic length lc of the material (see Pijaudier et al 1987). The boundary conditions chosen for these computations are embedded in the foundation. The mesh uses 777 elements (Q4) and the rock is modelled as perfectly elastic. Numerical results of these computations are given for the steps of time 39 days and 187 days (figure 1b and 1c). They can be compared to the real cracks observed in the dam after 6 months

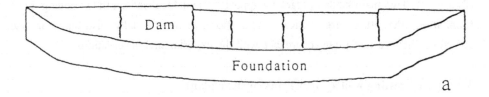

Damage: From 0.80 cracks occur

parameters values
ultimate shrinkage : 138 μm/m
activation energy : 71200 J/mol
damage law : At = 0.95
　　　　　Ac = 1.40
　　　　　Bc = 17000
　　　　　β = 1.06
　　　　　$ε_{do}$ = 1.10⁻⁴
ultimate Young modulus : 28774 MPa
maturity function : A = 5.60 10¹¹ days⁻¹

▦	de 0.000E+00 à 0.167E+00
▦	de 0.167E+00 à 0.333E+00
▦	de 0.333E+00 à 0.500E+00
▦	de 0.500E+00 à 0.667E+00
▦	de 0.667E+00 à 0.833E+00
▦	de 0.833E+00 à 0.100E+01

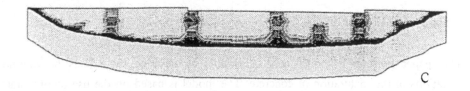

Figure 1 : **Cracking in a Roller Compacted Concrete dam**
a - Cracks observed after 6 months. b - c - Damaged zone obtained by calculation at
times 39 days and 187 days.

(figure 1a). Two main kinds of cracking appear : one is localized on the bounding dam-foundation, the others are vertical cracks more or less equally distributed in the structure. these results are in accordance with the experimental observations.

3.2 Case of testing wall at the Civaux nuclear plant

This wall was built up in order to evaluate the risk of cracking at early age in the case of various kinds of concrete that may be used to construct the Civaux nuclear plant. We present hereafter the case of a normal concrete wall (Ithurralde 1990) for which experimental results are available.

The testing wall is 20 meter long, 2 meter high and 1.2 meter thick. It was equiped with 8 thermocouples which gave the history of temperature during maturation at different locations in the center of different cross sections (figure 2). A part of these results was used to identify some thermal parameters such as exchange coefficients. The calculations assumed that it is a plane problem. However figure 2a shows that simulations of the evolution of temperature are in good accordance with the experimental data.

The thermomechanical modelling leads to damaged zones which show that there is a localization of damage in the lower part of the wall, which was assumed to be perfectly bound to the foundation. Cracks appear from the upper part to the lower part of the wall (see figure 2c). The comparison with in situ observations after 3 days (figure 2b) shows that, for the half length of the wall (symmetry was used), the prediction of the location of cracks are good in the center of the wall. The difficulty in simulating the real limit conditions at edges of the wall in contact with the mould obliged us to consider limit conditions completely free for calculations.

4 Conclusions

This paper presents a modelling process able to describe the effects of volumic variations in the maturation of concrete The model is based on the use of two state variables : maturity M and damage D The thermomechanical problem is solved within two separate steps : the first determinates the history of temperature and the second concerns the mechanical response which leads to strain, stress and damage.

The numerical results presented in this paper show that this kind of modelling leads to realistic prediction of damage induced by cement hydration in a gravity dam. This model can be extended to other types of concrete like HPC and concrete structure.

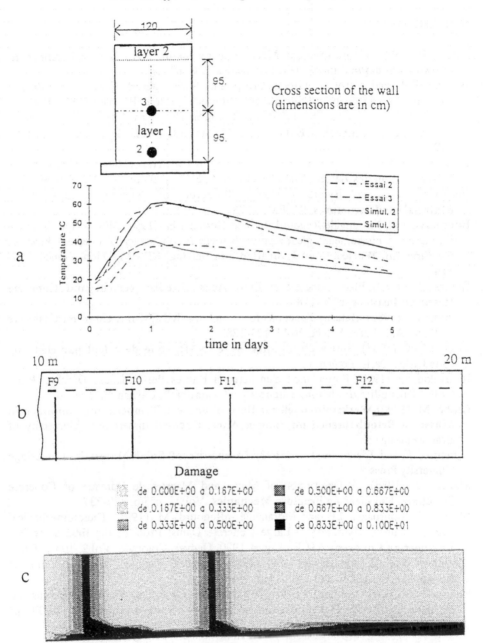

Figure 2 : Test on a large plain concrete wall
a- Evolution of temperature at points 2 &3, comparison between measures (essai) and calculations (simul.).
b- Location of the cracks on the second part of the wall at demoulding at 3 days
c- The prevision of damaged zones after 3 days (axis of symmetry is on the left)

5 References

Acker, P. (1988) **Comportement Mécanique du Béton : Apport de l' Approche Physico-chimique** Rapport de recherche des LPC, n° 152.

Bazant, Z.P. and Carol, I. (1993) Creep and Shrinkage of Concrete. **5th int. Symposium on Creep and Shrinkage. Concreep5**, Rilem. Barcelona, 177-189

Bazant, Z.P. and Prasannan, S. (1989) "Solidification Theory for Concrete Creep, Part I : Formulation". **journal of Engineering Mechanics,** ASCE, Vol 115, n° 8, 1691-1725

Bournazel, J.P. (1992) **Contribution à l'Etude du Caractère Thermomécanique de la Maturation des Bétons**. Thèse de doctorat de l' Université Paris VI, France.

Bergstrom, S.G. and Byfors, J. (1980) "Properties of Concrete at Early Ages" **Materials and Structures**, RILEM, n° 75

Bournazel, J.P., Moranville-Regourd, M. and Hornain, H. (1991) "Early Age Concrete Strength : a Phenomenological Approach to Physico-chemical Processes" Proc. of the **First Int. Workshop on Hydration and Setting,** RILEM, Dijon, France, 307-314

Byfors, J. (1980) Plain Concrete at Early Ages. **Swedish cement and Concrete Research Institute**, n° 3, n° 80.

Carino, N.J. (1982) Maturity Functions for Concretes, RILEM **int. Conf. on Concrete at Early Age**, Paris, France, Vol I, 123-128.

Cesar-LCPC (1987) **Notice d'utilisation de CESAR, code de calcul par éléments finis** LCPC, Paris, France.

Ithurralde, G. (1990) Formulation d'un béton à Hautes Performances Durable. Proc. **Coll. "Durabilité des bétons à hautes performances"**, Cachan, France

Laube, M. (1990) **Werkstoffmodell zur Berechnung von Temperaturspannungen in Massigen Betonbauteilen im Jungen Alter.** Doctoral dissertation, University of Braunschweig.

Lemaitre, J. and Chaboche J.L. (1990) **Mechanics of Solid Materials**. Cambridge University Press.

Mazars, J. (1986). A Description of Micro and Macroscale Damage of Concrete Structures. **Engineering Fracture Mechanics**, Vol 25, n°5/6, 729-737

Mazars, J., Bournazel, J.P. and Moranville - Regourd, M. (1991) Thermomechanical Damage Due to Hydration in Large Concrete Dams. Proc. of the **first materials Engineering Congress**, ASCE, August 1990, Denver, Colorado, Vol 2, 1061-1070

Pijaudier-Cabot, G. and Bazant, Z.P. (1987) Nonlocal Damge Theory. **Journal of Engrg. Mech.**, ASCE, 113 (10), 1512-1513

Regourd, M. and Gautier, E. (1980) Comportement des Ciments Soumis au Durcissement Accéléré". Durcissement accéléré des bétons, **Annales de l' ITBTP**, n° 387, 83-96

41 TEMPERATURE FIELD AND CONCRETE STRESSES IN A FOUNDATION PLATE

P. PAULINI and D. BILEWICZ
Institut für Baustofflehre and Materialprüfung, Universität Innsbruck,
Innsbruck, Austria

Abstract
Hydration temperatures can cause dividing cracks in thickwalled concrete building
members. The temperature field in a foundation plate was measured in the first week
of concrete hardening. Hydration temperatures are clearly influenced by day-night
changes of air temperatures. Calculations of concrete stresses showed that external
restraint stresses principally are responsible for thermal cracking. These can been
reduced with slide sheets to the extent that the occurrence of dividing cracks can be
avoided.
Keywords: Cracking, Heat of Hydration, Restraint Stresses, Structural Members
(Foundation Plate), Temperature Fields.

1 Introduction

Concrete temperature stresses resulting from hydration heat are significant for thick-
walled watertight building members such as foundation plates and walls or tunnel
linings in ground water. In such building components thermal stresses can lead to
cracks in early age concrete and put at risk the component's performance. Reinforcement cannot prevent the occurrence of cracks, it can only limit crack widths and
crack distribution. The aim of the engineer is therefore to avoid temperature stresses
to a large extent by e.g. the following methods
- reduction of fresh concrete temperature
- active cooling of the building component during hardening time
- reduction of cement content or use of low heat cement
- reduction of restraint.

The cement type as the source of hydration temperatures has an important influence
on the cracking behaviour. Springenschmid (1991,1992) shows that concretes with the
same cement type but from different producers may have considerable differences in
their cracking temperatures. The lowest tendency to cracking was found in cements
with a low hydration temperature rise and low alkali content. This varying reaction
behaviour of cements in the early phase was also pointed out by Paulini (1992). In
national standards no specifications exist for achievable hydration levels for the early
cement reaction. Requirements of high demould strength (e.g. for tunnel lining
concrete 5 MPa after 12 hrs) often conflict with a low hydration temperature rise. Site

Thermal Cracking in Concrete at Early Ages. Edited by R. Springenschmid. Published 1994
by E & FN Spon, 2–6 Boundary Row, London SE1 8HN, UK. ISBN: 0 419 18710 3.

management often has to conduct its own trials to choose a suitable cement type. Slide sheets are therefore increasingly being used, where it is statically possible, to avoid thermal cracking of building members, Simons (1988), Schade (1989).

2   Building site and laboratory trials

During concreting work for the extension of the Innsbruck sewerage plant an SRPC cement and slide sheets were used to avoid thermal cracking in a foundation plate. The plate was a 55 cm thick, 10 by 10 m, reinforced, watertight foundation plate. The concrete used was a B300 K3 GK22 WU according to the Austrian standard ÖN B 4200/T10. The following mix design was used:

Cement PZ 375 HS (SRPC)	340 kg/m^3
Water	156 kg/m^3
Aggregates 0/22 mm	1952 kg/m^3
Plasticiser BV1	2.04 kg/m^3

In a field measurement programme concrete temperatures were to be recorded for the first week of hardening. The measurement were made in summer 1993 in a period in which day-night temperatures fluctuated considerably. Hydration temperatures were measured by means of NiCr-Ni thermocouples (TC). A thermistor signal served as the reference temperature. Five TC's were placed in the concrete plate symmetrically to the neutral axis. In addition the air temperature was recorded 5 cm above the surface. The TC's were placed in a plastic tube, which was then filled with PU foam. The completed sensor was fixed to the reinforcement mesh and connected to a data logger. Fig.1 shows a diagram of the sensor.

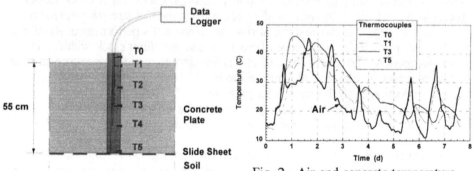

Fig. 1. Thermocouple sensor unit

Fig. 2. Air and concrete temperature readings

Fig. 2 shows the recorded air temperature T0, concrete temperature T1 at the surface, T5 at the bottom and the plate mid temperature T3. The time scale was chosen in such a way that day numbers refer to midnight. Concrete work took place from 7.00-10.30 a.m. with a fresh temperature of 21 °C. Approximately 5 hours after casting the hydration gain starts with a maximum temperature increase of 25 °C. 18 hours after casting the maximum core temperature of 46 °C is reached and the cooling period is initiated. The air temperature T0 is clearly influenced by the hydration peak. Air temperature changes of 15 - 20 °C between day and night were observed. On the 4th and 5th day air temperature fell suddenly, then increased again up to the 8th day. The air temperature cycles propagate into the concrete plate (Fig.3). During short daytime temperature peaks, heat is transferred to the plate while during the night the gradient is reversed and heat dissipates from the plate. The temperature wave reaches the bottom only after 8-10 hrs. Daily temperature gain on the surface thus partly coincides with temperature loss at the bottom. This explaines the oscillating stresses in the plate calculated subsequently.

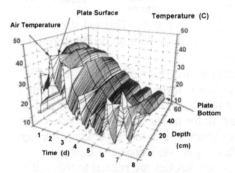

Fig. 3. Concrete temperature field

Age (h)	Compr. Strength (MPa)	Flexural Strength (MPa)	Splitting Strength (Mpa)	E-Mod. (GPa)
24	19.6	3.75	1.99	28.0
33	20.8	3.82	1.92	29.9
49	24.7	3.15	2.26	34.1
96	25.2	3.35	2.59	32.0

Tab. 1. Concrete properties

Development of concrete strength and elastic modulus were measured for the first four days. Concrete prisms (12/12/36 cm) were stored from 6 hours onwards in a climatic cabinet at constant 30 °C and 90% r.h. The measured values are shown in table 1.

3 Temperature stresses

Temperature stresses in building members consist of three different fractions which are shown in Fig. 4. Internal stresses σ_i arise from nonlinear temperature distribution over the cross section and are calculated so as to not give rise to any normal force and moment in the cross section. External restraint stresses arise only when temperature deformations of the building member are prevented. Warping stresses σ_w result from a linear temperature distribution between opposite surfaces while restraint normal stresses σ_n arise from uniform temperature differences relative to a stress free reference state T_0. This means for the stress state of Fig. 4 that the whole cross section is subjected to a deformation restraint. In practice such conditions seldom exist. Generally there is only a friction restraint in the border plane to soil or to the adjacent component, while the opposite surface remains free of restraint.

The further calculation is based on the principle shown in Fig.5. Here an eccentric friction force F_c acts on the plate bottom and produces a stress state σ_f. The friction force F_c can be determined by requiring that thermal deformations of the unrestrained plate and elastic deformations resulting from friction force on the bottom side must vanish. The assumption of a linear elastic stress-strain behaviour for concrete is true

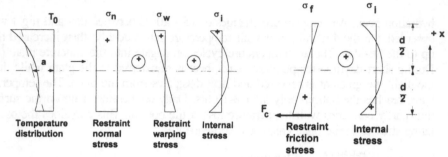

Fig. 4. Thermal stress fractions
with a centric restraint

Fig. 5. Thermal stress fractions
with an eccentric restaint

for small compression stresses and also for tension stresses, Heilmann et al (1969), Reinhardt (1984). Relaxation of early age concrete is taken into account according to Rostásy and Henning (1990) by introducing a relaxation factor k_r and a reduced elastic modulus E_r. The reference temperature T_0 for a state without any stress was found to be some degrees lower than the maximum hydration temperature, Springen-schmid (1991). For simplicity in this calculation T_0 is considered as maximum hydration temperature (max a_j Eq. 9) and restraint stresses are related to this state.

3.1 Internal stresses
Internal stresses arise from a nonlinear temperature distribution. First we describe the temperature distribution $T(x)$ in the plate as a polynome of coordinate x (Eq.1). The origin of x is set in the neutral axis of the plate (Fig. 5).

$$T(x) = \sum_{i=0}^{n} C_i \cdot x^i \tag{1}$$

Coefficients C_i are determined in such a way that the measured temperatures Tm in the support points s_i are matched exactly by the polynom. The five TC positions s_i, i=0..4, were arranged symmetrically to the neutral axis at 0, ±100 and ±260 mm defining a temperature polynom $T(x)$ of 4th degree. With the inverse of the polynom matrix M (Eq.2) coefficients C_j for each hourly measurement time j=1..183 can be calculated (Eq. 3).

$$M_i = s^i \tag{2}$$

$$C_j = M^{-1} \cdot Tm_j \tag{3}$$

The polynom $T(x)$ represents the temperature field for each depth in the plate and each of the measured times j. Now we look for a linear temperature distribution $G(x)$ whereby stresses from the temperature difference $T(x)-G(x)$ do not result in normal forces or moments in the plate. We assume the cross section to be a rectangle of depth d.

$$G_j(x) = a_j + b_j \cdot x \tag{4}$$

First we find the integrals of temperature TI and temperature moments TM over the plate depth d (Eq.5,6).

$$TI_j = \int_{-d/2}^{d/2} T(x)_j \cdot dx \tag{5}$$

$$TM_j = \int_{-d/2}^{d/2} T(x)_j \cdot x \cdot dx \tag{6}$$

The same values for the linear temperature distribution G(x) become (Eq. 7,8)

$$GI_j = a_j \cdot d \tag{7}$$

$$GM_j = b_j \cdot \frac{d^3}{12} \tag{8}$$

The coefficients a and b of the linear temperature distribution G(x) are calculated by equalizing both the temperature and the temperature moment integrals (Eq.9,10)

$$a_j = \frac{1}{d} \cdot \int_{-d/2}^{d/2} T(x)_j \cdot dx \tag{9}$$

$$b_j = \frac{12}{d^3} \cdot \int_{-d/2}^{d/2} T(x)_j \cdot x \cdot dx \tag{10}$$

Coefficient a represents the average temperature and coefficient b the slope of the linear distribution G(x). Fig. 6 is a plot of the coefficients over time, while Fig. 7 gives temperature distributions T(x) and G(x) over depth as well as the measured temperatures Tm at 38 hours after casting.

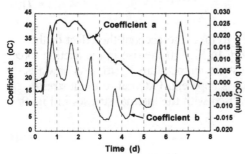

Fig. 6. Coefficients of linear temperatur distribution G(x)

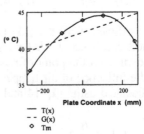

Fig. 7. Calculated and measured plate temperature distribution

Internal stresses σ_i are found with Eq. 11 from the difference of both temperature distributions T and G.

$$\sigma_i(x)_j = (T(x) - G(x))_j \cdot E_r \cdot \alpha_t \tag{11}$$

$$E_r = k_r \cdot E \tag{12}$$

E_r is a reduced elastic modulus of concrete according to Eq. 12. The relaxation factor k_r for an effective concrete age of 2-3 days is given as 0.65 and for more then

5 days as 0.85 , Rostásy and Henning (1990) . In this calculation k_r was set at an average value of 0.75. The elastic modulus was assumed at 32 GPa (Tab. 1) and the coefficient of thermal deformation α_t at 9.10^{-6}. Fig. 8 represents calculated internal stresses which are clearly influenced by daily air temperature changes.

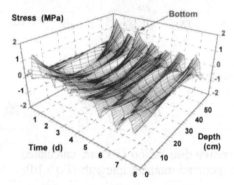 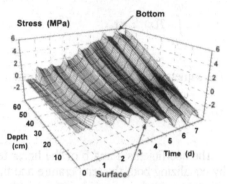

Fig. 8. Internal stress field

Fig. 9. Thermal stress field, combind internal and external restraint stresses

3.2 External restraint stresses

External restraint stresses in the plate arise only from the effect of friction forces F_c (Fig. 5). By requiring that elastic and thermic deformations in the restraint plane must vanish, we get for a rectangular section the friction force F_c at time j to

$$F_{c,j} = [(T_0 - a_j) + b_j \cdot x] f E_r \alpha_t \frac{d}{4} \tag{13}$$

The coefficient f $(0 \leq f \leq 1)$ is introduced to take into account the effect of the slide sheet and is set at 1.0 for a rigid support. In Fig. 10 the rise in the friction force F_c is shown from the point of reaching zero stress temperature T_0 (max a_j) onwards. Again the influence of air temperature cycles is apparent. External restraint stresses σ_f were calculated using Eq. 14 and are summarised with the internal stresses in Fig. 9.

$$\sigma_f = \frac{F_c}{d} \left(1 - \frac{6}{d} x \right) \tag{14}$$

In Fig. 11 plate edge stresses and stresses in the neutral axis are represented together with measured flexural strength $f_{c,fl}$ and uniaxial tensile strength $f_{c,t}$. Tensile strength $f_{c,t}$ was not measured and set for comparison at $f_{c,fl} / 2$ according EC2. Concrete strengths were measured for the first four days (Tab. 1) and were subsequently drawn constant. Restraint stresses exceed internal edge stresses from approximately the second day on and act mainly as a bending load on the plate. The criterion for the occurrence of a dividing crack through the whole plate is the attainment of uniaxial concrete tension strength $f_{c,t}$ in the neutral axis of the plate. This occurs for a short time on the 6th and 7th day in the early afternoon. While cracks on the bottom side can propagate once flexural strength $f_{c,fl}$ is exceeded on the

4th day, they do not occur on the surface because of flexural compression stresses.

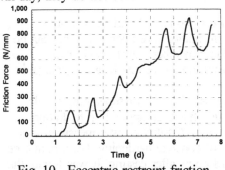

Fig. 10. Eccentric restraint friction force F_c, $f=1$

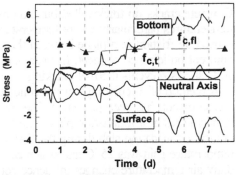

Fig. 11. Concrete stresses and strength, $f=1$

3.3 Effect of slide sheets

Slide sheets are being used increasingly to ensure the avoidance of cracks in thick-walled and long concrete members, Simons (1988), Schade (1989). They reduce external restraint stress which in our case make up appr. 70% of total thermal stress. Bitumen sheeting has proven particularly effective because hydration temperatures improve its viscous deformation behaviour.

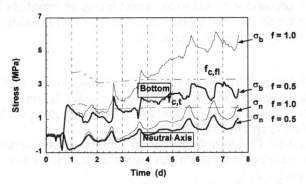

Fig. 12. Concrete stresses with varying slide sheet influence f, n = plate neutral axis, b = plate bottom

In the previous calculation the coefficient f was set at 1.0 assuming a rigid support. This value will mainly be influenced by the support conditions and the elasticity of soil. At the moment an exact estimation of external restraint stress reduction of slide sheets is not possible because suitable trials have yet to be carried out. Assuming a reduction of 50% (f=0.50), plate edge stress becomes lower than concrete flexural strength and neutral axis stress lower then uniaxial tensile strength. Fig. 12 shows this comparison between stresses in the plate bottom (σ_b) and the neutral axis (σ_n) for f=1.0 and f=0.50. However, in the foundation plate we studied, no dividing cracks actually occured.

4 Conclusion

Concrete cracks in thickwalled building members can be induced by hydration temperatures particularly where temperature deformations are restricted. This can be caused by friction in a plane between the building member and the soil or the adjacent building component. In many such cases the use of low heat cements and slide sheets prevents thermal cracking.

A method of calculating thermal concrete stresses based on temperature profile measurements has been presented. The calculation has been applied on hydration temperatures of a foundation plate. The temperature distribution was measured in a 55 cm thick foundation slab during the first week of concrete hardening. Maximum core temperature gain was 25 °C and was reached 18 hours after casting. It was found, that daily air temperature changes influence considerably concrete temperatures and stresses. The night cooling period particularly propagates into the concrete and reaches the plate bottom after a time lag of 8-10 hours. Concrete stresses were calculated from the temperature field assuming linear elastic material behaviour. With rigid support conditions, oscillating stresses should lead to dividing cracks 6 to 7 days after concrete casting. Thermal dividing cracks were avoided using slide sheets between foundation plate and soil. The degree of stress reduction of slide sheets is not yet completely understood. Further experiments will clarify this influence.

The method can be applied to building members which have a friction restraint on one surface like foundation slabs and walls, tunnel linings or watertight building members in contact with soil. The application to free walls on foundations has not been proved but should be possible in principle.

References

Heilmann H., Hilsdorf H., Finsterwalder K.: (1969), *Festigkeit und Verformung von Beton unter Zugspannung*, DAfStb. H.203, Ernst & Sohn, Berlin

Paulini P.: (1992), *A weighing method for cement hydration*. Proc. 9th ICCC., New Delhi, Vol.IV, pp.248-254

Reinhardt H.W.: (1984),*Verhalten des Betons im verformungsgesteuerten axialen Zugversuch*. Fortschritte im konstruktiven Ingenieurbau, Rehm-Festschrift, Ernst & Sohn, Berlin, pp.221-227

Rostásy F.S., Henning W.: *(1984), Zwang und Rißbildung in Wänden auf Fundamenten*. DAfStb. H.407, Beuth Verlag, Berlin

Schade D.: (1989), *Tiefgarage der Kongreßhalle Böblingen - eine weiße Wanne 10 m tief im Grundwasser*. Bautechnik, 66, pp.217-219

Simons H.J.: (1988), *Konstruktive Gesichtspunkte beim Entwurf weißer Wannen*. Bauingenieur, 63, pp.429-437

Springenschmid R.: (1991), *Risse im Beton infolge Hydratationswärme*. Zement-Kalk-Gips, 44.Jg., H.3, pp.132-138

Springenschmid R, Breitenbücher R.: (1992), *Influence of cement on thermal cracking of concrete at early ages*. Proc. 9th ICCC., New Delhi, Vol.V, pp.122-128

PART EIGHT

PRACTICAL MEASURES FOR AVOIDANCE OF CRACKING – CASE RECORDS

(Méthodes pratiques de prévention
de la fissuration –
Exemples d'applications)

42 REPORT ON CONSTRUCTION OF WATER-IMPERMEABLE CONCRETE STRUCTURES WITH HIGH-LEVEL GROUND-WATER ("WEISSE WANNEN" - "WHITE TROUGHS") IN BAVARIA

J.-St. KREUTZ
fink + kreutz, Beratende Ingenieure VBI, Nürnberg, Germany

Abstract
Perhaps the amount of reinforcement (load and restrain) has no major influence on the water-permeability of concrete structures. Reports about design and construction experience point to other influences: Specification of fresh concrete, handling of concrete after mixing, during and after construction, climate influences. Attack by ground water and soils. Influences of the reinforcement beside design for load-stress and restrain-stress, knowledge of the design-engineer about concrete, lack of communication between design-engineers and concrete-technology-engineers (Personal knowledge and coded rules). Influences of quality control, organisation, law and money. Suggestions for designing and constructing of watertight (water-impermeable) concrete components. Questions to the concrete scientists.
Keywords: Construction of Reinforced Concrete, Water-tight Concrete.

1 Introduction

The water-permeability of a structure is influenced by its design, materials and construction. In Germany a great effort in design work is done in calculating the amount of reinforcement in order to assure a maximum crack-width. A lot of discussion resulting in an enormous amount of papers deal with this special design problem.

By all new design rules in Germany the water-tightness of the lower floors of concrete buildings does not seem to increase. There is a lot of discussion going on before, during and after constructing of basements, mostly dealing with the amount of reinforcement of the groundslabs and the walls.

The simplified problem is: The concrete construction of a building is finished, but there is still a lot of water coming into the ground-floors on 'dry' days. After locating where the water comes into the building some questions about the concrete and its construction will be discussed. These discussions are attended by: Owner of the building or/and his attorney at law, architect, consulting structural engineer, engineers from the construction companies, sometimes also engineers from the concrete delivery company. The consulting structural engineer very often is in a bad position: The mistakes of the structural engineer can easily be proved because they are all in his plans and calculations. The mistakes from the construction can rarely be proved or cannot be seperated from one another. When some mistakes can be easily described, nobody takes care of the more complex mistakes and therefore nobody takes care of the real responsibility.

This paper wants to discuss some responsibilities for water permeabilities in concrete basements in Germany.

Thermal Cracking in Concrete at Early Ages. Edited by R. Springenschmid. Published 1994 by E & FN Spon, 2–6 Boundary Row, London SE1 8HN, UK. ISBN: 0 419 18710 3.

The building companies quite often try to avoid the contact between the structural engineer and the concrete delivery company because there are some corresponding interests. This report wants to show these corresponding interests to concrete technology engineers.

2 Planning and Calculating White Troughs in Germany

The normal problems with White Troughs were not dealt with in the last change of German code DIN 1045 in 1988. Also the additional rules in Heft 400 of DAfStb (German Council for Reinforced Concrete) created further confusion for constructing White Troughs. DIN 1045 and Heft 400 only deal with crack-width, not with any complete construction problem. But they are both rules which are strictly to be observed in Germany for every law problem within concrete constructions.

The calculation of average crack-width is the aim of different formulas and other rules in DIN 1045 (1988) and Heft 400 (1988). The idea is that with increasing reinforcement, decreasing bar diameters of the reinforcement, decreasing bar intervals and decreasing stress in the bars, the crack-width of the concrete wall or slab will be decreased. This is true and complete for slabs and beams and so on. But in my opinion it is not in White Troughs.

The dimensioning idea of DIN 1045 and Heft 400 (1988) for calculating the amount of reinforcement deals mostly with restraint of components. This restraint differs between longitudinal and bending. The calculation in DIN 1045 and Heft 400 also give very high amounts of reinforcement. Nevertheless there is some progress from the former DIN 1045 (1978 and older) which gave too thick concrete slabs, walls and beams for White Troughs because of the calculated design-stress.

3 Structural engineers Experience with Constructing White Troughs

The scepticism with calculating crack-width in order to assure watertightness of White Troughs has resulted from practical experience: There are very thin cracks, which can not be seen and their width cannot be measured, it is close to zero. Nevertheless considerable amounts of water can be observerd coming through these cracks. On the other hand there are cracks with widths over 0,35 mm and they bring no water at all. Also the costs of injection of cracks has no connection with the crack-width. And the cheapest repairing of cracks is the pasting over of wide cracks on the water-side using strips or coats.

Finite Element Calculations of White Troughs can calculate different stages of construction for the concrete parts, different shrinkage parameters, different curing and different stress history. Some consultants make a lot of these calculations, and because of this the software companies are developing these sorts of programs for civil engineers. For the prediction of the place of leakage in the White Troughs these calculations are of no help. The calculation of stresses resulting from shrinkage and other restraints is easy in theory, but difficult for a real construction especially for reinforced concrete constructions.

Summarizing the practical experience of 8 German consulting companies with White Troughs the amount of reinforcement (load and restraint) has no major influence on the water-permeability of concrete structures. There is a 30 year old white trough, which has less than 15% of the necessary amount of reinforcement in the water-side walls and slabs required by DIN 1045 nowadays. But even with different water-levels it is completely water-impermeable.

Reports about design and construction experience with White Troughs point to other influences:

3.1 Specification of fresh concrete
Most contractors buy concrete by their merchant or another person with no technical knowledge. They read for example B 25 on a drawing from the consultant and then they order the cheapest B 25 from a concrete delivery company.

Mostly the concrete-engineers do not know what sort of construction is to be built from their concrete.

There will be other authors in this RILEM symposium to talk about concrete mixing. Inspite of this symposium and other efforts there is an increasing difference in the knowledge about normal concrete items of the concrete-engineers and the site-engineers, contractors and consultant offices. Sometimes there is also not the same language in their talking together.

There are several important characteristics in White Troughs influenced or determined by effects of the specification of the fresh concrete. One of the most important influences is the strength of the concrete. In Germany for a B 25, a B 35 has to be taken into account for calculations of crack width, shrinkage and so on. Their is no discussion about avoiding or minimizing these superior strengths. Perhaps a B 25 can always be a B 25 or some B 25 must be B 45.

3.2 Handling of concrete after mixing, during construction and afterwards
The high amount of reinforcement very often gives no chance for normal methods of placing the concrete. Of course all these slabs and walls with problems placing concrete have problems with water-tightness.

The mistakes in the concrete construction often depend on the expansion joints and the joints between different construction stages. The handling of concrete of different ages needs much care and knowledge.

Perhaps expansion joints should be avoided in White Troughs. All the experience with repairing not watertight expansion joints ends up to more than 1000 German Marks for every meter. So every other construction is cheaper, especially when expansion joints have watertightness problems also after years.

The curing of the concrete in the walls and slabs is a problem in Bavaria for the last 15 years. And there has been no improvement in all these years.

3.3 Climate influences, especially the difference between summer and winter
There are differences in concrete construction in summer and in winter, which can be calculated by FE-analysis. And there are differences between summer and winter, which can be influenced by the handling of concrete: Same temperature of the fresh concrete, raining (which means high humidity) during the placing of the concrete and for some other days, great efforts of all people in constructing especially for curing.

Even if all these parameters are the same in summer and in winter and would bring the same thermal and shrinkage history for the concrete elements there are still differences between the constructions built in summer and built in winter: The winter constructions are all watertight, the summer constructions have a lot of wide cracks and other mistakes.

With the winter constructions there might be a problem with the high water levels during construction. But there are only 2 constructions which give evidence for this suspicion.

3.4 Attack by ground water and soils
Some mistakes with constructing White Troughs deal with chemical influences on the concrete by ground water and soils. In Germany there are some coded rules for these sorts of attacks. In general the problem has to be thought about during planning and construction.

3.5 Influences of the reinforcement beside design for load-stress and restrain-stress
According to the practical experience of 8 German consultants the amount of reinforcement has no major influence on the water-tightness of concrete structures. Of course there must be a true design for load-stress and restraint-stress. For the rest of the design it would be useful to know an overall amount of reinforcement which has to be under every surface of the concrete structures in both directions. The total amount of this reinforcement without load-stress and restrain-stress should not exceed a volume-percentage of 1.25. In some cases also 0.75 volume-percent may be enough.

All additional reinforcement should be as short as possible in order to assure the best distribution of load-tension cracks.

According to experience on construction sites the amount of reinforcement has various detrimental influences on water-impermeability, not to talk about the placing of the concrete. Perhaps the concrete gets cracked during early age shrinking according to the amount of reinforcement (Reinforcement under pressure).

3.6 Knowledge of the design-engineer about concrete
Its quite sad to see how many design-engineers in the consultant offices are completely lacking in knowledge of concrete technology. When there is knowledge about concrete then there is often not enough courage to be satisfied with reinforcement far below DIN 1045 and Heft 400. And when there is knowledge and courage, then a Bavarian checking engineer will specify much more reinforcement.

3.7 Lack of communication between design-engineers and concrete-engineers (Personal knowledge and coded rules)
The lack of communication and information between design-engineers and concrete-engineers starts after graduation. The civil engineering journals seldom try to bring knowledge "to the other side". There are journals for structural engineering and there are journals for concrete engineering. And they all are very scientific, of course.

3.8 Influences of quality control, organisation, law and money
Quality control of contractors never deals directly with the quality of the structures. This sort of quality management always thinks about processes and functions. And the auditors for the quality controlling system do not care about constructions and concrete. So it is evident that in Germany quality control influences water-tightness of concrete walls and slabs only indirectly.

When the discussions about the reasons for the water penetration into the building are going on, the mistakes of the contractors can rarely be proved or they cannot be seperated from the mistakes of the consultants. So for the contractors solicitor it is quite easy to do his work.

If in the last 20 years in Germany 10 percent of the amount and costs of the reinforcements of water-impermeable concrete constructions would be saved and invested in better curing and controlling of the concrete, what would be different now ? The water-tightness would be perhaps no major question in Germany.

3.9 Effects arising from the Owner of the Building

Most owners of the building don't think enough about the usage of building in planning or construction: He or she should give information about the real requirements for the water tightness: A garage needs less care than a computer room in the ground floors. In a garage the quantity of the incoming water could be restricted to a certain amount saving money in the construction. Of course in a computer room the quantity of incoming water must be restricted as much as possible.

Every contractor and consultant has great time pressure in the office as well as on the construction site. Every owner should know that more time for planning and constructing saves money and brings better results in the end.

Some aspects of time planning should also be taken into account: The pumps should stop as early as possible in order to bring back the normal level of ground water to initiate the inversion of shrinkage by bringing water/humidity in the ground slabs and walls. On the other hand the injection of cracks should be done as late as possible. Every service load of the building and all kinds of creep of the concrete and soil influences cracks and their width. There are many more influences from the water level and the times of its changing without going deeper into this subject.

4 Suggestions for designing and constructing of watertight concrete components.

Consultants feel not responsible for the details of concrete technology. But all consulting engineers as well as contractor engineers should know about the main parameters of concrete technology. Most of all this work is done by civil engineers, so it might be no structural problem, to discuss the problems of the construction and scientific work. The consultant should always try to talk to the concrete-engineer in charge when constructing White Troughs.

High amounts of cement as well as reinforcement must be avoided. Perhaps special concrete mix or special cement sorts can help to solve or reduce problems. All the items about cement and concrete and mixing and so on may be discussed by other authors on this symposium or elsewhere.

Codes and construction rules for White Troughs must depend on the experience with concrete structures and its elements concrete and steel. If the injection of cracks is taken into account there must be different design rules in Germany and perhaps also in Eurocode 2 (1992) here to be modified.

Calculating maximum crack-width and a minimum-reinforcement is not enough for the water-tightness of concrete structures. There must be more help for simple construction and more control on the construction site. This controlling should be done by independent engineers. They can also help with the responsibility of mistakes.

5 Questions to the concrete scientists.

In Germany a lot of scientific work is done with concrete and its behaviour. For a normal practical constructing engineer the knowledge of the scientists can not be used. So every scientist should think about the application of his knowledge by other civil engineers.

The influence of reinforcement on the water-tightness of the concrete constructions seems different if there is not enough load-stress or restrain-stress in the bars.

There is no complete conclusion about the difference in constructing White Troughs in summer and in winter.

The amount of cement, the sort of cement and the compensation of parts of cement by blast-furnace slag materials for White Troughs are discussed by concrete scientists. The results are not clear and simple for the public.

Prescriptions for typical concrete for White Troughs according to the sort of cement and the other ingredients of concrete should be published. Every "Designer" of a German WU-Concrete (water-impermeable concrete) should have the possibility of predicting the amount of shrinkage under site conditions.

43 ON THE RELIABILITY OF TEMPERATURE DIFFERENTIALS AS A CRITERION FOR THE RISK OF EARLY-AGE THERMAL CRACKING

M. EBERHARDT
Technische Hochschule Darmstadt, Darmstadt, Germany
S.J. LOKHORST and K. van BREUGEL
Delft University of Technology, Delft, The Netherlands

Abstract
Many decades already temperature differentials in hardening concrete
have been used as a criterion for the risk of early-age thermal
cracking. Practical experience and theoretical considerations, however,
have given rise to doubt the validity of such a criterion. In this paper
the results are discussed of a parameter study on the risk of thermal
cracking in a wall and a wall-slab element. For this study a numerical
simulation program was used. The basic structure of this program is
briefly outlined. The most important factors affecting the risk of
cracking are discussed and conclusions are drawn as regards the validity
of temperature differentials as a criterion for early age thermal
cracking.
Keywords: Crack Criterion, Cracking, Numerical Simulation, Relaxation,
Temperature Differentials.

1 Introduction

Early-age thermal cracking of concrete may jeopardize the functionality
and durability of a structure. Criteria for prevention of thermal
cracking have been specified in terms of allowable temperature
differentials. Temperature differentials in a cross section or between
different parts of a concrete structure should not exceed 15 to 20°C,
respectively. As early as in 1945 it has been noticed already by Rawhouser that "The number of degrees of change in temperature is only one
of the several factors which determine the cracking tendency of mass
concrete". He then continues with saying: "It is almost meaningless to
determine the number of degrees of temperature drop a concrete can withstand without considering these other factors". Rawhouser's statement
was mainly based on practical experiences. Almost half a century after
this experience-born statement Emborg concluded from his laboratory
tests and theoretical considerations in a similar way that: "It is of
crucial importance to realize that the complexity of thermal stress
analysis in early age concrete is of such a dignity that the monitoring
of temperatures as such may never constitute anything but a crude way of
controlling cracking in concrete structures". Factors which complicate
the judgement of early-age thermal stresses and the risk of cracking
are, among other things, the degree of restraint experienced by the
concrete, the temperature history and the relaxation properties of the
concrete. For an unfavourable combination of these factors cracking may
occur notwithstanding the fact that afore mentioned temperature criteria
are met. The reason why the temperature criterion in not a viable in all

Thermal Cracking in Concrete at Early Ages. Edited by R. Springenschmid. Published 1994
by E & FN Spon, 2–6 Boundary Row, London SE1 8HN, UK. ISBN: 0 419 18710 3.

cases originate from the fact there is no linear relationship between
temperature and stresses in early-age concrete. It are the non-linear
deformational properties of early-age concrete which substantially and
fundamentally affect the temperature-stress relationship. The
consequences of non-linearity between temperature and stress development
have been emphasized several times by, for example, Bernander. With the
help of today's powerful hardware numerical simulation of the complex
behaviour of early-age concrete becomes possible. For the prediction of
thermal stresses and the risk of cracking these simulation programs
should be able to determine the temperature distributions, development
of strength and stiffness and the creep, c.q. relaxation behaviour. The
question is as to whether the temperatures and the relevant materials
properties, which are continuously changing during hydration, can be
described sufficiently accurate so as to guarantee reliable predictions
of the thermal stresses and risk of cracking.

2 Structure and aim of the research project

In order to get a better understanding of the development of both
strength, thermal stresses and risk of cracking as a function of
internal and *external* influencing factors an extensive parameter study
was conducted. The study was carried out with the computer codes
"TEMPSPAN" and "CRACK". With TEMPSPAN the development of both the
hydration process and the temperature is calculated. With CRACK the
development of strength, thermal stresses and the risk of early-age
cracking is determined. In these codes the (calculated) degree of
hydration is used for the description of the development of materials
properties, including early-age relaxation.

   In the parameter study two types of structures were considered, viz.
a wall and a T-shaped element (wall-slab configuration). The parameters
dealt with in the study are the type of cement (slow, ordinary, rapid
hardening), initial temperature of the mix, the thickness of the wall,
time of demoulding, ambient temperature, wind velocity and degree of
restraint. The final goal is to show that the resolution of modern simulation programs has reached a level that enables the prediction of the
effect of the major influencing factors with such an accuracy that these
programs can be judged as reliable tools for making engineering
decisions in the design stage of a project as regards different types of
measurements which aim to prevent thermal cracking.

3 Numerical simulation of the risk of thermal cracking

3.1 Basic structure of the numerical simulations program

Concrete will crack when the imposed strains exceed the ultimate strain
or when the stresses exceed the ultimate strength. The question as to
whether a strain criterion or a strength criterion must be adopted, is -
to a certain extent - not relevant since for numerical modelling
stresses and strains must be considered as interrelated quantities.
Working with a strength criterion the probability of cracking can be
determined from the expression:

$$P_{crack} = P\{\sigma(t) > f_{ct}(t)\}$$

in which $\sigma(t)$ is the concrete stress at time t and $f_{ct}(t)$ the tensile strength at that time. Pending better procedures, the superposition method (Fig. 1) is adopted for determination of the thermal stresses. In case of full restraint thermal stresses can be calculated with (van Breugel):

$$\sigma(t) = \sum_{i=1}^{n} \Delta\sigma(t,\tau_i) = \sum_{i=1}^{n} \Delta T_i * \alpha_c * E(\tau_i) * \psi(t,\tau_i)$$

with: ΔT_i temperature increment in time step $\Delta\tau_i$
        α_c coefficient of thermal expansion
        $E_c(\tau_i)$ modulus of elasticity of concrete at time τ_i
        $\varphi(t,\tau_i)$ relaxation factor pertaining to a particular stress
                  increment $\Delta\sigma(\tau_i)$

The degree of restraint may vary between 0 and 100%, depending on the physical boundary conditions, i.e. the geometry of the structure. The degree of restraint is not known at the onset of a calculation as it is a function of the overall deformational behaviour of the structure. It has, therefore, to be adjusted during the calculation, taking account of the overall structural deformations.

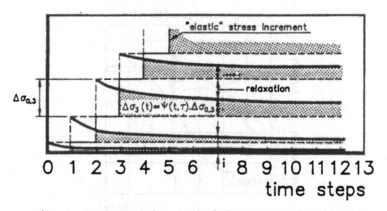

Fig. 1: Superposition method for stress development

3.2 Prediction of temperature and degree of hydration
Temperatures are predicted taking account of external factors of influence and the mix properties. The actual rate of hydration, i.e. of heat evolution, in an arbitrary point of the structure is calculated as a function of the local temperature and the momentary degree of hydration. In this procedure the adiabatic hydration curve of the mix was used as a reference curve. From this curve the actual rate of hydration is calculated in a step wise procedure. For adjustment of the rate of hydration an Arrhenius-like expression is used in which the apparent activation energy is considered to be a function of the temperature, the degree of hydration and the type of cement (see van Breugel).

3.3 Strength and modulus of elasticity
The compressive strength is related to the calculated degree of hydration $\alpha(\tau)$ making use of a relationship proposed by Fagerlund. The

tensile strength $f_{ct}(\tau)$, in turn, is related to the compressive strength $f_c(\alpha(\tau))$ according to:

$$f_{ct}(\tau) = c * [f_c(\alpha_h(\tau))]^{2/3} \qquad [MPa]$$

in which c is a constant. For the modulus of elasticity a three-phase model (aggregate, unhydrated cement and gel) was used. By differenting between hydrated and unhydrated cement it is possible to write the E-modulus as a function of the degree of hydration $\alpha(\tau)$.

3.4 Relaxation of early-age concrete

For the relaxation of early-age concrete an expression is used in which the effect of the increase of the degree of hydration, i.e. of micro-structural development, is allowed for explicitly, viz. (van Breugel):

$$\psi(t,\tau) = \exp\left(-\left[\frac{\alpha_h(t)}{\alpha_h(\tau_i)} - 1 + \omega_0^{1.65} \cdot \tau_i^{-d} \cdot (t-\tau_i)^n \cdot \frac{\alpha_h(t)}{\alpha_h(\tau_i)}\right]\right)$$

in which the factor n varies between 0.3 for compression to 0.6 in tension and d from 0.3 to 0.4 for a low heat and a rapid hardening cement respectively. ω_0 is the w/c ratio of the mix.

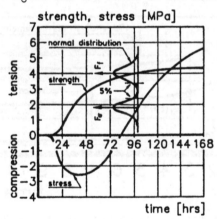

Fig 2: Probability of cracking - schematic representation of the calculation procedure

3.5 Probability of cracking

The calculated tensile strength and thermal stresses were assumed to be mean values of normally distributed quantities. The standard deviation of the tensile strength was estimated at 0.5 MPa, whereas the standard deviation of the stresses were assumed 20% higher, viz. 0.6 MPa. The procedure for the calculation of the probability of cracking is sche-matically shown in Fig. 2. For the sake of simplicity and in the absence of reliable data on this point, post-peak softening of young concrete has not been considered in the calculations. The effect of autogenous shrinkage has been included in the calculation (see Tab.1).

4   Probability of cracking - Parameter studies

4.1   List of parameters

In order to trace the conditions under which thermal cracking is most likely to occur a parameter study was carried out. Two types of structures were considered, viz. a wall and a wall-slab structure as indicated in Fig. 3. Parameter values have been summarized in Table 1 (see also Eberhardt).

In order to limit the number of calculations only axissymmetric conditions, i.e. symmetric temperature distributions, were considered. The calculations were limited to the first 168 hours after casting.

Table 1: Parameters considered in cracking sensitivity study

Parameter	Wall	Wall-slab structure
Cement type	Ordinary (PC-A) Rapid hardening (PC-C) Low heat (BFS-cement)	Ordinary (PC-A) Rapid hardening (PC-C)
Cement content	350 kg	350 kg
Water/cement ratio	0.5	0.5
Autogenous shrinkage	$20 \ 10^{-6}$ at t = 20 hrs $65 \ 10^{-6}$ at t = 80 hrs	$20 \ 10^{-6}$ at t = 20 hrs $65 \ 10^{-6}$ at t = 80 hrs
Ambient temperature	0, 10, 20, 30°C	10, 20°C
Wind velocity	1, 2, 3, 4, 5 m/s (for PC-A and PC-C)	0, 5 m/s
Wall thickness	0.5, 0.8, 1.0, 1.5, 2.0 m	0.25, 0.5, 1.0 m
Slab temperature	20°C (constant)	20°C (constant)
Mix temperature	15°C	10, 15, 20°C

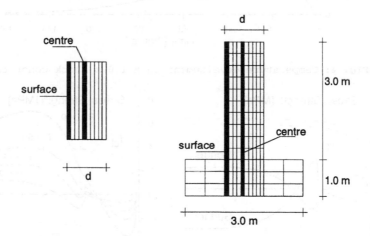

Fig. 3: Structural elements considered in parameter studies

4.2   Probability of cracking in concrete wall

General trends found in the parameter study of the wall were all in agreement with the expectations. Self-equilibrating stresses increased with:

a. decreasing ambient temperature
b. increasing wall thickness
c. early removal of formwork
d. reactivity of the cement (i.e. increasing fineness)
e. increasing wind velocity

An example of a temperature calculation is presented in Fig. 4. In this calculation a rapid hardening cement (PC-C) was used. The tensile strengths in the centre and at the surface of the wall are presented in Fig. 5 together with the self-equilibrating stresses. Even though the maximum temperature differential in the cross section reached up to 19°C, the stresses remained substantially below the tensile strength values. This result is in good agreement with literature data and practical experience.

Of major importance in view of cracking is the degree of restraint, as can be deduced from Fig. 5. Other parameter values being the same as in Fig. 5, cracking is likely to occur for a degree of restraint of 50% and higher.

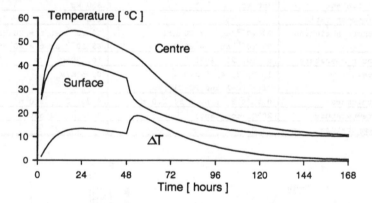

Fig. 4: Temperature development in a 1.0 m thick concrete wall

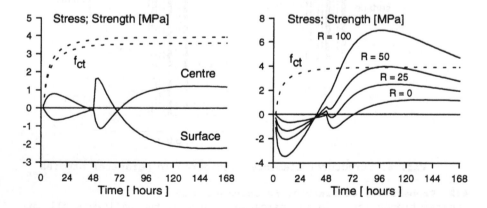

Fig. 5: Strength and stress in a concrete wall. Left: only Eigenstresses. Right: effect of degree of restraint (R).

4.3 Probability of cracking in a T-shaped wall-slab element

In this study a practical situation is simulated of a wall cast on a concrete slab (Fig. 3). For a degree of restraint of 0% the element is free to bend and deform in axial direction. At 100% restraint both bending and axial deformations are completely prevented.

In contrast with the findings of the preceding case, eigen-stresses could now lead to cracking. The highest stresses emerge in a section just above the wall-to-slab joint. Walls thicker than 0.5 m were likely to crack in most cases. In thin walls wind may have a positive effect since it reduces the peak temperatures and hence the temperature differentials between wall and slab. The risk of cracking increased for higher initial temperatures of the mix.

In all cases where the temperature differential between wall and slab was less than 10...15°C, cracking did not occur. Hence, for judgement of the risk of cracking caused by eigen-stresses the 20°C-rule appeared to be applicable.

The risk of cracking increased substantially with increasing the degree of restraint. For a 0.25 m thick wall a temperature differential between wall and slab of 12...15°C was found with a corresponding probability of cracking of 32% for a degree of restraint of only 25%. The development of temperatures, strength and stresses and the probability of cracking in the 0.25 m thick wall are presented in Fig. 6 and 7.

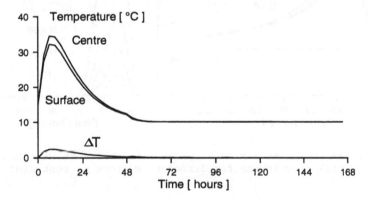

Fig. 6: Temperature in a 0.25 m thick wall

5 Conclusions

A study was carried out with the aim to get a better insight in the (absence of a) correlation between temperature differentials in hardening concrete and the probability of cracking. For this purpose an extensive parameter study was performed from which the following conclusions can be drawn:

    a. Wind velocity and degree of restraint are most important factors in view of thermal cracking.

    b. In a hardening concrete wall, which is free to deform, thermal eigen-stresses are not likely to cause cracks as long as the temperature differentials remain below 15°C.

    c. In hardening concrete elements which are cast against a hardened or stiff element thermal eigen-stresses in the composite element, e.g. a T-shaped wall-slab beam, may give rise to cracking even though the temperature differentials are less than 15°C.

The general conclusion of this parameter study is that the risk of thermal cracking in early-age concrete cannot merely be related to the temperature differentials which may occur. This finding is in agreement with earlier statements on this point of Rawhouser and Emborg. The studies show that the resolution of today simulation programs for prediction of the risk of cracking is sufficiently high to serve as a basis for decisions as regards the most suitable mixes, casting sequences and curing conditions.

It is noticed that in this study only the risk of temperature-induced cracking has been considered. In practice the effect of drying shrinkage may affect the findings of this study. Also the effect of the type of concrete, i.e. its relaxation properties, require further research. This, however, shall not be deducted from the potential of numerical simulations of the behaviour of early-age concrete.

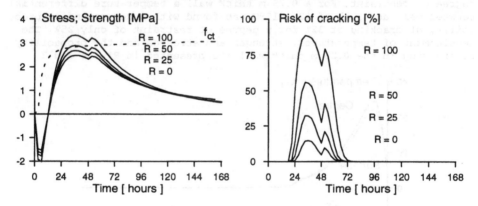

Fig. 7: Strength and stress development in a 0.25 m thick concrete wall (left) and risk of cracking for different degrees of restraint (right).

References

Bernander, M. (1982) International conference on concrete at early ages, Paris, Vol.2, pp.218-221.
Eberhardt, M. (1993) **Rißgefahr in jungem Beton** (Probability of cracking in early-age concrete), Msc-Thesis, Delft.
Emborg, M. (1989) **Thermal stresses in concrete structures at early ages.** PhD-Thesis, Luleå Univ. of Technology, 285 p.
Fagerlund, G. (1987) Relations between the strength and the degree of hydration or porosity of cement paste, cement mortar and concrete, in **Sem. on Hydration of Cement**, Copenhagen, 57 p.
Rawhouser, C. (1945) Cracking and temperature control of mass concrete, in **J. American Concrete Institute**, Vol.16, No.4, pp.306-346.
Van Breugel K. (1991) **Simulation of hydration and formation of structure in hardening cement-based materials**, PhD, TU-Delft, 295 p.

44 WHY ARE TEMPERATURE-RELATED CRITERIA SO UNRELIABLE FOR PREDICTING THERMAL CRACKING AT EARLY AGES?

M. MANGOLD and R. SPRINGENSCHMID
Institute of Building Materials, Technical University of Munich,
Germany

Abstract
Temperature-related criteria are commonly used to control the risk of
cracking on site. However, practical experience proves that these
criteria are often unreliable. An important reason for this uncertainty is due to the fact, that the zero-stress temperature (i.e. the
temperature at which stress of the restrained concrete is zero) is
usually different over the cross-section of a member ("zero-stress
temperature gradient"). Restraint and intrinsic stresses of the member
are zero only if the actual temperature gradient is the same as the
zero-stress temperature gradient. For an arbitrary state of concrete
temperatures the cracking risk may therefore be high or low as well,
depending on the shape of the zero-stress temperature gradient. For
the latter, the hardening temperatures during the first day of hydration (i.e. when the degree of hydration is still low) are decisive.
Keywords: Zero-Stress Temperature, Intrinsic Stresses, Crack
Criterion.

1 Introduction

When thermal stress analysis is carried out at the design stage of a
structure, the stress computations are based on code rules, assumptions and/or results from material tests (e.g. creep tests). From the
magnitude of calculated stress the risk of cracking is estimated in
advance. However, it is difficult to verify stress calculations and
the risk of cracking on site. On the one hand, in situ conditions do
not necessarily agree with the usually simplified assumptions made for
stress calculations. On the other hand, in situ stress measurement
requires know-how, time and money. Because of this in situ stress
measurement is usually carried out only within the scope of research
projects.

Unlike stresses, the in situ measurement of concrete temperatures
is simple and requires only little expenditure. Therefore, temperature
measurements are a common practical tool to estimate thermal stresses
and to assess the risk of cracking on site. For example, in order to
avoid splitting cracks it is often specified that the difference
between the mean temperatures of the restrained and the restraining
concrete member should be less than 20 K or 25 K. To avoid surface

Thermal Cracking in Concrete at Early Ages. Edited by R. Springenschmid. Published 1994
by E & FN Spon, 2–6 Boundary Row, London SE1 8HN, UK. ISBN: 0 419 18710 3.

cracks which are due to intrinsic stresses (or also called "self-equilibrating stresses" or "eigenstresses") it is mostly required that the difference of concrete temperatures at the surface and in the core of the member should be 20 K at most. For massive structures of more than 2 m in thickness it is recommended that this temperature difference should be less than 15 K, see Wischers (1965).

Practical experience proves, however, that these temperature-related criteria are often only a crude method of assessing the cracking risk. Even if temperature requirements are met, systematic thermal cracking of the young concrete may occur, see e.g. Dierks (1982). In other cases, only few thermal cracks are observed even though specified temperature differences are not met by far, see e.g. Hampe (1944). To find out the reasons for this, a closer look has to be taken at the prerequisites of temperature related criteria. This paper emphasizes that the magnitude of tolerable temperature differences depends very much on the temperature development which occurs at a very early age of hydration, in particular during the first day.

2 Zero-Stress Temperature and Stress-Inducing Temperature Differential

Concrete members with a thickness of 30 cm or more heat up considerably due to the heat of hydration of the cement. During the heating phase the concrete's modulus of elasticity still increases and relaxation is high. If the concrete hardens under restraint, the hindered expansion during the temperature increase is largely transformed into plastic deformation of the concrete and only in little compressive stresses. As the degree of hydration and thus the stiffness of the concrete increases, more and more of the hindered deformation is transformed into stresses ("elastic deformation") and not into plastic deformation any more. When the temperature maximum is exceeded and the concrete cools down again considerable tensile stresses are produced so that the compressive stresses rapidly decrease. The concrete is stress-relieved at the zero-stress temperature $T_{Z,2}$ (Fig. 1). In cracking frame tests it is possible to quantify zero-stress temperatures from the onset of hydration. To the rule $T_{Z,2}$ is considerably higher than the initial temperature and often only a few degrees lower than the maximum temperature. However, $T_{Z,2}$ as shown in Fig. 1 is only a single point within the whole range of the temperature development, at which stress of the restrained concrete is actually zero. In fact, corresponding to any arbitrary state of stress $\sigma(t)$ there is also a value $T_Z(t)$, the "actual zero-stress temperature", at which the actual stress would be zero under the given degree of restraint (Fig. 1). If the actual **concrete temperature T(t)** is higher than the actual **zero-stress temperature $T_Z(t)$** compressive stresses arise in the case of restraint. If T(t) is lower than $T_Z(t)$ tensile stresses arise. During the first 3 to 6 hours the hindered expansion is completely transformed into plastic deformation of the concrete. During this period the actual zero-stress temperature is the same as the actual concrete temperature (Fig. 1). $T_{Z,1}$ is the transition from plastic to viscoelastic behaviour of the concrete (Fig. 1). From then on one portion of the hindered deformation results in stress ("elastic deformation"), the other part results in inelastic deformation of the

concrete (relaxation, microcracking) and not in stress. As the degree
of hydration, and thus the concrete's stiffness increases, the more of
the hindered expansion is transformed into stress and the smaller is
the portion of the inelastic deformation. Therefore, the course of the
zero-stress temperature T_z deviates more and more from the course of
the concrete temperature and approaches an almost constant value.

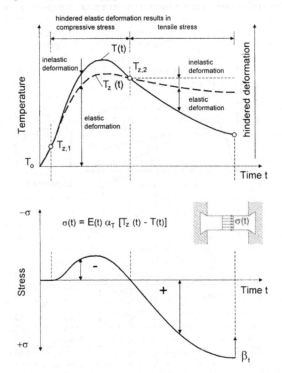

Fig. 1.   Development of concrete temperature $T(t)$, restraint stress
$\sigma(t)$ and zero-stress temperature $T_z(t)$ if the concrete hardens under complete or a high degree of restraint. Stress
arises due to the elastic portion of the hindered deformation. The inelastic portion does not result in stress.

In concrete members the temperature development at the surface and in
the center is different due to the heat dissipation. Therefore also
the stress development and the zero-stress temperatures at the surface
and in the center of the restrained member are different. Considering
the whole cross-section, the member hardens with a curved zero-stress
temperature gradient (Fig. 2). The temperature difference between the
actual temperature gradient and the actual zero-stress temperature
gradient leads to thermal stresses. Because the zero-stress temperatures at the surface and in the center of the member still develop
during hydration (see Fig. 1) also the zero-stress temperature gradient is not constant during the early stage of hydration ("actual
zero-stress temperature gradient"). Concrete temperatures during the
first day of hardening are therefore decisive for the magnitude of the

zero-stress temperature, see Mangold (1994). If the concrete hardens
at high zero-stress temperatures (e.g. on hot summer days) the thermal
tensile stresses and thus also the crack sensitivity of the member
will be high. If zero-stress temperatures are low, also thermal tensile stresses and the crack sensitivity are low. So, when judging
thermal stresses and the cracking risk from temperature measurements
one has to take into account the zero-stress temperature gradient.

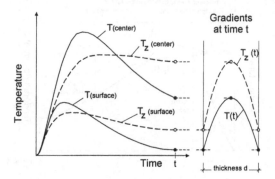

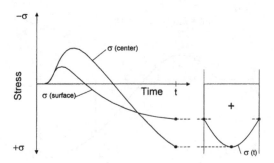

Fig. 2.   Development of the concrete temperature (T), zero-stress
temperature (T_Z) and stress (σ) at the surface and in the
center of a restrained member. Gradient of T, T_Z and σ over
the cross-section of the member at time t.

Commonly it is assumed that there is a constant zero-stress
temperature over the cross-section of a member (Fig. 3a). If the
temperature at the surface is lower than in the center, tensile
intrinsic stresses at the surface and compressive intrinsic stresses
in the center are determined from this assumption.

In fact, however, the zero-stress temperature gradient is curved.
If the curvature of the zero-stress temperature gradient is more
convex than the curvature of the temperature gradient, compressive
intrinsic stresses are present at the surface even though the temperature at the surface is lower than in the center (Fig. 3b). If the
ambient temperature is reached again compressive intrinsic stresses
remain at the surface in that case (Fig. 3c). Such a favourable zero-
stress gradient occurs e.g. if the surface of the member is cooled
from a very early age on, see Mangold (1994).

However, the zero-stress temperature at the surface may also be
higher than in the center (concave curvature of the zero-stress temperature gradient). This may happen if the young concrete surface heats
up due to solar radiation, e.g. in the case of a concrete slab or
pavement which is placed in the morning of a hot summer day. In this
unfavourable case tensile intrinsic stresses are present at the surface even if the heat of hydration has dissipated (Fig. 3d). Then the
crack sensitivity of the member is high.

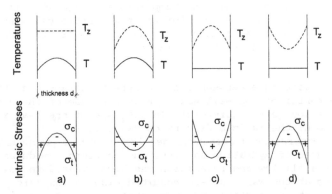

Fig. 3.   Intrinsic stresses over the cross-section of a concrete wall
(thickness d) in the case of
a)   equal zero-stress temperature over the cross-section
(simplified assumption) and convex temperature gradient
b)   convex zero-stress temperature gradient and convex
temperature gradient
c)   convex zero-stress temperature gradient after the heat
of hydration has dissipated
d)   concave zero-stress temperature gradient after the hest
of hydration has dissipated

3 Case Study - Influence of the Thermal Properties of the Formwork on Zero-Stress Temperatures and Thermal Stresses

3.1 Laboratory Tests
Zero-stress temperatures and thermal stresses in concrete members can
be quantified by laboratory tests with the cracking frame, see Mangold
(1994). At the Institute of Building Materials (Technical University
Munich) it was investigated how the thermal properties of the formwork
affect the magnitude of thermal stresses and zero-stress temperatures
in concrete walls of 1 m thickness during hydration. To simulate the
hardening process of the concrete wall in the cracking frame tests,
each beam specimen was heated according to the temperature development
in one portion of the cross-section of the wall, e.g. as in the center
or at the surface of the wall (The temperature development was calculated in advance with a finite-element program). The development of
stresses within the beam specimens was measured and from that the
zero-stress temperatures as well as the development of restraint and
intrinsic stresses within the wall could be calculated (a detailed

description is given in Mangold (1994)). The following cases were investigated:

Case 1: The heat transfer at the wall surface is inhibited by a heat insulating formwork. The coefficient of heat transfer compares to a 2 cm thick wooden formwork. After 48 hours the formwork is stripped.

Case 2: The heat transfer is almost unhindered as in the case of a steel formwork.

In both cases a constant ambient temperature of $T_a=20°C$ and a wind velocity of $v=2$ m/s was assumed for the temperature calculations. The temperature of the fresh concrete was assumed to be 20°C. The wall was protected from the influence of solar radiation.

3.2 Results

The maximum difference between the concrete temperature at the surface and in the center of the wall was 20 K in the case of the unhindered heat dissipation (steel formwork). In the case of the heat insulating formwork (wood) the maximum difference was 13.5 K.

These temperature differences developed from a very early age on when the degree of hydration and thus the concrete's modulus of elasticity was still low and relaxation was high. Therefore low concrete temperatures at the wall surface resulted also in low zero-stress temperatures at the surface and in a strongly convex curvature of the zero-stress temperature gradient. When the maximum difference between the concrete temperature at the surface and in the center occured, the zero-stress temperature at the wall surface was 6 K lower than in the center in case 1 and 13.5 K lower in case 2 (Fig. 4).

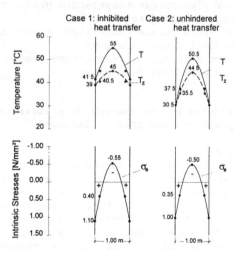

Fig. 4. Heat dissipation from a 1 m thick wall which hardens under restraint: case 1 in the case of a heat insulating (wood) and case 2 non insulating (steel) formwork. Gradient of concrete temperatures T and zero-stress temperatures T_z when tensile intrinsic stresses σ_e at the surface reach their maximum values.

This had consequences for the magnitude of thermal stresses:
The magnitude of intrinsic stresses is determined by the temperature
differential $\Delta\Delta T$ (t)

$$\Delta\Delta T \text{ (t)} = \Delta T \text{ (t)} - \Delta T_Z \text{ (t)} \tag{1}$$

where

ΔT difference between the actual concrete temperatures T at the surface and in the center of the member

ΔT_Z difference between the actual zero-stress temperatures T_Z at the
surface and in the center.

In the case of the heat insulating formwork (case 1) max $\Delta\Delta T$ was 13.5
K - 6 K = 7.5 K . In the case of the non-insulating formwork (case 2)
max $\Delta\Delta T$ was 20 K - 14 K = 6 K. Because of this in case 2 the maximum
tensile intrinsic stress was even slightly lower (1.00 N/mm²) than in
case 1 (1.10 N/mm²), although the difference between the concrete
temperatures at the surface and in the center of the wall was 6.5 K
higher in case 2. In both cases tensile intrinsic stresses were about
40 % of the actual concrete tensile strength.

The intrinsic stresses were zero if the curvature of the temperature
gradient and the zero-stress temperature gradient were the same
(Fig. 5). Then only longitudinal restraint stresses were present, even
though concrete temperatures at the wall surface were 8 K (case 1) and
12.5 K (case 2) lower than in the center.

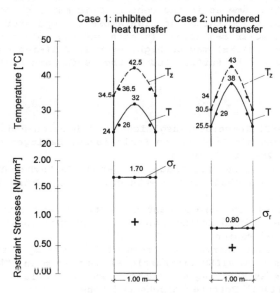

Fig. 5.   1 m thick wall as shown in Fig. 4 - Gradient of concrete
           temperatures T, zero-stress temperatures T_Z and restraint
           stresses σ_r at the time when intrinsic stresses are zero
           (after 4 days in case 1 resp. after 2.5 days in case 2)

4 Conclusions

An important reason for the uncertainty of temperature related criteria is due to the fact that the zero-stress temperature gradient is
not taken into account.

The cracking frame tests carried out at the Institute of Building
Materials revealed the following:
- The magnitude of the thermal stress gradient is determined by the
  difference between the actual temperature gradient and the zero-
  stress temperature gradient. If the curvature of the temperature
  gradient and the zero-stress temperature gradient are the same, no
  intrinsic stresses are present. Thus, the risk of surface cracks may
  be zero even if the measured concrete temperature at the surface is
  lower than the temperature in the center of the member.
- Zero-stress temperatures and the shape of the zero-stress temperature gradient depend mainly on the hardening temperatures of the
  concrete at a very early age (predominantly on the temperature development during the first day).
- It is favourable if the concrete surface hardens at a low zero-
  stress temperature (convex zero-stress temperature gradient). The
  tolerable difference between the concrete temperature at the surface
  and the center is much higher then as has been estimated up to now.
  Surface cooling to achieve such a favourable zero-stress temperature
  gradient is described by Mangold (1994a).
- If the zero-stress temperature at the surface is higher than in the
  center (concave zero-stress temperature gradient), tensile intrinsic
  stresses are present at the surface even if the temperature at the
  surface is the same as in the center. This may e.g. occur with slabs
  or pavements, which harden at high ambient temperatures under the
  influence of solar radiation. In this unfavourable case the cracking
  risk (surface cracks) may be high, even if differences in concrete
  temperatures at the surface and in the center are small.

5 References

Dierks, K. (1982) Temperaturausgleich in Stahlbetonwänden - ein
    Versuch gegen die Bildung von Spaltrissen, **Bauingenieur**, 57, Vol.
    3, pp. 257-264 (in German)
Hampe, B. (1944) Temperaturschäden im Beton, Deutscher Ausschuß für
    Stahlbeton. Ausschuß für Massenbeton, Vol. 1, Ernst & Sohn, Berlin,
    (in German).
Mangold, M. (1994) The development of restraint and intrinsic stresses
    in concrete members during hydration, Doctoral thesis, Technical
    University of Munich.
Mangold, M. (1994a) Thermal Prestress of Concrete by Surface Cooling,
    **Proceedings from RILEM International Symposium on "Thermal Cracking
    in Concrete at Early Ages"**, Munich, October 10-12.
Wischers, G. (1965) Betontechnische und konstruktive Maßnahmen gegen
    Temperaturrisse in massigen Bauteilen. **Betontechnische Berichte
    1964**, Beton-Verlag, Düsseldorf, pp. 21-58 (in German).

45 INHERENT THERMAL STRESS DISTRIBUTIONS IN CONCRETE STRUCTURES AND METHOD FOR THEIR CONTROL

A.R. SOLOVYANCHIK
Institute of Transport Construction, Moscow, Russia
B.A. KRYLOV and E.N. MALINSKY
Institute of Concrete and Reinforced Concrete, Moscow, Russia

Abstract
Results of scientific research on the inherent thermostress formation peculiarities, are presented along with the practical recommendations of data obtained by casting in situ structures and the manufacturing of precast reinforced concrete members to provide an increased crack resistance.

"Inherent" or "residual" stresses in concrete structures means the stresses existing in the bulk of the structures under homogeneous temperature distribution. At a uniform temperature distribution several other types of residual temperature stresses also exist in mass concrete bodies. Those which are called structural micro and submicrostresses are caused by various thermophysical and physico-mechanical properties of the constituents of the material down to the level of crystals and molecules. In this presentation this kind of temperature stresses will not be considered.

The temperature distribution in the body, or cross-section, of a structure at which temperature stresses are absent is called the temperature field (or curve) of zero residual stresses. The temperature field may be concave or convex, favourable or unfavourable. The essence of these concepts will be demonstrated below.

Favourable (efficient) inherent thermostressed state means an inherent thermo-stressed state which increases crack resistance of structures, while an unfavourable (non-efficient) state is one which reduces the crack resistance.

Keywords: Thermal Stress Distribution, Effect of Temperature on Curing, Temperature Fields.

1 Introduction

Since long ago, it was known that in many products, elements, and large scale structures made of hardening materials, a peculiar consequence due to the transformation process from a plastic to hardened material and therefore the presence of a temperature gradient along the body of the structure resulted in residual (or inherent) thermal microstresses. Existence of residual thermal micro stresses ("stresses" in the following text) were discovered in the former USSR by Professor Lukianov about 50 years ago when studying the thermostressed state of solid concrete bearings for bridges and examining the reasons of crack formation in them.

Thermal Cracking in Concrete at Early Ages. Edited by R. Springenschmid. Published 1994 by E & FN Spon, 2–6 Boundary Row, London SE1 8HN, UK. ISBN: 0 419 18710 3.

At Kuibyshev and later at the Saratov hydroelectric power station construction which used solid concrete 50 t weight blocks, the existence of inherent thermal stresses in such concrete structures was not only documented, but the construction technology of such large blocks was also developed in a manner which provided increased stability against temperature influences. This was accomplished by the regulation of inherent thermal stresses. At that time the physical nature of the formation process of inherent thermostressed states and the influence of the concrete structure size on the creation of these formation processes in solid structures was not fully understood. This caused difficulties when making predictions concerning the values of inherent stresses and when researching the increased crack resistance of such structures by the creation of a favourable inherent thermal stress distribution.

2 Physical Foundation of Inherent Thermostress Formation in Concrete Structures

The prediction of values of inherent temperature stresses in structures induced by concrete hardening, to develop measures for the rational changes of temperature stress values, and a clear understanding of the nature of physical foundation of the inherent thermostress formation is very important. During the first stage of the inherent temperature stresses study, it was demonstrated by Professor Lukianov that the creation of physical foundations for inherent thermostressed state forecasting is related to the solution of a simple problem. This problem is defined as the transition time of cement paste into a cement stone, or in other words, the time of transformation of concrete mix into a hardened concrete which is able to respond elastically to applied temperature and other loads.

In those days Lukianov [1] considering the low-strength cements (M200, M300) used in construction, was convinced that in the process of concrete curing at temperature +15 to +20 °C, the material became elastic in 24 to 36 hours after placing.

Later, on the basis of experimental investigations through the construction of the Kuibyshev Hydro Power Station, Komzin [2] suggested the adaption of an elastic time of transition of concrete rather than the plastic state, as the interval of time during which the concrete reaches the $0.3 R_{28}$ strength value. But this value required confirmation. Therefore Solovyanchik investigated this problem on a base of structure formation examining processes in concrete [5].

The theories of Le Chatelier, Michaelis and Baikov are considered to be fundamental. Each researcher usually supports one or another theory of concrete hardening. Therefore at the same time, they are developing and improving separate aspects of the problem. When conducting our own studies we used theoretical prerequisites for structurization nature of the binding materials in the curing process developed by Rebinder. Lately, we have also applied the results of the explorations made by Shpynova [4], as well as the results of some other authors, to the nature of physico-chemical phenomena related to this process, and to explain in different ways the reasons of strength gain in hardening cement paste.

In the light of up-to-date considerations, the process of cement paste hardening may be conditionally divided into two stages: formation of primary low-stable structures and formation and consolidation of secondary structures from the finest crystals of calcium silicate hydrate. These processes determine strength and elastic properties of hardened cement paste.

During the first stage after the addition of water to a concrete mix, aluminates and calcium alumoferrites react much faster than the other minerals. Their hydration products in the presence of gypsum are high- and low-sulphate hydrosulphoalumoferrites and calcium hydroalumoferrites. If gypsum is lacking, mainly calcium hydroaluminates are formed.

In a hardening cement paste contacts occur between skeleton structures generated on the surface of the adjacent cement grains. This results in appearance of hydrosulphoalumoferrite and hydroalumoferrite structures. The silicate structure grows against a background of aluminate structure and a separate plates of $Ca(OH)_2$ crystals are formed. However, at that time the basic strength of cement stone structure is being provided by crystals of calcium hydrosulphoalumoferrite of ettringite type due to their size being considerably larger as compared to that of fibrous crystals of calcium hydrosilicates. At the end of that period primary structures disintegrate.

Sychev [8] discovered interaggregate contacts of electrostatic and electromagnetic nature in the primary structure. Such primary structures predetermine the presence of elastic properties in hardening cement paste and the ability of self-restoration of strength properties of material in the presence of some deformations.

The period of existence of the primary structure depends on the type and mineralogical composition of the cement and other factors. This structure may last from several hours up to 3 days.

The secondary structure is formed as a result of valence surface phenomena [8]. In such a structure shear caused by an imposing load may cause brittle failure. Thus, on the basis of the above stated, it is obvious that the period for the primary structure transformation into the secondary one may be considered the period for the conversion of concrete mix into hardened concrete.

3 Experimental Investigations

In order to determination of the time of transition of the primary structure into the second one, the authors participated in a study as early as 1970 about the process of structure formation in a cement paste by means of electron microscope in the Scientific Research Institute for Concrete and Reinforced Concrete (NHZHB) in Moscow.

Observation of the processes of cement hydration have shown that for sulphate resistant Portland cement M400 of Novoamvrosievsk Cement Plant (mineralogical composition: C_3S - 45%; C_2S - 35%: C_3A - 3.5%; C_4AF - 13%) in 29 to 30 hours the calcium hydrosilicate structures are in many areas so densified that separate crystals as well as crystalline joints were hardly visible. This is because in these parts formed consolidations are being created, which are able to respond resiliently to the loads imposed. Structures from calcium hydrosulphoaluminates by this time are disintegrated to a great extent.

Simultaneously conducted studies on the strength gain of hardened concrete and the heat release of the cement led to interesting results. It appears that the time of transition of the primary structure into the second one corresponds approximately with the concrete strength from 0.22 to 0.3 R_{28} or an average 0.25 R_{28}. At this time the most intensive concrete strength growth is monitored. The most intensive heat release was recorded earlier in the period and during the largest growth of hydrosulphoalumoferrite structures. The most intensified concrete strength gain, occured with the peak curve of cement heat release and was displaced approximately by 5 hours. This way the conducted investigations have shown that as a transition time from the primary structure to the secondary, the most intensified concrete strength gain could be adopted.

From the analysis of numerous experimental data on concrete strength gain with Portland cements of different compositions a cumulative curve for the secondary structure formation in accordance with C_3A content in the cement was obtained (Fig. 1).

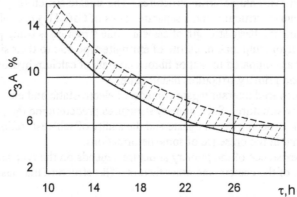

Fig. 1. Relationship between period for spatial crystalic structure formation and C_3A content in cement

With available data on peculiarities of structure formation in a cement paste it was possible to approach the development of theoretical prerequisites for formation of inherent thermostressed state in concrete and in reinforced concrete structures.

4 Formation of Temperature Field of Zero Stresses

For the determination of conditions and time of zero stresses temperature curve or field formation it was assumed that the time of its formation was connected with the peculiarities of structure formation during the cement hydration in the concrete structure and depends on temperature influence on concrete. In the case of uniform temperature distribution along the entire concrete body during the concrete hardening, the process of secondary crystal structure formation takes place with the equal rate along the whole concrete body, which therefore does not cause residual temperature stresses in already hardened structure.

In the conditions of non-uniform temperature distribution, as may be the case of periphery heating, the secondary calcium hydrosilicates crystallic structure in internal layers is formed later than that of external layers, and the connection of crystalline structures may take place providing significant temperature gradient along the structure thickness. With subsequent temperature equalising, for example due to cooling, external layers have to contract more than internal ones. But this would not take place as internal layers prevent external ones from deforming and cause occurrence of inherent tensile stresses and additional residual strains which reduce the structure's crack resistance. As for structure's internal layers, they become prestressed.

When internal electrical heating of a structure is used, a preheated concrete mix is used, or cement heat release is considered as a source of heating, the process of secondary crystal structure formation from calcium hydrosilicates is faster in internal layers than in external ones. If secondary crystals are formed in external layers later, structure thickness temperature difference being appreciable and internal layers temperature being higher than that of external ones, then following temperature equalisation over the entire concrete body (for example during cooling down), internal layers will contract more than external layers. But insofar as external layers would prevent internal ones from deformation, they become contracted and more crack resistant than usual, prerequisites for formation of inherent thermostressed state in concrete structures and occurrence of residual temperature stresses therein.

The analysis of conditions for the formation of inherent thermostressed state in concrete structures shows, that during peripheral heating the temperature distribution along the structure thickness, during the connection of calcium hydrosilicates crystallisation front-lines coming from the surface of the structure to its central layers may be adopted as a temperature curve or field of zero stresses. However during internal heating - the temperature distribution along the structure thickness during the secondary crystalline structure formation in external layers is present, its accuracy being sufficient for practical purposes. Consequently, in both cases the temperature distribution over the structure thickness during the formation of secondary crystalline structure from calcium hydrosilicates in the layer with the largest lag of cement hydration process may be accepted as the zero stresses temperature curve (temperature field).

5 Practical Checking of Theoretical Prerequisites

The precision of the theoretical prerequisites developed has been checked by the investigation of the effects of thermal treatment temperature regimes on temperature curve patterns for zero stresses in various large and small size bulk structures and on their crack resistance. The results of the investigations conducted on small and medium size structures are presented below.

Inherent thermostressed states in small and medium size structures were previously considered to be absent. In this experiment the thermostressed state of solid lightweight concrete panels having thicknesses of 35 to 40 cm was studied. The results of field inspections showed the required crack resistance to have been achieved when certain regimes were used for heat treating the panel, while other regimes have not yielded the same effect. To determine the causes of such phenomena, the dynamics of temperature change through the panel thickness during heat treatment has been studied and the magnitude of possible inherent temperature stresses in panels has been estimated. The investigation results have been compared with the results of field inspections carried out (Fig.2).

Analysis of panels thermostressed state showed that when equalising the temperature through the panel thickness, occurrence of inherent tensile stresses of 0.28-0.37 MPa is possible in surface layers (accounting for plastic properties of expanded-clay lightweight concrete). The presence of inherent tensile stresses would reduce panel resistance to temperature effects. The field investigations, conducted by Solovyanchik, have showed that multiple cracks had been observed in panels, both in the period of their cooling and during building maintenance. These were subjected by the precast concrete plants to heat treatment by the so-called standard regime. (Fig.2).

Indices	LARGE PANEL HOUSEBUILDING FACTORIES							
	Podolsk factory	Bratsk factory			Krasnoyarsk factory	Syktyvkar factory		Perruhov factory
		regime 1	regime 2	regime 3		steam curing	electric curing	
Concrete mix temperature	20°C	20°C	20°C	20°C	20°C	35-45°C		85°C
Thermal treatment regimes used	standard 	standard 	peak 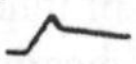	2-peaks	NIIZHB 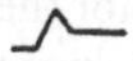	SKO 	WNIIST impulse curing	Curing at t = 65-75°c
Shape of temperature curves for zero stresses			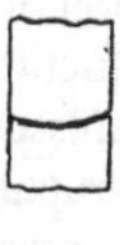		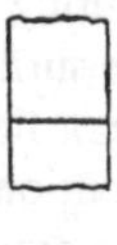	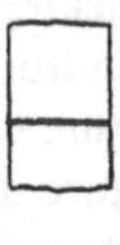		
Inherent stresses in surface layer	tension 0.28 MPa	tension 0.37 MPa	tension 0.03 MPa	tension 0.32 MPa	absent	absent	compression 0.38 MPa	compression 0.4 MPa
Cracks	present	present	absent	present	absent	absent	absent	absent

Fig 2. Formation of inherent thermostressed state in expanded clay lightweight concrete panels

In panels where the zero stresses temperature curve was convex, external layers appeared to be compressed and there was no cracking in buildings during their operation.

Thus, the studies conducted have confirmed the necessity of considering the inherent thermostressed states of structures when specifying regimes for concrete thermal treatment and estimating crack resistance of small and medium massive structures.

6 Determination of Zero Stress Temperature Curve Formation Time Under Variable Concrete Curing Temperature

When cast in situ structures are erected or precast structures are fabricated at plants, normally precasters or contractors have the concrete strength gain curves a tempature +20°C. Having such a curve makes is quite easy to determine when concrete achieves strength value of $0.25 R_{28}$, and at this time establish when the temperature curve of zero stresses is reached. In real structures temperature of concrete may be higher or lower than +20°C. In connection with this it is necessary to have the distribution of temperature in the concrete structure before the concrete strength, in layers with maximum delay of hydration process, reaches value of about $0.25 R_{28}$. Then the actual time of concrete hardening would be recalculated by means of temperature functions.

Data concerning temperature regime of hardening concrete is normally obtained on the base of thermophysical analysis of hardening concrete by means of computers. The influence of temperature on kinetics of heat release in concrete is taken into consideration by means of the above mentioned functions.

7 Utilisation of Inherent Thermostressed State for Improving Crack Resistance of Concrete Structures

Methods of creation of favourable inherent thermostressed state for large and small size concrete and reinforced concrete bulk structures are being developed in Russia. In particular when fabricating large size bulk concrete and reinforced concrete structures electro-heating of the core of structure is recommended [3]. When heat curing small mean size bulk structures using peripheral heaters at a certain stage of the technological process, it is recommended to create a convex temperature field in the hardened concrete of the structure by decreasing the ambient temperature.

Suggested methods of regulation of inherent thermostressed state have found application in practice, especially when erecting concrete piers for bridges, manufacturing massive blocks for power stations, large size structures in precasting plants, etc [7].

Regulation of inherent thermostressed state in concrete structures in many cases allow considerable reduction of concrete cooling expenses in massive structures, reduction of production area in precasting plants, improvement of the durability of structures and their external appearance, and the reduction of expenses on thermoinsulation of concrete when concreteing in cold weather.

8 References

Komzin B.N. (1959) Investigation of thermal stresses in the blocks of hydrotechnic structures, erected in cold weather, Thesises of dissertation. M. M1S1, pp. 29.

Krylov B.A., Lukianov V.S., Mironov S.A. et.al. (1972) Method of manufacturing of massive structures. USSR, Certificate of Invention, N 476991 Cl. C 04 B 1/30.

Lukianov V.S., Denisov I.J. (1959) Prevention of cracks in bridge piers. M.Transzheldorizdat, pp. 110.

Shpynova L.G. (1975) Formation and genesis of microstructure of cement paste. L'vov, Visha Shkola, pp. 246.

Solovianchik A.R. (1970) Prevention of crack formation form temperature influences in external keramzite concrete wall panels. Dissertation, M. TSNIIS, pp. 168.

Solovianchik A.R. (1972) Temperature influence on kinetics of heat release by cement. Proceedings TSNIIS, vol. 72, M.

Solovianchik A.R., Abromov V.P. et.al. (1981) Method of manufacturing of concrete structures. USSR. Certificate of Invention, N 808485,B.I.,N 8.

Sychev M.M. (1974) Hardening of binding materials. L. Stroyizdat.

46 PRACTICAL EXPERIENCE WITH CONCRETE TECHNOLOGICAL MEASURES TO AVOID CRACKING

R. SPRINGENSCHMID
Institute of Building Materials, Technical University of Munich, Germany
R. BREITENBÜCHER
Philipp Holzmann AG, Frankfurt, Germany
M. MANGOLD
Institute of Building Materials, Technical University of Munich, Germany

Abstract
Up to now temperature criteria have been used to control early age cracking. They neglect essential concrete properties. A method has been developed to measure the stress due to hydration heat and early age nonthermal effects. In the cracking frame a concrete specimen hardens under semi-adiabatic conditions at a high degree of restraint. Concrete mix compositions optimized in cracking frame tests were sucessfully employed on numerous large structures (e.g. lock walls or concrete linings for tunnels). Requirements on the cement, the aggregates, the fresh concrete temperature, the compressive strength, the use of entrained air concrete as well as on curing were made and described in the call for tender. This paper reports about the practical experiences with these measures.
Keywords: Cracking Frame, Cracking Tendency, Cement, Aggregates, Fresh Concrete Temperature, Fly Ash, Silica Fume, Entrained Air, Lock Walls, Tunnel Lining.

1 Introduction

Many structures should contain as few cracks as possible or no cracks at all, e.g. if the structure is to be impermeable to water (as in the case of reservoirs, watertight constructions, locks or dams) or because of the higher risk of corrosion when the cracks are water-bearing. Up to now a common measure to avoid cracking at early age was the limitation of the temperature increase of the concrete and of the temperature differences within the concrete structure. In this case, concrete properties which are important for crack resistance (coefficient of thermal expansion, tensile strength, early stress relaxation, etc.) are not considered.

In many cases supplementary steel reinforcement is provided to prevent wide cracks. The amount of reinforcement must be sufficiently large so that the hindered contraction of the concrete results in many narrow cracks instead of a few wide cracks. However, in many cases, cracks can be largely avoided by choosing an appropriate concrete with a low cracking tendency. This is better than keeping the width of the cracks small by using a high reinforcement content or injecting many cracks.

Thermal Cracking in Concrete at Early Ages. Edited by R. Springenschmid. Published 1994 by E & FN Spon, 2–6 Boundary Row, London SE1 8HN, UK. ISBN: 0 419 18710 3.

2 New Method to Define Concrete with Low Cracking Tendency

Extensive research work has made it possible to optimize a concrete in a way that its cracking tendency is low and that the other required properties, such as early age strength and watertightness, are obtained as well. The effect of technological measures on the cracking tendency of concrete can be quantified by measurements of the restraint stresses in cracking frame tests, see Springenschmid and Nischer (1973), Springenschmid and Breitenbücher (1990), Breitenbücher (1989), Breitenbücher and Mangold (1994). The so-called cracking temperature T_c is used as a standard of comparison for the cracking tendency. T_c is the temperature at which a concrete beam, with the high degree of restraint given by the cracking frame, breaks. The lower the cracking temperature is, the lower is the cracking tendency of the concrete.

Strictly speaking, it is irrelevant by which measures a low cracking tendency of the concrete is obtained. The local cost situation as well as the expected extent of cracking and the consequences of cracking are decisive for the concrete specification. For field work the following measures may be adopted:

(1) Low temperature of the fresh mix. The fresh concrete temperature has a strong effect on cracking, above all when concrete is placed in summer. A limit of up to 20 or 25°C can usually be fulfilled by simple means, such as spraying the coarse aggregate with water to generate cooling due to evaporation. Most favourable fresh concrete temperatures between 8 and 12°C in summer can be achieved by cooling with liquid nitrogen or by adding finely chipped ice to the mix.

(2) Selection of a cement with a low cracking temperature. The cracking temperature of a cement is determined in a standard test with the cracking frame using a well defined concrete composition and defined experimental conditions, see RILEM Technical Recommendation (1993). The cracking temperature of favourable cements should not exceed 10°C. There are many cements, which fulfill this requirement. However, quite often those are not the customary low heat cements.

(3) Selection of aggregates with a low coefficient of thermal expansion such as many limestones, basalt, diorite, granite or quartzporphyr. Furthermore the aggregates with a moderately rough surface give better adhesion and therefore a higher tensile strength. However, favourable aggregates may require transportation over long distances which gives rise to high costs.

(4) Use of air entraining admixtures to lower the cracking tendency of the concrete. Increasing the air content to a value between 4 and 6 vol.%, the cracking temperature is lowered by approximately 4 K.

(5) Partial replacement of cement by fly ash.

(6) Strengths which are higher than strictly necessary should not be specified, because high strengths often lead to a concrete with a higher cracking tendency. Instead of the 28 day strength the 56 or 90 day strength may be specified.

(7) In order to minimize the stresses caused by drying shrinkage, the water content should be as low as possible. For this a large maximum aggregate size and an optimized granulometric composition,

especially for the sand, are favourable. If necessary, the sand must be prepared in a classifier. Sands optimized in a classifier have a low water demand which results in better workability of the concrete. With plasticizer as well as with a low fresh concrete temperature, the water demand will be further reduced resulting in somewhat smaller shrinkage stresses.

Beyond these concrete technological measures, additional measures are often necessary to further reduce the cracking risk on site. Among these measures are
- the reduction of the degree of restraint and
- Appropriate curing with control of moisture and curing temperatures.For this, the surface of the concrete must be
  (1) moist cured, in order to prevent the concrete from drying, and
  (2) kept cool up to the time at which the Young's modulus reaches at least half of its 28 day value (usually for about 16 to 24 hours), see Mangold (1994). An advantageous curing is to cover the concrete surface with moist water absorbing blankets. As the water evaporates, a favourable cooling effect is produced. Thus, the unfavourable heating up of the concrete surface can be avoided on hot summer days. As soon as the concrete has reached a high Young's modulus, it must be protected against further cooling. A waterproof sheet can be laid on the moist blanket for this purpose.

3 Practical Procedure (Call for Tender)

In the invitation to tender for a construction project, the list of requirements must contain all the instructions which enable the bidder to calculate the supplementary costs, without the need of performing tests. Therefore it is unreasonable to specify a concrete with a low cracking temperature (e.g. lower than 5°C) in a call for tender. It is only to be specified, which of the above mentioned measures are required.

4 On Site Experience (Examples)

4.1 Berching lock of the Rhine Main Danube Corp.
In order to prevent cracking of massive structures such as lock walls, it was up to now only specified to keep the temperature increase in the concrete as low as possible. Therefore, low heat cements have traditionally been used for the concrete of massive structures. As could be shown with cracking frame tests, concretes with low heat cements heat up less, but the development of the Young's modulus and the tensile strength is generally slower, so that their cracking temperature may be distinctly above 10°C. Therefore, in order to ensure a low cracking tendency, cements with a low cracking temperature are specified now and not only low heat cements.

The Rhine Main Danube Corp. built three similar locks in South Germany. Each lock was approximately 240 m long, with a lift height of 17 m. In the Dietfurt and Bachhausen locks the usual low heat blast

furnace slag cement (CEM III/A 32.5) was used. For the last lock
Berching (built from 1990 to 1992), taking the newest knowledge into
account, an ordinary Portland cement with a cracking temperature of
10°C was used. To ensure a high frost resistance and a low cracking
tendency as well, an entrained air concrete was specified. An examination of the crack formation of the three locks (all locks with the
same reinforcement) along the whole length of the lock wall (154.2m)
showed the following results:
* Dietfurt lock (Low heat blast furnace cement), 64 cracks, mean crack
  width 0.20 mm.
* Bachhausen lock (Low heat blast furnace cement), 68 cracks, mean
  crack width 0.17 mm.
* Berching (Ordinary Portland cement with low cracking temperature),
  38 cracks, mean crack width 0.19 mm.
    This means that for a concrete containing ordinary Portland cement
of low cracking tendency, approximately 45% less cracks occured than
with the concretes containing low heat cements which were used previously.

4.2 Concrete Linings for Tunnels
The blocks of concrete linings for large railway tunnels are usually
11 m long and 30 to 60 cm thick, fig. 1. In order to concrete in a
daily cycle, the concrete must be stripped at latest after 12 hours.
Therefore the concrete must be optimized so that it reaches an adequate 12 hour strength and nevertheless has a low cracking tendency.

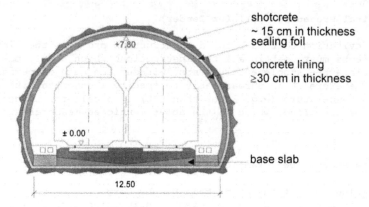

shotcrete
~ 15 cm in thickness
sealing foil

concrete lining
≥30 cm in thickness

base slab

Fig. 1. Cross-section of a railway tunnel.

    In 1982, when the first large tunnel for the new railway line
Hannover-Würzburg of the German Federal Railway Company was built, the
concrete had been optimized exclusively with regard to a high 12 hour
strength. An early age strength Portland cement (CEM I 42,5 R) was
used in order to obtain a compressive strength as high as possible at
the time of stripping. Only a few days after stripping, the first
radial and roof cracks (in the longitudinal direction of the lining)
appeared. The number of cracks and the total crack opening area
increased significantly in the first winter. Must a concrete with

sufficient high early age strength inevitably lead to excessive cracking ?

  To clarify this point, concretes with cements of different types,
strength classes, from differents plants, with partial replacement of
cement by fly ash and with different fresh concrete temperatures were
systematically investigated in comparative laboratory tests with the
cracking frame. Fig. 2 shows a comparison between the cracking temperatures and the 12 hour strength of thermally insulated cubes with a
15 cm edge. The following was established:

(1) A high 12 hour strength generally leads to a higher cracking
     temperature.
(2) The required 12 hour strength of 3 MPa may be reached with
     cracking temperatures between 2 and 18°C, fig. 2.
(3) Low fresh concrete temperatures (12°C), which still provide suf-
     ficient compressive strength, yield lower cracking temperatures.
(4) Particularly high cracking temperatures are produced by a fresh
     concrete temperature of 25°C. Cracking temperatures increase
     faster than the fresh concrete temperature, i.e. with a propor-
     tionality factor of 1.2 to 1.5.
(5) At a binder content of 280 kg/m³ ordinary Portland cement (OPC)
     plus 60 kg/m³ fly ash lower cracking temperatures are obtained
     than with 340 kg/m³ cement.

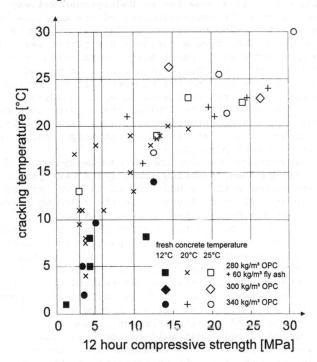

Fig. 2. Influence of the fresh concrete temperature, cement content,
         fly ash and compressive strength on the cracking temperature

In the period 1983 to 1987, for 16 new railway tunnels with a total total length of 33.3 km, built in the southern section of the new line Hannover-Würzburg (Germany), the technological requirements for the concrete were modified according to the above results of the cracking frame tests.

The examination of the tunnels after 5 years of service showed that long sections exhibited either no cracks at all or the crack formation was reduced considerably compared with the previous tunnels.

In the last period 1989 to 1992, another four railway tunnels were built. Due to further studies, the specifications were supplemented as follows:

(1) Limitation of the fresh concrete temperature to a maximum of 25°C.
(2) Cement with a maximum cracking temperature of 10°C.
(3) Use of a summer concrete mix design with 280 kg/m³ cement plus 60 kg/m³ fly ash and a winter mix design with 340 kg/m³ cement. The cement content has to be modified when the 12 hour compressive strength falls below 3 MPa or exceeds 6 MPa.
(4) Use of entrained air concrete also in sections which are not exposed to frost.
(5) Removal of the heat which accumulates in the roof of the tunnel.

4.3 Concrete with Silica Fume for an Underground Railway
Silica fume improves the impermeability of the concrete matrix. It is known from experiments on unrestrained laboratory samples that concretes with silica fume have a considerably better resistance against chemical attack and a water penetration depth which is up to 90% lower than without silica fume. Thus, for a tunnel in ground water, it was planned to add silica fume to the concrete for the lining. As the concrete lining had no external sealing and was placed directly against the external shotcrete lining, the requirements on the water impermeability of the construction were very high. A silica fume content of 25 kg/m³, i.e. 8% of the cement weight and a water cement ratio of w/c=0.50 were planned.

In order to estimate how the cracking tendency was influenced by the addition of silica fume, cracking frame tests were performed on the same concrete either with the addition of 25 kg/m³ silica fume or alternatively with 30 kg/m³ fly ash. In the cracking frame tests, the concrete with silica fume heated up about 3 K more than the concrete with fly ash, fig. 3. During the heating phase the concrete with fly ash converted the restrained thermal expansion into compressive stresses and could be cooled down after 4 days to 0°C without cracking. However, in the concrete containing silica fume tensile stresses arose even before the temperature maximum was reached (fig. 3). This must be attributed to self-desiccation which is particulary important for concretes with silica fume. At low w/c ratios < 0.40, it is well known that the development of tensile stresses is accelerated by self-desiccation. It was remarkable that self-desiccation occured to such a degree at a w/c ratio of 0.50. Corresponding to the high tensile stresses when reaching the ambient temperature, the cracking tendency of the concrete with silica fume was high (T_c=17.5°C). Therefore, the concrete with fly ash was chosen for the construction instead of the concrete with silica fume.

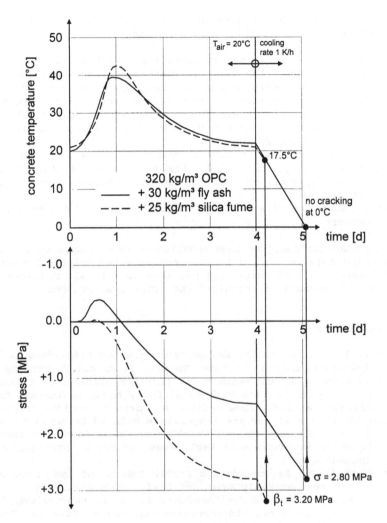

Fig. 3. Cracking frame tests on concretes containing fly ash and silica fume, aggregate with limestone.

5 Conclusions

Cracks in concrete structures do not occur because of high temperatures, but because of the thermal stresses caused by the hydration heat and the restrained deformations of the concrete. In addition, nonthermal effects (e.g. self-desiccation) often have an important influence. At the Institute of Building Materials of the Technical University of Munich these stresses are measured in cracking frame tests. The results of these tests have been applied in practice to numerous large structures.

The following measures have been successfully taken in pratice to ensure a low cracking tendency of the concrete.

(1) Low fresh concrete temperatures (optimum 8 to 12°C).
(2) Low strength of the concrete, particulary at early age.
(3) The use of cements with a low cracking temperature ($T_c \leq 10$°C). The cracking temperature of cements is determined in a standard test with the cracking frame. There are low heat cements with a slow development of Young's modulus and tensile strength which do not meet this requirement.
(4) Replacement of a part of the cement by fly ash of high quality.
(5) The use of aggregate with a low coefficient of thermal expansion.
(6) Air entrainment, so that the concrete contains about 4 vol.% entrained air.
(7) The addition of silica fume should be avoided for restrained members.

In numerous constructions troublesome cracks have been extensively or completely avoided with these concrete technological measures, although it was of course not possible to take all measures at the same time because of technical and economical reasons.

6 References

Breitenbücher, R. (1989) Zwangsspannungen und Rißbildung infolge Hydratationswärme. Doctoral Thesis, Technical University of Munich.

Breitenbücher, R. and Mangold, M. (1994) Minimization of Thermal Cracking in Concrete Members at Early Ages, in **Thermal Cracking in Concrete at Early Ages** (editor R. Springenschmid), Chapman & Hall.

Mangold, M. (1994) Why are Temperature-Related Criteria so Undependable to Predict Thermal Cracking at Early Ages?, in **Thermal Cracking in Concrete at Early Ages** (editor R. Springenschmid), Chapman & Hall.

RILEM Technical Recommendation (1993) Testing of the Cracking Tendency of Concrete at Early Ages, 2nd Draft, December 1993.

Springenschmid, R. and Breitenbücher R. (1990) Beurteilung der Rißneigung anhand der Rißtemperatur von jungem Beton bei Zwang. **Beton- und Stahlbetonbau**, 85, 29-33.

Springenschmid, R. and Nischer, P. (1973) Untersuchungen über die Ursache von Querrissen im jungen Beton, **Beton- und Stahlbetonbau**, Heft 9, pp. 221-226

47 RISK OF CRACKING IN MASSIVE CONCRETE STRUCTURES - NEW DEVELOPMENTS AND EXPERIENCES

S. BERNANDER and M. EMBORG
Department of Civil Engineering, Division of Structural Engineering,
Luleå University of Technology, Luleå, Sweden

Abstract
Ever since the dawn of concrete technology the risk of cracking in early age concrete has been almost exclusively based on temperature related criteria - a practice which - on an inadequate basis - presupposes a more or less direct proportionality between temperature and stress development in the concrete. It can be shown that this presumption is often grossly inaccurate. Research at the Luleå University of Technology demonstrates the possibilities in this context of considering also a number of other crucial factors in the assessment of temperature induced stresses in hardening concrete.
Keywords: Classification of Cracks, Cracking, Curing with Artificial Cooling/Heating, Maturity Development, Restraint, Thermal Stress Analysis.

1 General

The fact that, in particular, massive concrete structures tend to crack as a result of temperature induced stresses during hardening is a problem as old as the practice of using concrete as a construction material. The issue has thus been subject to the attention of civil engineers for almost a century.

Cracking in *plain unreinforced* concrete may have serious static implications. A far as reinforced concrete is concerned cracking mainly affects functional and durability requirements (tightness, chemical resistance, frost resistance, etc.). Yet it is only of late that concrete engineers begin to gain a reasonable understanding of the intricacies of these problems. Research at Skanska AB since the early sixties and at the Technical University of Luleå has now resulted in simulation procedures by which the problem of temperature induced stress build-up and the consequent possible cracking in concrete can be analyzed in a way, which - to a significantly higher degree than before - matches the enormous complexity of the issue. In the nineteen-fifties and earlier, cracking in maturing concrete was controlled mainly by simply trying to limit the temperature rise due to hydration. The degree of restraint to which the young concrete may be subjected, was considered in a qualitative manner by suitable arrangements of construction joints and construction sequences.

Thermal Cracking in Concrete at Early Ages. Edited by R. Springenschmid. Published 1994 by E & FN Spon, 2–6 Boundary Row, London SE1 8HN, UK. ISBN: 0 419 18710 3.

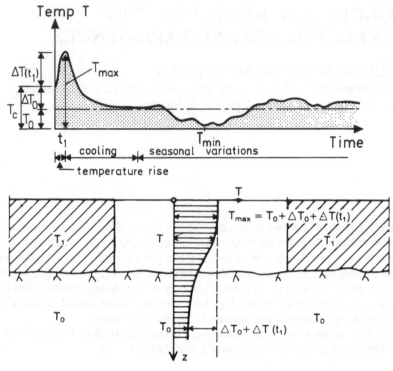

Fig.1   Example of temperature conditions in a newly cast concrete section and in adjoining structures

2 New development - stress analysis

In the last two decades it has become common practice to try to limit cracking by fulfilling certain more specific requirements with respect to temperature differences within the cast section as well as to temperature differences between new and old concrete, see Fig 1. It is e. g. customary to limit the value of $\Delta T = T_{max} - T_1$ to values between 12 and 20 °C.

However, in the author's opinion, it is astonishing that a problem, which has attracted the attention of concrete engineers for so long , has been offered so little qualified analysis. Any experienced engineer in structural mechanics should, after all, not have to ponder over this problem for very long in order to realize, that volume change due to temperature only constitutes one of the major agents affecting the risk of cracking - and that, therefore, temperature alone should never be permitted to be the sole basis on which the risk of cracking is made. Thus, a number of factors, other than natural temperature development, have decisive influence on the early age cracking in concrete (see Fig 2):

The degree of restraint on the cast section imposed by the stiffness of adjoining structures, geometry, length and nature of construction joints etc.
The transient mechanical properties of hardening concrete and the dependence of these properties of cement type, cement content and maturity development.

Non-uniform temperature and non-uniform maturity development - not least when
  temperature is governed by artificial means
Conditions related to fracture mechanics
Artificial temperature control - cooling, heating, insulation etc

For the tunnel at Tingstad, (Gothenburg) 1964 - 1966 and the bridge over Alssund 1978
- 1980 (Bernander and Gustafsson (1981a and 1981b)), methods were developed for crack
prediction considering the plasticity of the young concrete and in some measure, also
varying restraint. This approach has since been developed further by the authors at the
Luleå University of Technology particularly in conjunction with a doctoral research project
(Emborg (1990)). This work was focused on the rheological properties of early age
concrete and comprehensive tests have been carried out to define the most important of
these properties. These have then served as basis for stress analysis and mathematical
simulation. A computer program named TEMPSTRE-N has been developed by means of
which it is possible to analyze the development of temperature related stresses for a broad
range of concrete structures or structural components. The method permits consideration
of practically all the factors listed in Fig 2. It is thus now possible to assess the relative
efficiency of various practical measures to exercising crack control.

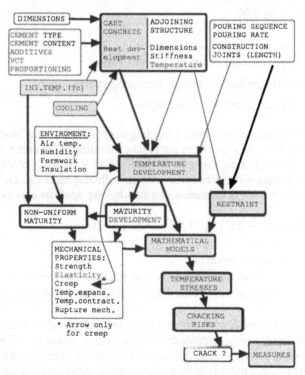

Fig. 2 Diagram illustrating interacting factors governing temperature-induced stresses
   and cracking in early age concrete

Hence the designer and the concrete engineer now have useful tools at their disposal for optimizing crack control with regard to the requirements that may be set up in various cases. This - should we say ? - advanced technique has been applied in connection with a number of projects, of which the following may be mentioned: (a) Railway box-section tunnels, Helsingborg and Stockholm, Sweden, (b) repair of bridge piers for the Öland and Smögen bridges, Sweden, (c) concrete lining in a head-race bifurcation tunnel at Carhuaquero Power plant, Peru, (d) the Storebaelt bridge, Denmark, (e) rock storage caverns in the Middle East, (f) the Igelsta railway bridge, Södertälje, Sweden, (g) basin walls purification plants, Stockholm

The TEMPSTRE-N model is indicated in Fig. 3. A material model according to Bazant and Chern (1985) has been modified for the behavior of early age concrete. The analysis of thermal stresses is evaluated adopting the so called Plane Surface Method (i. e. the cross section remains plane in bending according to the beam theory) where two equations of equilibrium are solved taking bending and translation of the structure into consideration. At a certain time step the increments of the elastic and creep deformation, fracturing behavior and thermal movements may be written as

$$\Delta\varepsilon = \Delta\varepsilon_{ec} + \Delta\xi + \Delta\varepsilon^{\circ} \qquad (1)$$

where $\Delta\varepsilon$ is the increment of total strain
$\Delta\varepsilon_{ec}$ is the increment of elastic strain and creep
$\Delta\xi$ is the increment of inelastic fracturing strain
$\Delta\varepsilon^{\circ}$ is the increment of thermal strain

The considered viscoelastic response according the Maxwell Chain Model (see Fig 3a) including evaluation of creep data from tests and calibration of the creep model as well as evaluation of so called relaxation spectra is described in Westman (1994). The behaviour of the fracturing element is described with a strain-softening modulus, D. The constitutive equation for the thermal stress increment in a step-by-step analysis may be expressed as

$$\Delta\sigma = B(\Delta\varepsilon - \Delta\varepsilon^{''} - \Delta\xi^{''}) \ ; \ 1/B = 1/E^{''} + 1/D \qquad (2)$$

where $E^{''}$ is the relaxation modulus
$\Delta\varepsilon^{''}$ is inelastic strain including relaxation strain and thermal strain
$\Delta\xi^{''}$ is inelastic strain increment, (Bazant and Chern (1985) and Emborg (1990))

The application of this more sophisticated stress analysis has shown, that temperature-related crack criteria may be acceptable only in such cases, where the effect of other important influencing parameters are defined and incorporated in the background experience. However, in a number of cases the result of stress analysis have disagreed markedly with assessments based on temperature related criteria. At times it was clearly not possible even to predict whether the concrete at a certain time was in tension or in compression. This applies especially to cracks occurring in the heating phase.

3 New experiences - New classification of early age thermal cracks

3.1 Cracks in the heating and cooling phases of the temperature cycle - surface cracks 'versus' through cracks

A traditional feature in literature and in the terminology among concrete engineers in this context, has been to distinguish sharply between

1) surface cracks (which are said to occur as a result of internal restraint)
2) through cracks (which are said to be the result of external restraint)

In the opinion of the authors this classification of temperature-induced early age cracks is inaccurate and not to the point for two reasons:

A) Internal restraint - e. g. from uneven temperature distribution seldom occurs in practice as an isolated phenomenon. In general, the effects of restraint from very early age temperature differences are to a varying degree, superimposed by simultaneous compressive stresses due to restraint from adjoining structures as indicated in Fig. 4 and 5. Internal restraint may - nota bene - also cause through cracks in many structural configurations.

B) Static analysis of stress development necessitates consideration of the stiffness characteristics of adjacent structural elements, which therefore must be part of the analytical model. The so called ' external' restraint from e. g. a previous cast then becomes an ' internal' restraint in the analyzed portion of the structure. Cf Fig 4 where e. g. this 'internal' restraint gives rise to early age *through cracks* in the old concrete.

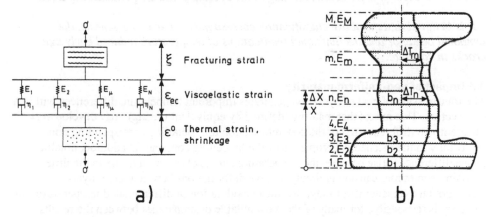

a)   b)

Fig. 3. Thermal stress analysis according to TEMPSTRE-N.
a) Rheologic model describing deformation due to fracturing at high tensile stresses, elastic and creep deformation and thermal deformation (after Bazant and Chern (1985) b) The construction is subdivided into a number of layers N with thickness Δx. Each layer may be attributed its own age, temperature development and maturity as well as separate mechanical properties and restraints.

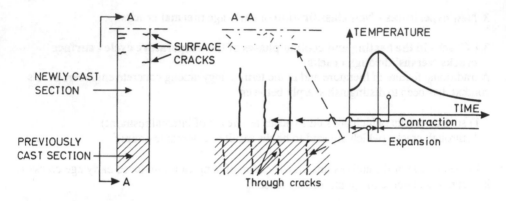

Fig. 4. Example of early age expansion cracks and contraction cracks

3.2 New classification of cracks

The authors therefore suggest that early age temperature cracks be classified as follows:

A) *Early age cracks in the expansion (or heating) phase.* These cracks appear in the expansion phase shortly (one to a few days) after pouring and tend to close at the end of the cooling phase. (Cf Fig 4). Their effects on static capacity, function and durability has to be deemed from one case to the other.

B) *Cracks occurring in the contraction (or cooling) phase.* (Cf Fig 4). Cracks appearing in the cooling phase are usually through cracks and may, depending on dimensions and other factors, show up weeks, months and in extreme cases even years after the section has been poured. Cracks forming in the cooling phase are permanent as a rule.

It is believed that the above classification corresponds better to the genesis, the temporal behavior and the functional implications of temperature induced early age cracks in concrete.

3.3 Impact of non-uniform maturity

The transmission of heat of hydration generates important temperature differences within a cast section. This means that maturity (defined by equivalent time t_e) may develop very differently in different parts of the concrete section thus entailing corresponding non-uniformity in the mechanical properties of the hardening concrete. In addition, notable phase differences arise between the temperature curves of different layers in the time domain - sometimes causing untimely relative deformations between these layers.

It is not an overstatement to say, that the consideration of differentiated temperatures and maturity is responsible for many of the most notable discrepancies between the results gained by the analysis applied by the authors and corresponding assessments based on conventional temperature-related criteria. Since the initial temperature of the concrete as well as the ambient temperature strongly affect the temperature distribution in the concrete, the relationship between these two parameters may be decisive for - in particular- early age cracking in the expansion phase, see Emborg and Bernander (1994).

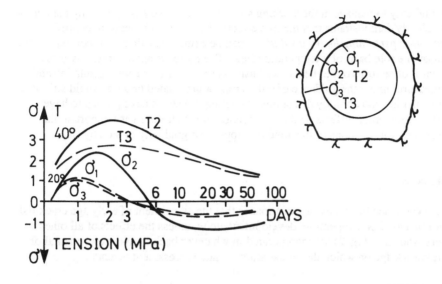

Fig. 5 Temperature and stress development in the concrete lining of a head-race tunnel at
Carhuaquero, Peru. The example illustrates how tension stresses, which would
have developed in the expansion phase as a result of uneven temperature, are
superimposed by tangential compression stresses due to restraint from the rock.
Concrete: cement content 300 kg/m^3, w/c=0.54, T_c=20 °C.

3.4 Restraint

The necessity of paying as much attention to the 'degree' of restraint as to temperature
itself, ought to go without saying. None the less we may recognize, that in practice there is
a widespread use of general purely temperature based criteria for assessing crack risk even
in major projects - e. g. such as the Storebaelt bridges. This implies that "the *degree of
restraint*"- however important it may be - is still vastly overlooked, or treated flimsily, in
concrete engineering of today. Often 100% restraint has been presupposed and
astonishingly little work has been dedicated to the study of the true degree of restraint,
which will depend on a variety of parameters such as stiffness of adjoining structures,
dimensions, geometry, configuration, length and nature of construction joints etc. For
instance, the capacity of a structure to rotate (or bend) often has a decisive effect on the
stress build-up in different layers.

In the many practical cases, where TEMPSTRE-N has been applied, the importance of
correctly modelling the freedom of movement as defined by adjoining structures, has been
clearly established, see e. g. Bernander and Emborg (1994).

3.5 Insulation and cooling

Current measures to controlling cracks in young concrete include insulation and internal
cooling. When temperature based crack criteria are applied, such measures appear both

effective and easy to manage at the working site. The stress analysis applied by the authors, however, clearly demonstrates that the true effects of artificial temperature control - *especially those pertaining heavy insulation* - can be extremely difficult, if not impossible, to predict on the sole basis of a temperature field. The more stringent stress analysis has shown, that it often turns out to be a risky business to *'fool around'* with insulation and cooling measures under the guidance of only temperature related criteria. Rapid stripping of heavy insulation and abruptly discontinued cooling processes have proved to have dramatically unfavorable effects on the stress fields. It is therefore very important that temperature control functions of this kind be suppressed gradually or step by step.

4 Conclusions

A general experience from these applications is that the assessment of early age crack risk should not be based on temperature development alone, unless the effects of all other parameters (shown in Fig. 2) are incorporated in whatever background experimental or empirical knowledge on which the temperature - related assessment is made.

Acknowledgment

This work has been supported by the Swedish Council for Building Research, the Foundation for Swedish Concrete Research, the Swedish National Road Administration, the Swedish State Power Board, Cementa AB, NCC AB and Skanska AB. The work was supervised by Professor Lennart Elfgren, head of the Division of Structural Engineering at Luleå University of Technology.

References

Bernander S (1981a): Temperature stresses in early age concrete. **Rilem Int. Symposium on Early Age Concrete,** Paris 1981

Bernander S, Gustafsson S (1981b): Temperature induced stresses in concrete due to hydration. **Nordisk Betong** , No 2, 1981 (in swedish)

Bernander S, Emborg Mats (1992): Temperature conditions and crack limitation in massive concrete structures. Chapter 27, **Handbook for Concrete Technology,** Stockholm, 1992 (in Swedish) 50 pp

Emborg M (1990): **Thermal stresses in concrete structures at early ages**, Doctoral Thesis 1989:73 D, Div of Struct Eng, Luleå Univ. of Technology, revised edition, Luleå 1990, 286 pp

Bernander S, Emborg M (1994) Avoidance of early age thermal cracking in concrete structures - predesign, measures, follow-up. **Proc. from Rilem Int. Symp. "Avoidance of Thermal Cracking in Concrete",** Munich Oct. 1994, 8 pp.

Emborg M, Bernander S (1994) Assessment of the risk of thermal cracking in hardening concrete, **Journ. of Struc,. Eng (ASCE),** (to be published 1994),

Westman G (1994) Basic of creep and relaxation of young concrete. **Proc. from Rilem Int. Symp. "Avoidance of Thermal Cracking in Concrete",** Munich Oct. 1994, 8 pp.

48 THERMAL CRACKING IN THE DIAPHRAGM-WALL CONCRETE OF KAWASAKI ISLAND

D.D. LIOU
Bechtel Corporation, San Francisco, California, USA

Abstract
Kawasaki Island is currently being built in the middle of
Tokyo Bay as a component of the Trans Tokyo Bay Highway
Project. The centerpiece of this island is a cylindrical-
shaped cast-in-place diaphragm wall, which serves as a
cofferdam during construction of the permanent structure. A
state-of-art instrumentation system is used to monitor the
behavior and safety of the man-made island during the long
construction period, with many strain gages, soil pressure
gages and thermocouples installed in the diaphragm wall.
This paper discusses the correlation between the stress data
recorded by the instrumentation system and the varying
thermal conducting environment around the wall, describes
cracking patterns in the concrete observed at the inside face
of the diaphragm wall, and suggests the existence of a
critical maximum wall thickness beyond which excessive
thermal cracking in concrete would be expected to occur.
Experience gained in integrating the design, instrumentation
and construction of a large-sized diaphragm wall is
discussed.
Keywords: Cracking, Monitoring Instrumentation, Diaphragm
Wall and Marine Construction, Superplasticized Concrete.

1 Introduction

When completed, Kawasaki Island will serve as a work base for
four slurry shield machines during the next phase of boring
two 10 km tunnels under the bay, and later as a ventilation
shaft for the entire Trans-Tokyo Bay Highway when the project
is completed. The island is being built at a site where the
original surface soil is extremely weak, and the average
water level is 28 m deep.

The island features a diaphragm wall of 56 panel elements,
with a 100 m outside diameter, a 119 m depth, a 2.8 m
thickness. 120,000 m³ of concrete is used to build this
gigantic wall. In designing the concrete mix, a combination
of fly ash and cement with low hydration heat is used in
place of normal concrete to reduce the cracking potential and

Thermal Cracking in Concrete at Early Ages. Edited by R. Springenschmid. Published 1994
by E & FN Spon, 2–6 Boundary Row, London SE1 8HN, UK. ISBN: 0 419 18710 3.

enhance the workability of the concrete. Superplasticizer is added plentifully to the concrete mix to provide a flowable concrete that can be continuously conveyed over a long distance.

2 Design and Construction of the Diaphragm Wall

Construction of the diaphragm wall is the fourth construction phase of the man-made island project. It was preceded by a soil improvement phase in which the original soil around the construction site was improved by sand pile compaction and deep mixing methods; a steel jacket phase in which steel jackets were fabricated, transported, positioned, and welded into two concentric steel jacket rings; and an embankment phase in which the annular space between the two steel jacket rings was filled with sandy fill material resulting in an artificial embankment with its top surface above the sea level. It is through this artificial embankment that the diaphragm wall was constructed.

The cofferdam-type diaphragm wall of Kawasaki Island is designed for both ground support and water cutoff purposes. The wall is constructed by using Electric Mill (EM) excavation equipment, which has a 2.88 m by 3.20 m cutting cross-section. The cylindrical-shaped wall is divided into alternate primary-panel elements and secondary-panel elements, with the primary panels constructed before the secondary panels. The excavation of a primary-panel trench, which has a same cross-sectional area as that of the EM excavation equipment, is achieved by one cut of the machine while the excavation of a secondary-panel trench, whose inside circumferential length is 7.8 m, is achieved by three cuts. During excavation, slurry containing both bentonite and CMC is used to maintain the stability of the trench.

Primary panels are more heavily reinforced than secondary panels. For both panel elements, reinforcement is provided by using six sections of pre-fabricated steel rebar cage, each approximately 20 m long. When excavation of a panel element shaft reaches the 119 m design depth, a new section of rebar cage is connected, one by one, to the finished portion of the rebar cage which is already in the trench, and the elongated rebar cage is then lowered into the shaft. Both welded and mechanical joints are used to join the sections together. In the top two sections connections are also provided between the rebar cage of a primary panel and those of the adjacent secondary panels by using pipe joints. In the top two sections, the rebar cage of a secondary panel is equipped with a movable portion that enables it to fit snugly into the neighboring cages of the primary panels.

Placement of superplasticized concrete forming the diaphragm wall is carried out by using tremie pipes. During the concreting of a primary panel, gravel is poured into the excavation as well. Placement of concrete and gravel is shown

in Figure 1. Concrete is poured into the central part of the
steel rebar cage, while gravel is poured into the two outer
·zones where the joints between the primary and the adjacent
secondary panels will be located, which makes it unnecessary
to cut the primary-panel concrete during excavation of the
secondary panels.

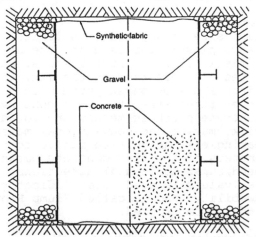

Fig.1. Plan View of Primary Panel
During Construction

3 Selection of Concrete

As with any diaphragm-wall project, a number of factors
influence the selection of a concrete mix, including
performance requirements, site location, and placement
conditions. Because of the distance to the shoreline (5 km),
concrete is produced from two batching plants on-board a
25,000 ton ship moored adjacent to the island. Concrete
pouring spans all four seasons, with summer being the most
difficult because of heat, humidity, and frequent typhoons.
    From the outset, it was clear that superplasticized
concrete would be used for the project. Superplasticizer
enhances reinforced concrete by providing better coverage of
steel as concrete flows around it and requires almost no
vibration, an important advantage when pours are made in
tight, inaccessible spaces.

3.1 Trial Mixes
The target compressive strength of superplasticized concrete
is 53 N/mm^2 at 91 days. Two rounds of trial mixes are carried
out in the laboratory to determine the best blend of cement
and the most suitable superplasticizer for that mix.
Superplasticizers ranging from 1.8 to 2.6% of the weight of
cement are added to all trial mixes. The cement selected is a
combination of 20% fly ash, 35% clinker, and 45% slag. This

combination has been found to have acceptable flowability, consistency, and low hydration heat, a desirable property that reduces the cracking potential in concrete. The superplasticizer selected is a product that contains cross-linked polymer and polycarboxylic ether, and a dosage of 1.8% by weight of cement is used.

3.2 Slump Flow

Since the maximum size of aggregate is limited to 20 mm and the dosage of superplasticizer is high, the selected concrete mix possesses drastic fluid characteristics. When the standard slump test is performed on the mix, the concrete quickly flows into a plano-convex patch with a ruggedly circular perimeter right after the test cone was lifted. The pizza-shaped concrete patch is merely one to three centimeters thick, making the conventional measurement of slump value meaningless. As an alternative to the slump value, two diameters of the flat concrete patch are measured along two perpendicular horizontal directions. The average of the two diameter values is used as an indicator of the concrete's flowability, and is called Slump Flow. A Slump Flow value of 65 ± 5 cm is used.

3.3 Field Adjustment of Concrete Mix

Because the concrete is placed year-round, the full range of weather conditions is encountered, requiring occasional adjustment of the final mix to accommodate actual conditions encountered in the field. For example, when the temperature of air is so hot that ice is required, the dosage of superplasticizer is adjusted upward to maintain the flowability of the mix.

4 Instrumentation

The designer of a man-made island faces more uncertainties than are encountered in the design of a more traditional land-based structure. Several mechanisms controlling the behavior of the structure at different construction stages must be managed, as discussed by the author in Reference 1.

The success of the design may ultimately be based on judgement in determining the most probable values for engineering properties of the materials. Additional difficulties are encountered in determining engineering properties of materials in a marine environment, and in assigning probable numerical values, not only to the naturally occurring heterogeneous materials, such as soil, but also to the manufactured materials. Laboratory and field tests can be performed on land, but taking field samples in the deep sea is manifestly more difficult. Certain materials also may display a wide range of values when handled in a wet work site. These uncertainties are also discussed in Reference 1.

Because of these uncertainties and other unknown factors, an instrumentation system that closely monitors conditions and changes is necessary in the construction of a man-made island. The instrumentation system for the Kawasaki Island project includes 800 instruments, many dedicated and flexible data loggers, cabling, personal computers, and numerous other peripheral equipment. The categories and quantities of instruments used are modified for each construction phase, as appropriate to the degree of design conservatism of that phase.

There are four control sections and four supplementary control sections in the wall (a supplementary control section contains fewer instruments than a control section). In a control section, sister bars are installed on the steel rebar cage of the wall to monitor stress levels in steel reinforcement in both vertical and horizontal directions. A thermocouple is installed in the vicinity of each sister bar to measure the temperature in steel rebar temperature. Other instruments installed in the diaphragm wall include concrete stress meters, contact earth pressure cells, piezometers, and inclinometers. Although concrete stress meters can measure stresses in concrete directly, they are used basically for checking the measurements recorded by sister bars. Inclinometer measures angle changes along the inclinometer's guide pipe directly, and measures horizontal deformation of the diaphragm wall indirectly. The guide pipes of the inclinometer are also used to measure the temperature profile in the center of the diaphragm wall.

5 Thermal Stresses in the Diaphragm Wall

The principal source of thermal stress in the diaphragm wall of Kawasaki Island is the heat generated during hydration of the constituents of the cement. Superplasticized concrete was selected in part because it generates relatively low heat during hydration. The following discussion describes the associated temperature changes and thermal stresses.

5.1 Temperature Changes in the Diaphragm Wall
Temperature changes in the diaphragm wall are continuously monitored by thermocouples installed in the wall. Following placement of concrete into a primary panel, the temperature curves of the concrete rise quickly toward a peak in the first 4 days following the pouring. Once a temperature curve passes its peak value, the concrete begins to cool at a rapidly decreasing rate. For the thermocouple near the outside face of the primary element, the peak cooling rate is 1.2°C/day at 7-day age. The cooling rate reduces to 0.7°C/day at 14-days age.

In a secondary panel following placement of concrete, the temperature curve at the center of the panel is quite similar to its primary panel counterpart, but the peak temperature is

higher and is attained faster. On the other hand, the
temperature curves at the outside face of the panel exhibit
different characteristics from their primary-panel
counterpart,cooling at smaller but rather stable rates. For
the curve near the center of extrados, the cooling rate is
about 2.0°C/day at 7-day age.

5.2 Steel Stress And Temperature Relationship

The measured stress curves in steel reinforcement generally
have a similar shape to their temperature curve counterparts,
strongly suggesting that temperature change is the major
cause of these stresses.

5.3 Steel Stress and Concrete Stress Relationship

The measured concrete stress generally has a linear
relationship with the corresponding measured steel stress.
However, the ratio between concrete and steel stress varies
with the location where the instruments are embedded,
suggesting a large variation in Young's Modulus for the
concrete in the diaphragm wall.

6 Observation of Cracking

The cracking patterns are readily observed on the inside
surface of the diaphragm wall following removal of the inner
steel jackets and embankment fill. Water leaking through the
cracks in the diaphragm wall moistens the concrete surface,
making it appear much darker than a dry surface. Figure 2
shows the wetness patterns on the inside surface of the
diaphragm wall.

Fig.2. View of Diaphragm Wall

The surfaces of the narrower primary panels are generally dry, while the lower portions of all secondary panels are wet. Some wet surfaces reach as high as Elevation TP-10 m., while the lowest ones are near sea bed level, Elevation TP-28.5 m.

Horizontal concrete cracks are clearly visible in all wetted areas of each secondary panel. No vertical cracks are visible. In general, there is considerable distance between two major neighboring horizontal cracks. The distances between horizontal cracks range in length from 2.5 m; to 6 m.

Horizontal drains installed near the elevation of the sea bed have been quite effective in collecting the water, which was initially substantial. Following installation of these drains, the lower surface of some panel elements actually have become drier than the portions above.

The speed of concrete placement does not appear to contribute substantially to thermal cracking in this type of concrete. No differences in cracking patterns were observed between portions of panels that were poured at differenct rates.

7 Discussion and Conclusions

It is observed that cracking occurs mainly in the concrete of the secondary panels of the diaphragm wall, and the cracks are predominantly horizontal. The patterns of cracking correspond closely to the conspicuous wetness patterns on the inside surfaces of the secondary panels.

There are essentially two possible forces at work that can move water through the diaphragm wall, causing it to wet the wall's surface: capillary force and gravitational force. The adhesion effect between water and concrete enables water to rise through vertical cracks in the concrete wall. The height of capillary rise of water depends upon the width of the concrete crack. Horizontal cracks, however, enlarge the path of the water and cut short the capillary rise. The other force, gravity, causes water to leak through both vertical and horizontal cracks in the wall as long as they are located below the phreatic line in the wall.

With this in mind, a number of inferences can be made related to the patterns of cracking observed at the inside face of the diaphragm wall.

The first inference is that temperature effects are the main cause of the observed cracking in secondary panels, and that secondary panels are more susceptible to thermal cracking. The primary panels, with a better heat transfer environment, show no major visible surface cracking. The primary panel is surrounded by cooler gravel, and then fill, allowing the hydration heat to radiate relatively freely through all surfaces of the primary panel element. During the period when the concrete is most susceptible to thermal effect cracking, no thermal gradient exists in the element that exceeds the critical thermal gradient. The heat transfer environment is less favorable for a secondary panel. Not only is the volume to surface area ratio of the secondary panel much smaller, it is also flanked by two primary elements which are much warmer than the embankment fill. As a result, concrete in the secondary panel has thermal gradients higher than the critical thermal gradient at the early age when it is inclined to experience thermal cracking. The effect of being sandwiched between two primary panels is predominantly thermal, rather than physical, or vertical cracking would be expected.

The second inference is that there does exist a critical volume-to-surface area ratio beyond which concrete in a diaphragm-wall panel is inclined to experience excessive thermal cracking. This critical volume-to-surface area ratio varies with the surrounding media. For the kind of superplasticized concrete used in Kawasaki Island, the critical volume-to-surface area ratio is somewhere between 0.7 m to 1.2 m, depending on whether the surrounding media is gravel, soil, improved soil, or finally, fill.

This observation has significant implication for designers and constructors of deep underground structures, such as LNG storage tanks and subway stations, in which the diaphragm wall construction method is employed. It would be possible to incorporate the influence of the surrounding media's thermal conducting properties into the design as some sort of weighted surface area requirement for the concrete element. In fact, for the secondary panels at Kawasaki Island thermal cracking in concrete at early age would be greatly reduced if excavation machines with a cutting width of around 2 m were used. Where conditions are such that a secondary panel has to be greater than 2 m thick, part of the secondary panel could be precast or else constructed later.

The fourth inference is that the phenomenon of thermal cracking at early age in the diaphragm wall can be adequately simulated by one-dimensional analytical models, which can take into account the temporal variations of the Young's Modulus and strength of concrete, temporal variation of thermal gradient, and the thermal conductivities of boundary media. With additional instrumentation, adequate field data could be accumulated for calibration of such an analytical model.

8 References

1. Liou, D.D. **Instrumenting a Man-Made Island Project**, Proceedings of ARECDAO 93, Barcelona, Spain (March, 1993).

49 MEASURES TO AVOID TEMPERATURE CRACKS IN CONCRETE FOR A BRIDGE DECK

W. FLEISCHER and R. SPRINGENSCHMID
Institute of Building Materials, Technical University of Munich,
Germany

Abstract
It is intended to concrete a 296 m long bridge deck, which is more than 1.60 m thick in the middle and only 0.40 m thick at the brackets in one continuous working operation without interruption. In order to avoid temperature cracks in the concrete, i. e. are essentially surface cracks due to intrinsic stresses (eigen stresses) and cracks due to longitudinal stresses, particular measures are necessary. Above all, these measures concern the mix design (using a cement with a cracking temperature below 10 °C as tested in the standard test in the cracking frame), the temperature of the fresh concrete (7 to 12 °C), tests in the laboratory (e. g. tests in the temperature-stress testing machine to estimate the development of the temperatures and stresses in the bridge deck), as well as the construction itself. The constructional measures involve influencing the degree of restraint by the process of concreting, wet-curing of the surface on the first day, which causes cooling of the surface, and thermal insulating the formwork of the brackets.
Keywords: Bridge Deck, Cracking Frame, Temperature-Stress Testing Machine, Cracking Temperature, Restraint, Intrinsic Stresses, Cracking, Curing.

1 Introduction

Next to an existing bridge of the German autobahn A 9 across the valley of the river Saale near the small town of Rudolphstein, close to Hof - i. e. the frontier bridge between Bavaria and the former German Democratic Republic - the construction of a new bridge is planned in conjunction with the widening of the autobahn from four to six traffic lanes. In order to adjust the shape of the new bridge to the existing bridge with 8 arcs made of natural stone, the new bridge will have 10 spans (Fig. 1).

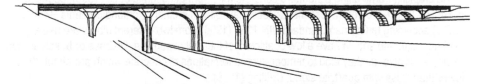

Fig. 1. View of the planned bridge near the small town of Rudolphstein in Bavaria and the existing bridge in the background (design: Autobahndirektion Nordbayern, Nürnberg and Ingenieurbüro Prof. Dr.-Ing. G. Scholz & Partner, München)

Thermal Cracking in Concrete at Early Ages. Edited by R. Springenschmid. Published 1994 by E & FN Spon, 2–6 Boundary Row, London SE1 8HN, UK. ISBN: 0 419 18710 3.

The bridge deck is 296 m long. The cross-section of the bridge deck can be seen in Fig. 2. The bridge deck will only be reinforced and not prestressed. The static system will be a continuous slab made of central-mixed concrete and not of precast slabs.

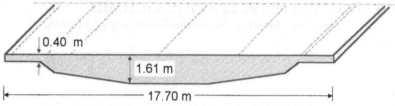

0.40 m

1.61 m

17.70 m

Fig. 2: Cross-section of the planned bridge deck

As joints always entail a maintenance problem, it is intended to concrete the bridge deck in one continuous working operation without interruption. This means that about 6,000 m^3 of concrete have to be placed without interruption. If it is possible to place 100 m^3 of concrete per hour, the total time required for concreting will be almost 3 days in shifts. Due to this demand and due to the large dimensions in the middle of the cross-section as compared to the thin brackets, a special concrete technology is necessary to avoid temperature cracks. Simply increasing the amount of reinforcement would not be sufficient, because this can not prevent cracking (re-inforced concrete needs cracks to fully activate the reinforcement), but can only influence the amount, distance, length and width of the cracks. To minimize the cracking risk of the concrete measures are taken concerning composition, temperature, curing and degree of restraint of the concrete.

2 Concrete Technology

To minimize the stresses due to heat of hydration, merely minimizing the heat development of the cement is not sufficient, Springenschmid et al. (1986), Mangold et. al (1994). The aim is that only small tensile stresses due to temperature changes are induced in the concrete. Therefore the following requirements have already been specified in the project tender specification:

(1) The cement should have a cracking temperature below 10 °C as tested in the standard test in the cracking-frame using a standard concrete mix, Springenschmid et al. (1990), because it is known, that very different restaint stresses may result if ordinary Portland cements PZ 35 F (OPC) according to German Standard DIN 1164 (1990) from two different plants are used.
(2) The aggregate should have a low coefficient of thermal expansion. Diabase or basalt are for example suitable, as they lead to temperature deformations of concrete which are about 30 % lower than those with quartzite stone, Dettling (1959).
(3) The cement content should be about 280 kg/m^3. Fly ash according to German Guideline (1990) should be added so that the content of cement and fly ash reaches 340 kg/m^3.
(4) The required concrete strength B 35 (which means an average compressive strength of cubes of at least 40 N/mm^2 according to German Standard DIN 1045 (1988) should be reached at an age of 90 days, instead of 28 days.

(5) The temperature of the fresh concrete immediately before compacting should be 7 to 12 °C. If necessary, the fresh concrete should be cooled using for example liquid nitrogen.

Tests in the temperature-stress testing machine are performed to estimate the temperature stresses which may develop in the bridge deck. This machine is similar to the cracking frame, Springenschmid et al. (1994). It allows simulation of temperature course of thick concrete layers and measurement of the longitudinal stresses under full restraint. For these tests concretes of the same mixes and temperatures as at the future job-site are used. During construction the stresses, which will actually appear in the bridge deck (i. e. stresses due to load, restraint and intrinsic stresses), will be measured by "stress-meters" according to Tanabe (1993).

3 Temperature Development and Stresses at Early Ages

3.1 Survey
Owing to the shape and dimensions of the bridge deck the following two sections are necessary for an investigation of the temperature changes and the resulting stresses:

(1) Longitudinal restraint stresses in the longitudinal direction of the bridge deck.
(2) Intrinsic stresses (eigen stresses) over the complete cross-section in Fig. 2.
Furthermore, for the cross-section it is distinguished between
(2a) the distribution of the stresses over the thickness of the cross-section (vertical direction),
(2b) the stresses which arise due to the different dimensions of the bridge deck at the thin brackets and in the thick central section (horizontal direction).

3.2 Restraint Stresses in the Bridge Deck in Longitudinal Direction
For the cracking susceptibility the temperature development itself of the concrete is not decisive, but the stresses which arise in the concrete. In order to estimate the stresses, tests in the temperature-stress testing machine are performed, which take the special circumstances of the bridge deck into account. These tests give the stress response due to the change in concrete temperature.

Because the concreting of the bridge deck will take nearly 3 days, the concrete temperatures will show large differences over the longitudinal direction of the bridge. The concrete which has been placed at first will already be cooling when the last concrete is being placed. For the assessment of the longitudinal stresses the second zero-stress temperature is decisive.

If the deformations in the longitudinal direction are restrained, the younger concrete will show favourable compressive stresses, which decrease due to decreasing temperature. After passing the second zero-stress temperature cooling results in tensile stresses (Fig. 3, 4). If the deformations are not restrained, expansions instead of compressive stresses and contractions instead of tensile stresses will arise in the longitudinal direction. Therefore in order to minimize tensile stresses, the concrete should be able to contract wherever tensile stresses may occur.

In the present case longitudinal deformations of the concrete are restrained by the neighbouring concrete and to a small degree by the formwork. The reinforcement and the piers do not provide considerable restraint on the deformations of the bridge deck. However, favourable compressive stresses arise to a small degree if the expansion of a concrete section due to heating is partly restrained. These compressive stresses are reduced when the concrete is cooled down to the second zero-stress temperature (Fig. 3, 4).

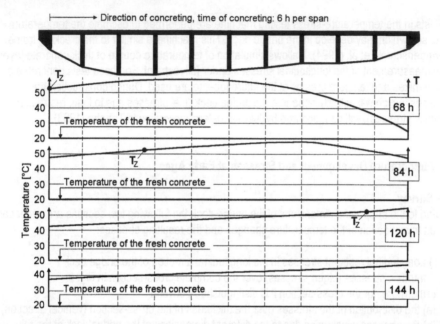

Fig. 3: *Temperature in the centre of the cross-section of the bridge deck between 68 and 144 hours after start of concreting (T_Z: temperature is equal to actual zero-stress temperature)*

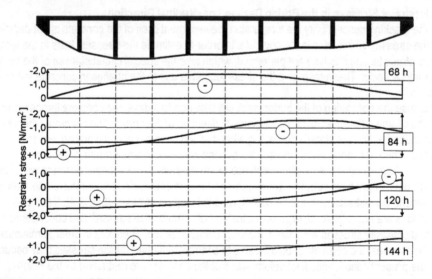

Fig. 4: *Longitudinal stresses if full restraint of deformations in the centre of the cross-section of the bridge deck is assumed (according to tests in the temperature-stress testing machine)*

If the expansions of the young concrete are not restrained tensile stresses would occur as soon as the temperature of the concrete decreases. In this case the second zero-stress temperature is equivalent to the highest temperature of the concrete, which means an unfavourably very high second zero-stress temperature.

As soon as tensile stresses occur due to restrained deformations, contractions of the concrete section should be allowed in the longitudinal direction. The neighbouring concrete sections should prevent the contractions as little as possible. In this way the tensile stresses in the concrete section are reduced. To enable this the formwork should have joints for example.

Summarized, this means that the degree of restraint can be influenced by the concreting process. For example the concreting should be finished in about 60 to 70 hours, if it is intended to avoid that the concrete placed last, will crack due to tensile stresses resulting from contractions which will occur in the older concrete, because the older concrete has been reached the second zero-stress temperature.

3.3 Intrinsic Stresses over the Thickness of the Cross-Section
Present practice in most cases, is to cover the concrete with thermal insulating mats after compaction. Nevertheless, surface cracks appear often as soon as 2 to 3 days after concreting especially in concrete members with a thick cross-section. Such surface cracks are only a few centimetres deep and often do not close completely. They originate in the intrinsic stresses (eigen stresses) which remain in the concrete member after the concrete has cooled and are caused by the gradient of the zero-stress temperature. The intrinsic stresses can be minimized by influencing the temperature of the concrete during the first days.

Artificial cooling of the surface concrete is favourable according to Mangold (1994),

(1) if cooling starts as soon as possible after concreting, i. e. the concrete still shows plastic deformations. This causes a significantly lower zero-stress temperature T_Z at the surface than in the middle of the cross-section.
(2) if the expansion of the concrete member is restrained during the surface cooling procedure. If there is no restraint, compressive stresses can not occur at the surface in spite of heating due to hydration. In this case only moderate cooling should be used to avoid early cracks.

How may these considerations be applied to the thick cross-section of the present bridge deck? Because of the large thickness of the cross-section curing can only influence the surface of the concrete member. With the help of curing a negative gradient of zero-stress temperature T_Z (T_Z lower at the surface than in the middle) can be created, which is favourable for the time of construction, i. e. before epoxy-sealing of the surface and laying of the asphalt pavement. In this case only small tensile intrinsic stresses occur, if the uncovered concrete surface is cooled suddenly (e. g. due to a thunder storm combined with a large decrease in temperature; a not uncommon occurrence in Central Europe), the risk is low that surface cracks occur in the concrete before the sealing. For this the curing should be done in two steps.

(1) Cooling of the concrete surface. For example with thin felts on the surface which is either already wet or is wetted immediately after placing. Afterwards care should be taken that the felts do not dry out. Due to the evaporation of the water from the felts cooling occurs which causes moderate cooling of the concrete surface.

(2) After about one day the felts are covered with mats, preventing further evaporation of water and cooling of the surface down to the ambient temperature, while the concrete is still being heated due to hydration. At the same time the concrete can take up water from the felts, which is still wet. This minimizes shrinkage of the concrete.

Why should cooling stop after about one day? For an explanation the development of the modulus of elasticty, of the temperatures and the longitudinal restraint stresses in the centre and at the surface of the cross-sections may be looked at in four different cases at least (Fig. 5):

(1) Centre of the cross-section. Here curing by cooling of the surface plays only a minor part.
(2) Upper surface of the cross-section cured by covering with thermally insulating mats.
(3) Upper surface of the cross-section cured by cooling with water (cooling due to evaporation) for 5 days.
(4) Upper surface of the cross-section cured by cooling with water on the first day and then covered with mats which prevent further evaporation and cause a moderate cooling of the concrete temperature to the ambient temperature.

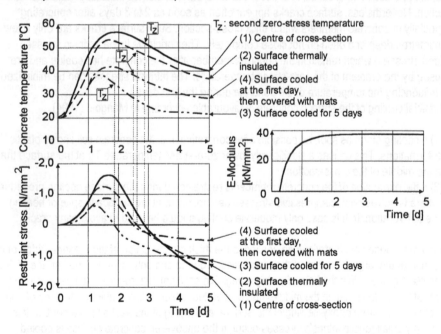

Fig. 5: Temperature and stress curves (restrained deformations) of the concrete (at the surface of the bridge deck under different curing conditions and in the centre of the thick cross-section) as well as the development of the modulus of elasticity (according to tests in the temperature-stress testing machine, curves (2) to (4) are estimations according to Mangold (1994))

The development of the temperatures and stresses (Fig. 5) show that curing by covering with thermally insulating mats causes a second zero-stress temperature which is up to 15 K higher than when the concrete surface is cooled with water. Curing by cooling the surface results in a favourable negative gradient of the zero-stress temperature. Furthermore, it can be seen that the concrete surface should not be cooled too long. Cooling should only last until the surface concrete reaches the second zero-stress temperature. If cooling lasts longer additional tensile stresses result at the surface, which are nearly as large as those stresses which arise due to covering with thermally insulating mats.

However, the investigations should not be confined to the surface region and middle section, but also the development of temperature and stress in the whole cross-section should be taken into consideration. After nearly one day the modulus of elasticity shows only very small increase. At the same time the centre of the cross-section with the large thickness is still heating up (Fig. 5) and expanding. Therefore tensile stresses arise at the surface. If at the same time a thin surface area would be cooled, additional tensile stresses would occur which would not relax because of the high modulus of elasticity and therefore surface cracks could arise.

What is the main effect of a negative gradient of the zero-stress temperature at the surface area, due to curing the surface with water-cooling on the first day as compared to covering the surface with thermally insulating mats? Rapid cooling of the surface after stripping the insulation and before sealing and laying the asphalt pavement, is critical. Because if insulating mats are used for curing, larger tensile intrinsic stresses occur at the surface due to rapid cooling, than if the surface is cured by cooling already on the first day.

3.4 Tensile Stresses in the Brackets in the Longitudinal Direction
The cross-section of the bridge deck is up to four times thicker in the middle than at the brackets (Fig. 2). In order to avoid transverse cracks in the brackets due to different temperatures between the brackets and the rest of the bridge deck, it would be advantageous to insulate the formwork of the brackets against temperature loss only before the second zero-stress temperature is reached. Then the concrete of the thin brackets would not be cooled quicker than the thicker parts of the cross-section. However this is not possible in practice. The only possibility is to secure thermal insulation to the formwork of the brackets before concreting and remove it after hardening. Thus the concrete in the brackets will have nearly the same temperature development as the concrete of the thicker sections. That means the development of the deformations and the stresses of the concrete of the brackets will not differ significantly from the concrete of the thicker parts of the cross-section.

4 Conclusions

It is planned to concrete a 296 m long bridge deck of a German autobahn in one continous working operation without interruption. The deck is more than 1.60 m thick in the middle and only 0.40 m thick at the brackets. A special mix-design, additional tests in the laboratory, and special measures during construction are necessary to minimize the cracking risk of this concrete member. The most important are as follows.

(1) It is not sufficient only to limit the heat development of the cement. In addition the cement should have a cracking temperature below 10 °C, tested with the standard test in the cracking frame.

(2) Tests in the temperature-stress testing machine will be performed with concretes of the same mixes and temperatures as at the job-site, to estimate the temperature stresses which

may develop in the concrete member.

(3) Due to the great length of the bridge deck without joints and due to the time required for concreting, great differences in the stresses in the longitudinal direction occur already during concreting (e. g. favourable compressive stresses in the concrete which is only a few hours old and unfavourable tensile stresses in the older concrete). The process of concreting should be carried out in such way that the restraint of deformations, which mostly occur because of neighbouring concrete sections, results initially in compressive stresses which lead to subsequent tensile stresses which are as low as possible.

(4) Wet-curing of the concrete surface on the first day by covering with wet felts, which cools the surface, results in favourable low zero-stress temperatures at the surface and a favourable negative gradient of the zero-stress temperature. This minimizes the danger, that surface cracks arise due to a rapid cooling of the surface before sealing.

(5) After one day wet-curing the felt should be covered with mats to avoid a large temperature difference between the surface and the middle of the concrete, which is still heating up due to hydration, leading to tensile stresses at the surface.

(6) The formwork of the brackets should be thermally insulated to minimize the temperature differences between the thin brackets and the thick middle section of the bridge deck, thus surface and transverse cracks are avoided in the brackets.

5 References

Dettling, H. (1959) Die Wärmedehnung des Zementsteins, der Gesteine und der Betone. **Doctoral thesis**, Technische Hochschule Stuttgart.

German Guideline (1990) Richtlinien für die Zuteilung von Prüfzeichen für anorganische Bindemittel - Prüfrichtlinien. **Mitteilung IfBt**, 4, 131-140.

German Standard DIN 1045 (1988) Reinforced concrete structures; design and construction.

German Standard DIN 1164 part 1 (1990) Portland cement, Portland blast furnace cement, blast furnace slag cement and trass cement; terminology, constituents, requirements, delivery.

Mangold, M. (1994) Thermal prestress of concrete by surface cooling, in **Proceedings of the international symposium, Thermal cracking of concrete at early ages**, Munich, October, 10-12, 1994.

Mangold, M. and Springenschmid, R. (1994) Why are temperature-related criteria so unde-pendable to predict thermal cracking at early ages? in **Proceedings of the international symposium, Thermal cracking of concrete at early ages**, Munich, October, 10-12, 1994.

Springenschmid, R. and Breitenbücher, R. (1986) Are low-heat cements the most favourable cements for the prevention of cracks due to heat of hydration? **Concrete Precasting Plant Technology**, 11, 704-711.

Springenschmid, R. and Breitenbücher, R. (1990) Beurteilung der Reißneigung anhand der Rißtemperatur von jungem Beton bei Zwang. **Beton- und Stahlbetonbau**, 2, 29-33.

Springenschmid, R., Breitenbücher, R., and Mangold, M. (1994) Development of the cracking frame and the temperature-stress testing machine, in **Proceedings of the international symposium, Thermal cracking of concrete at early ages**, Munich, October, 10-12, 1994.

Tanabe, T. (1993) Method for in-situ measurement of thermal stress in concrete using a special developed stress-meter. **Rilem Technical Recommendation**, TC 119, 2nd Draft.

50 AVOIDANCE OF EARLY AGE THERMAL CRACKING IN CONCRETE STRUCTURES - PREDESIGN, MEASURES, FOLLOW-UP

M. EMBORG and S. BERNANDER
Department of Civil Engineering, Division of Structural Engineering,
Luleå University of Technology, Luleå, Sweden

Abstract
By means of newly developed methods for thermal stress analysis of concrete elements, it is possible to compute the risk of cracking for different structural scenarios. This has resulted in simulation procedures by which the problem of temperature induced stress build-up, and the consequent possible cracking can be analyzed in ways that to a significantly higher degree than before correspond to the enormous complexity of the issue.

The simulation procedures have frequently been used in Sweden for obtaining crack-free concrete in massive structures (such as bridge piers, tunnels, foundations) as well as thinner structures (basin walls, bridge girders etc.). Precalculations of cracking risks are performed for structures during design and planning of construction.
Keywords: Computer Simulation, Measures for Avoiding Cracking, Risk of Cracking, Structural Members.

1 General

Newly developed methods for computer aided simulations of stress development induced by temperature volume change in early age concrete have been developed at the Luleå University of Technology. The modelling used in the computer programs permits consideration of - besides temperature - restraint, transient mechanical properties, non-uniform maturity development, construction sequence and various environmental aspects.

This new approach is very useful in the predesign, design and construction of structural members in which hydration stresses may become critical. By means of precalculations the relative efficiency with regard to cracking of different constructional measures can be quantified and optimized (see e. g. Emborg and Byfors (1993)). The approach will be demonstrated in this paper by means of some examples.

2 Example 1: Risk of very early surface cracks due to differential temperatures over a cross section - impact of non-uniform maturity

As been emphasized by Bernander and Emborg (1994), the consideration of differentiated temperatures and maturity is responsible for the most notable discrepancies between the

Thermal Cracking in Concrete at Early Ages. Edited by R. Springenschmid. Published 1994
by E & FN Spon, 2–6 Boundary Row, London SE1 8HN, UK. ISBN: 0 419 18710 3.

results gained by the new analysis applied and corresponding assessments based on conventional temperature-related criteria. A characteristic feature of the thermal stress distribution in the cross section is the presence of high early-age tensile stresses at the surface, changing into compression later in the hydration temperature cycle, see Fig 1a. Possible cracks tend to close.

Fig 1b illustrates maximum early age tensile stress levels in walls of different depths - 0.5 - 2.0 m at different ambient air temperatures, T_{air}, and for varying initial concrete temperatures T_i. The parameter on the horizontal axis indicates corresponding maximum temperature differences ΔT_{max} over the cross section. No external restraint is present. In the diagram, the lines B-B and C-C may be said to represent the relationship between stress levels and temperature differences under respectively favorable and unfavorable conditions. If for instance the stress level $\sigma(t)/f_{ct}(t) = 0.7$ was to be chosen as a limit criterion, then - depending on initial concrete and ambient temperatures - the acceptable differential temperature over a 2m thick wall section would vary from 13 °C to 34 °C.

An important finding, that has emerged with increasing experience gained by the thermal stress analysis practiced by the authors, is that the relationship between ambient air temperature and initial concrete temperature has a vital impact on the very early age stresses and susceptibility to surface cracking due to temperature differentials. This phenomenon is mainly due to the fact that temperature-time development may vary widely within a concrete section giving rise to corresponding differences in maturity, strength, creep and other mechanical properties. The effect is demonstrated in Example 1 involving stress analysis of a 1.5 m thick concrete wall without external restraint.Three scenarios with regard to initial and ambient air temperature are investigated, the stress levels of which are shown in Table 1 derived from the graph in Fig 1b. As may be seen the stress levels in Case I   are in fact lower than in Case II and Case III. Yet the temperature difference is markedly greater in Case I (24.5 °C) than in Case II and III (18 °C and 14.5 °C respectively): Thus, the correlation factors in Table 1 are not only very different from unity but differ in the three cases by a factor up to 2 allowing for very little confidence in temperature-related criteria.

In Case II, with the low air temperature and the higher initial pouring temperature, concrete maturity near the surfaces will be effectively retarded by heat losses to the cold environment while the maturity development in the core will be enhanced by the higher temperatures there. Hence, much of the volume expansion in the central parts of the wall will take place while the concrete at the surfaces is in a more plastic state. This condition is favorable as it effectively decreases the risk of early age surface cracks.

In Case III - on the contrary - the surface concrete will be leading the strength and maturity development across the section. Hence, when, eventually, the delayed rapid volume expansion takes place in the interior, high stresses develop in the stiffer surface concrete layers. As a result of advanced maturity, the stiffness of the surface is already high and its capacity to deform plastically in the limited time interval of maximum rate of interior expansion is very restricted. Therefore, the risk of surface cracks increases appreciably.

It may be deduced from Fig 1 and Table 1 that the winter situation is more favorable than summer situations despite the fact that the temperature difference ΔT may be more than 70 % higher in the winter. It may also be deduced from the figure, that for the studied type of concrete and form, there is a very low risk of surface cracks for wall thicknesses less than about 1 m.

Diagrams of the kind shown in Fig. 1 form an excellent tool for early planning of crack-free structures. By recording the temperature differences ΔT and maximum temperatures in centre of the structure the 'real' cracking risk may readily be checked. Graphs similar to Fig 1. are under evaluation in a code-related project for different cement types and concrete mixtures as well as for various types of form, insulation, wind speeds etc. The diagrams will constitute a base for future Swedish regulations and for optimization of measures against cracking.

Table 1 Three cases of differential temperature over a cross section (d=1.5m)

	Parameter	Case I	Case II	Case III
Initial concrete temperature	T_i	20 °C	20 °C	10 °C
Ambient air temperature	T_{air}	5 °C	20 °C	20 °C
Temperature differentials across the section	ΔT	24.5 °C	18 °C	14.5 °C
Tensional stress levels with TEMPSTRE-N	$\eta_{cr} = [\sigma(t)/f_{ct}(t)]^{max}$	0.45	0.51	0.52
Early age cracking risk based on temperature criterion $\Delta T_{cr} = 20°C$	$\Delta T / \Delta T_{cr}$	1.225	0.90	0.752
Correlation factor	$\dfrac{\Delta T / \Delta T_{cr}}{\eta_{cr}}$	2.72	1.76	1.39

3 Example 2: Wall cast on a rigid sub-base - restraint effects within section

When planning for crack-free concrete it is extremely important to make realistic assessments with respect to the degrees of restraints. Fig 2 illustrates a 2 m thick wall on a stiff sub-base, where the external restraint at the base is 100 % while the corresponding restraint in the upper part is negligible. As may be seen from the graph, the stress development in the two locations are radically different. At location (2) early age tensile stresses occur at the surface in analogy with Example 1 above indicating a peak stress at about 3 days. In contrast at the base of the wall (location (1)), the restraint from the slab dominates inducing a stress development characterized by early age compressive stresses and subsequent tensile stresses. Further, if the wall is free to translate on the sub-base, the influence of rotation of the wall and slab must be considered, see e. g. Bernander and Emborg (1994).

This example demonstrates the crucial influence of the degree and nature of restraint - a parameter that is still vastly overlooked even in concrete engineering of today.

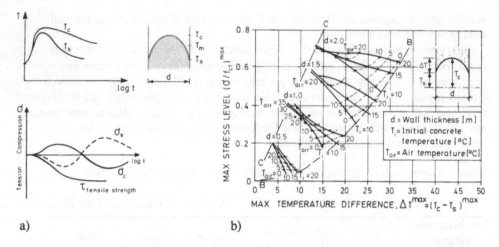

a)   b)

Fig 1. a) Temperature differences over a wall section causing early age tensile stresses
at surface.
b) Maximum tensile stress level as a function of maximum temperature
difference between core and surface for different placing temperatures and air
temperatures. Computations by menas of TEMPSTRE-N. Concrete: Swedish
Standard Portland Cement, type Slite, cement content 331 kg/m^3 w/c=0.55,
$f_{cc}(28)$ = 53.8 MPa (cube strength). Form stripped a the age of 4 days

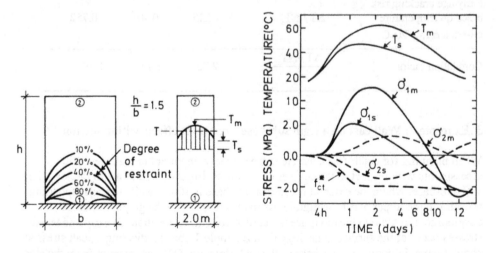

Fig 2.   Wall cast on a stiff foundation. Thermal stresses in the upper and lower part of a
wall. T_s and T_m denote surface and mid section temperatures respectively. σ_{1s},
σ_{1m}, σ_{2s}, and σ_{2m} are corresponding stresses in locations (1) and (2).
f_{ct} denotes the tensile strength. $T_i = T_{air}$ = 15 °C. Cement: Std Portland
350 kg/m^3. Form :12 mm plywood.

4 Example 3: Effects of varying restraint from sub-base

In a major bridge project a large foundation slab was to be cast on a rough concrete sub-base. The preevaluations of thermal stresses demonstrated, as in Example 2 above, the importance of considering restraint correctly. However, in this case the restraint could be changed and optimized. The temperature field in the foundation is shown in Fig. 3a, indicating a maximum mean temperature of about 60 °C and a temperature difference between mean temperature of the interior parts and the rims of the foundation of $\Delta T_{max} \leq 15$ °C - this being a restriction set by the client for this particular object.

By varying the degree of restraint (R), modifying the geometry of the foundation (i. e. L/B in Fig. 3b) and studying different locations on the edges of the slab (i. e. midpoint on side (1) and corner (2)) completely different thermal stresses were obtained. Curve A corresponds to location (1), restraint R = 100% and L/B = ∞, leading to compressive stresses in the heating phase and tensile stresses later on. For R= 0, a quite different stress-time curve is obtained (curve B) showing high very early age peripheral tensions which gradually change into compression. For location (2) other levels of stress were evaluated.

By optimizing the restraint R, e. g. by introducing a partly flexible material, reinforcement etc. at the interface between the foundation and subbase, the tensile stresses could be minimized. For instance, increasing the degree of external restraint provokes early age compression which neutralizes peripheral tensions.

5 Example 4: Efficiency of measures against cracking of a tunnel element

Wall and roof elements of typical tunnel sections, Fig 4a, are during hydration subjected to restraints both related to different temperatures within the wall (and the roof) and to the stiffness of the existing slab. In evaluations of crack risk of such sections, various scenarios with regard to pouring must be studied e. g. different air and placing temperatures, effects of rate of pouring the concrete, artificial heating of the slab, cooling of the concrete, effects of bending and translation.

In Helsingborg, Sweden, a long railway tunnel was constructed. It was found in the predesign phase that the risk of cracking was too high for nearly all pouring scenarios. An effective way of reducing high tensile stresses was to cool the hardening concrete by means of embedded cooling pipes placed in the lower part of the wall see Figs 4a and 4b. Hydration temperatures on the building site were recorded (Fig 4c) constituting a basis for renewed stress evaluations showing maximum tensile stress levels less than 0.6, i.e. the risks of cracking were not too high. The final experiences of casting the tunnel section in Helsingborg with internal cooling were very positive - more than 2000 m tunnel walls were cast without any thermal cracking.

For some other tunnel sections in Stockholm, the crack risk was reduced by heating the existing slab by means of resistance cables, see Fig 5a. The objective of this method is to synchronize - as far a possible - the temperature cycle of both wall and slab, thus reducing the effective restraint. Fig 5b exemplifies temperatures in a slab and a wall under very similar conditions leading to a minimum tensile stresses. The crack risk, in terms of stress levels, was in some cases reduced from about 0.8 for non-heated sections to less than 0.2 for sections equipped with heating cables.

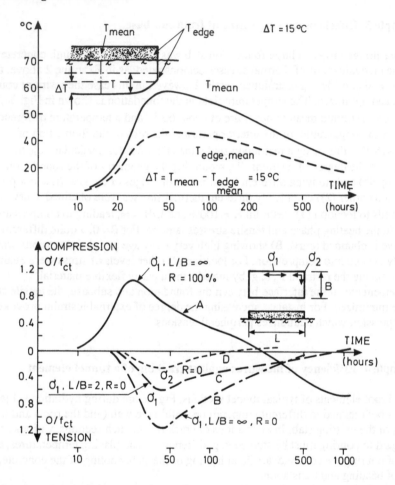

Fig 3   a) Mean temperatures in a foundation slab cast on a tremie concrete subbase.
b) Thermal stresses for different restraints from the sub-base, different geometry
of the foundation and at two locations. Standard Portland cement 314 kg/m³ , fly
ash 70.9 kg/m³, silica fume 20.2 kg/m³, thickness of the slab: 2.5 m

6 Conclusions

The application of thermal stress analysis in early age concrete according to the Luleå
approach has not only proved to be a powerful practical engineering tool, but has in many
ways modified - in som aspects radically changed - our understanding of hydration
temperature cracking phenomena.

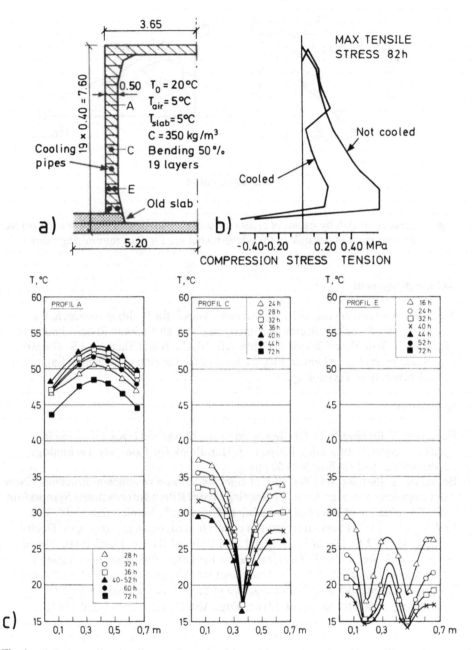

Fig 4   a) Box tunnel element cast on a cold slab

b) Thermal stress analysis indicating the reduction of cracking risk for a cooled section compared to a non-cooled section.

c) Examples of recorded temperatures at three levels of the wall with different degrees of cooling.

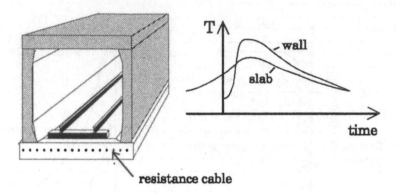

Fig 5.   Heating of slab by means of resistance cables rendering as uniform a temperature
development as possible in the whole tunnel section thus minimizing crack risk.

Acknowledgement

This work has been supported by the Swedish Council for Building Research, the
Foundation for Swedish Concrete Research, the Swedish National Road Administration,
the Swedish State Power Board, Cementa AB, NCC AB and Skanska AB. The work was
supervised by Professor Lennart Elfgren, head of the Divsion of Structural Engineering at
Luleå University of Technology.

References

Bernander S, Emborg M (1992) Temperature conditions and crack limitation in
   massive concrete structures. Chapter 27, **Handbook for Concrete Technology,**
   Stockholm, 1992 (in Swedish) 50 pp
Bernander S, Emborg M (1994) Risk of cracking in massive concrete structures - New
   developments and experience. **Proceedings from Rilem International Symposium
   on "Avoidance of Thermal Cracking in Concrete"**, Munich Oct. 1994, 8 pp.
Emborg M (1990): **Thermal stresses in concrete structures at early ages**, Doctoral
   Thesis 1989:73 D, Div of Struct Eng, Luleå Univ. of Techn, Luleå 1990, 286 pp
Emborg M, Byfors J (1993) **The Crack-Free Program**. Procedure of assessing a
   concrete without thermal cracks. Betongindustri AB, Stockholm 1992 (in Swedish)
Emborg M, Bernander S (1994) Assessment of the risk of thermal cracking in
   hardening concrete, **Journ. of Struc. Eng (ASCE),** (to be published 1994)

51 WATER TIGHT DESIGN, ARTIFICIAL COOLING OR EXTRA REINFORCEMENT

W.G.L. WAGENAARS
Hollandsche Beton Groep (HBG), Gouda, The Netherlands
K. van BREUGEL
Delft University of Technology, Delft, The Netherlands

Abstract
For a typical floor-wall-roof construction with variable dimensions it has been investigated which method of avoidance or preventing of early age thermal cracks, could be applied most economically if water tightness of the structure would be the main design criterion. Tightness of the structure is considered to be achievable by either prevention of cracking by cooling during the hardening process or stringent crack width control by application of extra reinforcement. Conclusions were formulated from the relationship between the length of the structure and the thickness of the walls on the one hand and on the other hand the most economical way to assure water tightness.
Keywords: Cooling Systems, Curing by Artificial Cooling/Heating, Extra Reinforcement, Self Healing Cracks, Water-tight Concrete, Width of Cracks.

1 Introduction

When water comes in contact with cement, a chemical reaction takes places by which heat is being generated. Mainly for structures with thick walls, where the resulting heat cannot dissipate rapidly to the environment, the temperatures in a structures can rise considerably. If the corresponding volume variation is prevented, the structure may crack. Depending on the kind of structure, this early age thermal cracking can or cannot be allowed. Criteria for controling or preventing early age thermal cracking can for instance be judged on base of water tightness or preventing corrosion of reinforcement. In this present investigation water tightness of a structure has been the main design criterion.

There are several possibilities to control or avoid early age thermal cracking. Two reasonable options are:

- internal cooling with cooling pipes to prevent cracking
- extra reinforcement to control the occurring cracks

Thermal Cracking in Concrete at Early Ages. Edited by R. Springenschmid. Published 1994 by E & FN Spon, 2–6 Boundary Row, London SE1 8HN, UK. ISBN: 0 419 18710 3.

Both measures escalate costs. Consequently the question rises, which control measurement leads to a water tight structure at the lowest costs.

A typical floor-wall-roof construction with variable dimensions has been investigated (fig. 1). The floor of this typical construction is cast first, thereafter the wall and the roof are casted simultaneously on the already hardened structure.

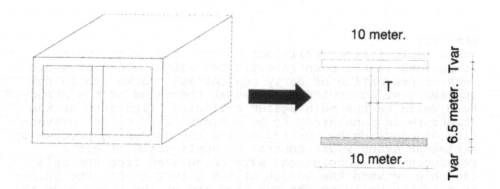

Fig.1   Typical floor-wall-roof construction, with variable wall thickness(t_{var}) and length (L_{constr})

2   Cracking

Prior to discussing measures to control or avoid early age thermal cracking, it is necessary to consider the negative effects of cracking. Problems raised by cracking can be divided into three categories:

- loss of durability
- decrease of the functionality of the structure (for example decrease of water tightness)
- decrease of the esthetic value

2.1 Loss of durability
Main purpose of durability requirements is to reduce the risk of corrosion of the reinforcement. Corrosion of the reinforcement, with the exception of the crack width, also dependents upon the climatic environment, the quality of the concrete and the concrete cover. In several international codes a minimum concrete cover of approximately 50 mm is demanded for these kind of structures. For this thickness of the cover a maximum crack width of 0.3 - 0.5 mm is allowed. The decisive maximum crack width needed for water tightness of a structure is lower. Therefore, if the crack width is reduced for purposes of water tightness, the risk of corrosion of the reinforcement will be acceptable.

2.2 Decrease of water tightness

Practice teaches, and test results have confirmed this, that penetrating tension cracks, if not too large, may heal. This information can be of use when permiting early age thermal cracks.

This self healing capacity of concrete structures consists of the possibility to narrow the flowing canals (continuous cracks). Prior conditions for self healing are, among other things;

- the crack width and the surface roughness
- water current is low
- the crack is stable

The main mechanism responsible for the self healing process are;

- narrowing of the flowing canals by inorganic parts carried along with the water flow
- blockage by small crumbling concrete parts
- formation or depositing of chalk
- continuous hydratation of cement

It is difficult to explain to what extent each phenomenon is responsible for the total self healing process. Restrictive information on this subject comes mainly from German laboratories. Results from experimental research and observations in practice show, that stable cracks with a crack width of 0.20 to 0.25 mm were almost completely healed within approximately one week. It is however, not acceptable to translate these research results directly in practical design criteria. There are some uncertainties, i.e:

- the type of groundwater, pollution etc.
- weak points in the structure
- concrete quality, roughness, kind of granulate
- uncertainties in the maximum occurring water pressure
- peripheral disturbance, interfaces etc.

Taking into consideration the above mentioned uncertainties, it has been assumed that cracks resulting from a singly enforced deformation, with a maximum crack width of 0.15 mm, are capable to heal.

2.3 Cracking leading to a decrease of the esthetics value

The acceptance of visual cracks is mainly to be decided by the client. Practice teaches that there are great differences between clients. Because of the subjective character of this criteria, it has been decided to judge the structure on its functionality. In almost all cases the durability criteria will then be satisfied implicitly.

3 Artificial Cooling

For the determination of the required cooling capacity two
computer programmes were used, viz. the programme "VBS" and
"MANDRY" [1,8]. The " VBS " (concrete curing system) is a
finite element package, which calculates for a time step
(normally 15 minutes), the development in temperature,
hydration, strength and stresses (fig 2).

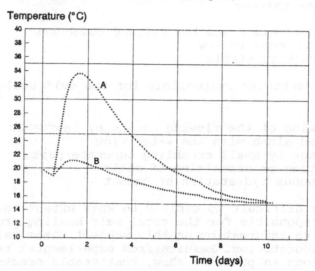

Fig.2   Example of temperture development in the center of
         a concrete wall, wall thickness is 750 mm. A = wall
         without cooling, B = artificially cooled.

For the determination of the risk of cracking a probabilistic approach has been followed. The required reduction of
the peak temperatures was determined so as to prevent the
maximum thermal stresses to exceed the (calculated) actual
minimal tensile strength of the concrete (fig 3).
    The required reduction of the peak tempertures has to be
translated into a number of cooling pipes. The " MANDRY "
theory was used for designing the number, position and
diameter, of the cooling pipes. For the design of the
required cooling capacity three different cooling systems
were considered:

- a closed cooling system with a cooling aggregate
- an open system using a flow cooler
- an open system using uncooled surface water

    The various cooling systems can be characterized as
follows. In the event of a low temperature of the incoming
water flow, less cooling pipes are needed. On the other
hand, the initial costs of a cooling system increases with
the required temperature reduction.

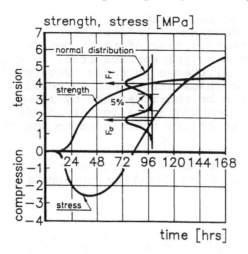

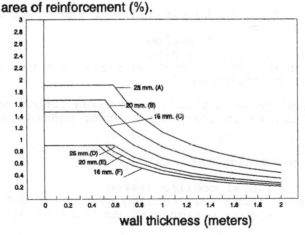

Fig.3   Example of development in; average and 5% upper
limit of the thermal stresses, average and 5% lower
limit of the tensile strength.

4   Extra Reinforcement

Fig.4   Extra reinforcement as a function of the wall thicknesses. Maximum crack width is 0.15 mm, concrete
quality is B35. A,B,C = total of extra reinforcement. D,E,F = minimum reinforcement.

The type of structure as described in this article (see
figure 1), loaded by a temperature difference as a result
of the hydratation process, can be schematized as a tension
bar exposed to a imposed deformation. The following remarks
should be made on the response of the structure to this
load case.

- The crack pattern, caused by the above described temperature load will almost never be completed.
- The reinforcement in this kind of thick concrete structures is concentrated at the surface area. As a consequence, there is a difference in crack pattern of a tension bar and, for thick structures, a tension disk. The crack pattern can be characterized by a few continuing cracks (primary cracks) and a lot of smaller cracks (secondary cracks) which are restricted to the reinforced edge area.

Both positive effects can be used to reduce the amount of extra reinforcement needed for limiting the maximum crack width.

5   Costs

The costs of extra reinforcement were simple to determine. However, to determine the costs of artificial cooling it was necessary to analyse the total cooling procedure. This was done on the basis of data collected in practice. The costs for artificial cooling are divided into four main categories:

- once-only costs, to be divided over the total length of the construction
- costs per meter construction
- costs per meter cooling pipe
- costs depending on the casting cycles and the casting length of the structure

note: length of construction = summarized length of walls, c.q. floors, cooled in the project under consideration.

6   Results and conclusions

6.1 Optimal artificial cooling system
Firstly the most economical cooling system was determined. Comparisons are shown in figure 5. Conclusions are;

- The once-only costs have a great influence on the total costs for short construction lengths. The system, for which uncooled surface water is used, has the lowest once-only cost. Therefore this system is the most economical cooling system for short construction lengths.
- The costs of the cooling pipes are the most important cost component for large construction lengths. Less cooling pipes are needed for a system with a lowerincoming water flow temperature. As a result a system cooled by a cooling aggregate in a closed system is a economical optimum for large construction lengths.
- A system cooled by a flow cooler in an open system is,

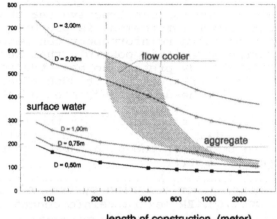

Fig.5   Optimum cooling system for different wall thicknes-
         ses and construction lengths.

from a cost point of view, medial.
-    The costs of artificial cooling varies between the 0.6%
     and 5% of the total concrete costs.

6.2 Artificial cooling versus extra reinforcement

After the optimum artificial cooling system was determined,
the costs of both cooling and extra reinforcement were
compared. A choice can be made between artificial cooling
to prevent cracking and extra reinforcement to control the
occurring cracks. Conclusions for the most economical way
to assure tightness are presented in table 1.
    In this Table a lower-upper structure can be recognized.
In the section left-below, artificial cooling is the most
economical solution. In the section top-right, extra reinforcement is the best solution. As already mentioned, the
once-only costs have a great influence on the total costs
for short construction lengths. The cost of extra reinforcement are, on the other hand, independent of the length
of the structure. Because of this, extra reinforcement is a
more economic solution in case of structures with a short
length. Furthermore the required reinforcement was calculated making use of a tension disk model. As a result, the
amount of extra reinforcement (in mm²) is independent,
beyond a certain thickness, of the thickness of the wall/-
floor. On the other hand, many more cooling pipes are
required to reduce the peak temperatures. The costs of
artificial cooling will increase for structures with thicker walls. Conclusion, extra reinforcement is advantageous
in case of constructions with thick walls.
    Finally three general notes;
    It should be remarked that the research results here
presented are based on an arbitrary construction. In practice each construction will be different. The result of

this research is therefore restricted to a few general
tendencies.

It is possible that for constructive reasons more than
the minimum required reinforcement is present. Because of
this the option of extra reinforcement can become attracti-
ve earlier. Furthermore it is possible that sufficient
reinforcement is already present to meet the required crack
width criteria in a construction. In such a case no extra
reinforcement will be needed to guarantee a water tight
construction.

It is emphasized that in this paper the focus of attention has been on water tightness, which is only one criterion that has to be met by the structure. Other criteria
may be important as well.

Table 1.   Most economic curing system, **AC** = artificial
             cooling, **ER** =extra reinforcement

Wall thickness ============ construction length	0.50 m	0.75 m	1.00 m	2.00 m	3.00 m
20 m	ER	ER	ER	ER	ER
50 m	ER	ER	ER	ER	ER
100 m	AC	AC	AC	ER	ER
250 m	AC	AC	AC	ER	ER
500 m	AC	AC	AC	ER	ER
700 m	AC	AC	AC	ER	ER
800 m	AC	AC	AC	AC	ER
1000 m	AC	AC	AC	AC	ER
1500 m	AC	AC	AC	AC	ER
2000 m	AC	AC	AC	AC	ER

7 References

Bakker, R.F.M.,Ramler, J.P.G. Sarneel, R.G.M. (1992)
    Theoretische Achtergronden V.B.S. HBG.
Braam, C.R. (1989) Control of Crack Width in Deep
    Reinforced Concrete Beams. TU Delft.
Breugel, K. van. (1991) Simulation of Hydration and Forma
    tion of Structure in Hardening Cement-Based Materials.
    TU Delft
Cordes,H. Trost,H. Ripphausen,B. (1989) Zur Wasserdurch
    lässigkeit von Stahlbetonbauteilen mit Trenrissen.
    Beton- und stahlbetonbau, nr 3.
Fehlhaben, König, Poll e.a.(1991) Betonbau Beim Umgang mit
    Wasser Gefährdenden Stoffen. Beuth Berlin.
Mandry, W. (1961) Über das Kühlen von Beton. Berlijn.
Wagenaars, W.G.L.(1992) Kostenaspecten van Verhardings-
    beheersmethoden. TU Delft.

52 A LARGE BEAM COOLED WITH WATER SHOWER TO PREVENT CRACKING

M. YAMAZAKI
Kajima Technical Research Institute, Chofu, Tokyo, Japan

Abstract
A large specimen beam 20 meter long, 1.5 meter high, 1 meter thick was cast on a rigid concrete bed. On half of the specimen, aluminum forms were used and the surfaces were cooled by water shower. For the other half of the specimen ordinary plywood forms were used and the surfaces were left to free air cooling. This test aimed at trying to prevent thermal cracking of concrete by water shower.

The temperatures of the sections, both water cooled and air cooled, were measured at two sections for about fourteen days from the beginning of concrete placement.

Simulation analysis was carried out to get the temperature and stresses of the concrete, and to verify the effects of the surface cooling method on prevention of concrete cracking.

Keywords: Cooling, Cracking, Curing with a Water Shower, Mass Concrete.

1 Preface

Sizes of the members of concrete structures are growing greater in these days, and the sections of the reinforced concrete members are large enough to cause cracking by the thermal stresses caused by the heat of hydration of the cement in the concrete.

To eliminate or suppress the thermal stresses, various methods are tried in various stages of construction. They are for example:

to use less quantity of cement,

to use low-heat cement or pozzolanic additives,

to cool down the mixing materials or the mixed concrete,

to cool down the concrete by embedded cooling pipes,

or to insulate the surfaces of concrete structure.

But this last method is effective when the massive concrete is free from the other restraining structures.

This paper reports a trial method of suppressing concrete cracking implying cooling of the concrete from the outside of the form. Historically speaking, cooling outside the mass

Thermal Cracking in Concrete at Early Ages. Edited by R. Springenschmid. Published 1994 by E & FN Spon, 2–6 Boundary Row, London SE1 8HN, UK. ISBN: 0 419 18710 3.

concrete was considered as a taboo because of the thermal stress due to the increase of the temperature difference between surface and the midst of the section. But this thermal gradient has no harmful effect when the structure is strongly restrained by an outer structure such as a foundation slab or a rock foundation.

2 Specimen

The specimen was a 20 meter long, 1.5 meter high, 1 meter thick concrete beam cast on a rigid concrete bed. The concrete bed was 60 cm high, 2 meter wide and was rigidly fixed by prestressing bars to the reinforced concrete testing floor 2 meter thick. The section of the specimen is shown in Fig. 1. The vertical reinforcement consisted of D13 bars, which connected the specimen to the concrete bed. No other reinforcement was used except one horizontal re-bar mentioned later in section 4. The cement was ordinary portland cement. The admixture was an air-entraining and water-reducing agent. Coarse aggregate was crushed stones. The concrete mix used is shown in Table 1.

3 Test method

Two kinds of forms were used. One was aluminum form that was selected for its high thermal conductivity, and the other was a plywood form that is usually used at construction sites.

Water shower was applied on the aluminum surfaces to cool down the concrete caused by hydration heat. The shower was maintained for three days from just after the concrete casting.

On the other half of the specimen, the surfaces of the shuttering and top surface of the concrete were left free for comparison.

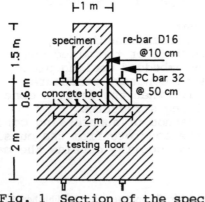

Fig. 1 Section of the specimen

Table 1　Concrete mix	
Gmax	20 mm
Slump	15 cm
Air content	4 %
W/C	56.5 %
s/a	45.0 %
Cement	300 kg/m3
Water	172 kg/m3
Sand	809 kg/m3
Coarse agg.	1015 kg/m3
Addmix.	0.762 kg/m3

4 Measurements

4.1 Temperature
The temperatures of the concrete were measured at several
points of the cooled section and the non-cooled section.
Copper constantan thermo-couples were used to measure the
temperature of concrete at the two sections. One of which is
at the aluminum form and the other is at the plywood form.
At both sections, the alignment of the sensors were as shown
in Fig. 2. Temperature of air and water were also measured.

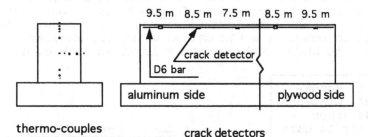

thermo-couples   crack detectors
Fig. 2 Alignment of thermo-couples and crack detectors

4.2 Crack Detector
A D6 re-bar with wire strain gages 1 meter apart was placed
longitudinally just below the top surface of the specimen to
detect cracks which may occur in the concrete. The gages
were attached to the bar, temperature compensated, and water
proofed at the factory of the strain gage fabricator. The
bar was placed in its final position before casting. Fig. 2.

4.3 Separation Detectors and Slip Detectors
To detect separation between the beam and the concrete bed,
strain gages were attached on the vertical re-bars at both
ends of the specimen. These gages were similar to the crack
detectors. The position of the detectors are shown in Fig. 3
    Embedded strain gages were placed on the top surfaces of
the concrete bed as the slip detectors. One end of the gage
was anchored to concrete bed, and the other was attached to
the specimen. See Fig. 3.

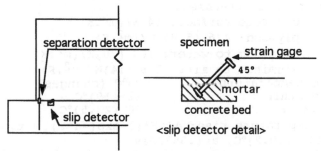

Fig. 3 Separation detector and slip detector

5 Anticipatory Pre-test Analysis

A presumption analysis on the thermal stress in the concrete
was studied through the CP Method, which was one of the
computer programs released by the JCI (Japan Concrete
Institute) as a powerful tool to analyze thermal stresses of
mass concrete. The computer program is named as "The Program
for Calculation of Thermal Stress of Mass Concrete, Version
2., Japan Concrete Institute, 1989".

The CP Method is based on the principle of "plane section
remain plane" for concrete structure at young age. External
restraint is presumed into axial and bending restraint.
Equation of equilibrium is made for both conditions. The CP
Method analyzes the longitudinal stress of the beam. The
program is composed by a FEM method using tri-angular
elements. The model mesh used here is shown in Fig. 4.

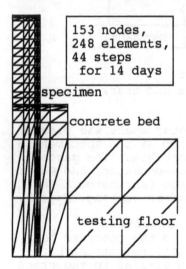

153 nodes,
248 elements,
44 steps
   for 14 days
specimen

concrete bed

testing floor

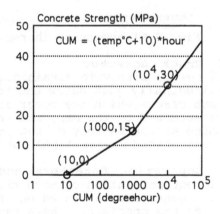

Concrete Strength (MPa)

CUM = (temp°C+10)*hour

$(10^4,30)$

$(1000,15)$

$(10,0)$

CUM (degreehour)

Fig. 4 FEM mesh used   Fig. 5 Concrete strength and maturity

Table 2 Input data for temperature computation

Thermal conductivity of concrete: 2.9 W/(m.K)
Coefficient of heat transfer:
            concrete surface: 14 W/(m^2.K)
            plywood: 7.6 (5.8) W/(m^2.K)
            aluminum to water: 400 (70)W/(m^2.K)
            aluminum to air: 18.6 (14)W/(m^2.K)
Atmospheric temperature :   25°C (changes)
Water temperature :   22 (19)°C
Concrete cast :   30 (23.5)°C
Adiabatic temperature rise: $\Delta T = K*(1-\exp(-a*t))$ t:days
        K=45.6(49.2)°C, a=1.557/day
                numbers in () are mentioned in Sec. 7.

Table 3   Input data for concrete stress computation

Relation between maturity(CUM) and strength :Fig. 5
Relation between strength Fc and Young's modulus Ec:

Ec=α*4695*SQRT(Fc)   Ec,Fc: (MPa)

α :depreciation factor = 0.73 (t < 3 days),
= 1 (t > 5 days),   linearly interpolated (3 < t < 5)
Restraint factors :Rn =1.6, Rm1 =1.0, Rm2 =1.65.
These three factors are calculated from the charts in the
Commentary of the Japan Society of Civil Enginees(JSCE)
Standard Specification for Design and Construction of
Concrete Structures 1986, Part 2 (Construction).

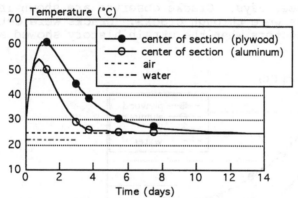

Fig. 6 Temperatures anticipated

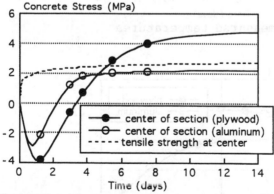

Fig. 7 Concrete stresses and strength anticipated

The input data, anticipated and/or used, are shown in Table 2
and Table 3.   Restraint factors used in the program are not
explained in this report.   Readers may obtain details in the
Commentary of the JSCE Standard Spacification.

The result of the pre-test analysis is shown in Fig. 6 and Fig. 7. By these results of the analysis, it was deduced that cooled section would not crack.

6 Test Results

Temperature histories of the concrete measured at the test are shown in Fig. 8. Temperature distributions in the concrete section are shown in Fig. 9. As a result, the peak temperature of the center-point of the concrete section at the aluminum form showed ten degrees Centigrade lower than that of the competitive section. This was the effect of the water shower and high thermal conductivity of the form material.

Cracks were observed after the forms were removed when the concrete temperature had stabilized to room temperature at the age of fourteen days. Cracks observed are shown in Fig. 10. These cracks were through cracks. Cracks were detected by the crack detectors when the strain history showed sharp

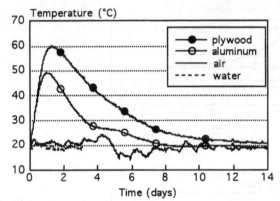

Fig. 8 History of measured temperatures

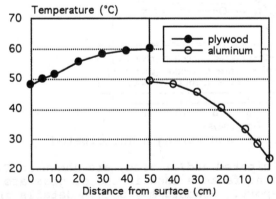

Fig. 9 Temperature distributions

gaps in Fig. 11. With these figures, it is deduced that the cracks occurred when the nearest detector showed the largest gap. According to the strain histories, three cracks must occurred at 3.1 day (Cr-1a), 5.56 day (Cr-2a) and 5.69 day (Cr-1p).

The separation detectors and slip detectors showed no slippage and no separation at both ends of the beam.

The crack width measured at the mid height of the beam are also shown in Fig. 10.

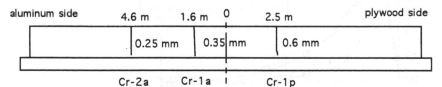

Fig. 10 Cracks observed

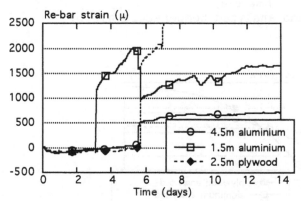

Fig. 11 Strain history of crack detectors

7 Post Test Analysis

The pre-test analysis failed in predicting cracking on the aluminum form section. This was due to miss-anticipation of initial temperatures.

A post test analysis was tried after the test based on the measured environmental data. The measured air and water temperatures were used as input data but modified a little simplification. The temperature of the fresh concrete was 23.5 °C. The air temperature dropped at the age of 5 days and it is presumed that it affected an increase of the concrete tensile stress. New input data are shown in Table 2 in parenthesis. Analyzed results of temperatures are shown in Fig. 12 as compared with measured data. They show good coincidence.

Analyzed concrete stresses are shown in fig. 13. When analyzing the concrete stresses the input data in Table 3

were not changed at all. This figure shows the cracks at
both sections.

The conditions of (Cr-1p) and (Cr-2a) are elucidated, but
that of (Cr-1a) coukd not be explained. This is left for
future work.

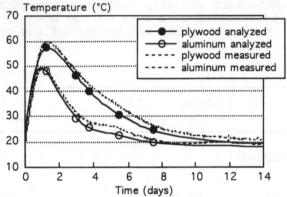

Fig. 12 Measured and analyzed temperature history

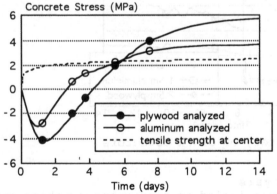

Fig. 13 Analyzed results of concrete stresses

8 Conclusion
The water sprinkling method in conjunction with aluminum form
is an effective way to cool down the concrete section. But
cracks were observed in the specimen in both the cooled
section and non-cooled section.

The temperature prediction of air and water came out
wrong, which was the main reason of the failure of the pre-
test analysis. It is necessary to estimate temperature
accurately when the water sprinkling method is used as a
method for prevention of cracking.

53 REDUCTION OF THERMAL STRESSES IN STRUCTURES WITH AIR-COOLING

H. HEDLUND, P. GROTH and J.E. JONASSON
Division of Structural Engineering, Luleå University of Technology,
Luleå, Sweden

Abstract
The risk of thermal cracking depends apart from temperature itself on a number of parameters such as e.g. restraint conditions, mechanical behaviour in the young concrete, and temperature development. During the last decade studies of thermal behaviour in concrete have been carried out at Luleå University of Technology. In the massive pier shafts at the Igelsta Bridge, Sweden, several in-situ studies have been made for both non-cooled and air-cooled sections. Early experiences from the building site as well as results from measurements and simulations had shown unacceptably high tension levels in the piers unless remediating measures were taken. A system of air-cooling with embedded pipes was chosen. The positive effects of this approach have been confirmed by further measurements. In order to design an air-cooling system the required cooling effect must be determined. The computed cooling effects based on measured changes in air temperature along the cooling-pipes have been compared with the corresponding effects calculated from general heat transfer equations. The results imply that a feasible model of designing a cooling system was found.
Keywords: Cooling Systems, Heat Transfer, Risk of Cracking, Structural Members (Concrete Pier Shafts).

1 Background

The last decade, studies of thermal stresses and problems related to thermal behaviour of concrete due to hydration have been performed at Luleå University of Technology. In laboratory tests thermal- and mechanical properties have been studied on small concrete specimens. Theoretical models have been developed and calibrated to the results from the laboratory tests and implemented into two computer programs for structural analyses, HETT (Jonasson) and TEMPSTRE-N (Emborg, 1989, Bernander et al, 1986).

Thus, it is now possible to compute the risk of thermal cracking due to hydration in newly cast concrete constructions considering the properties of young concrete and the restraint conditions for the structure.(Bernander, Emborg, 1994)

2 Pilot case study

One of Europe's longest railway bridges, the Igelsta Bridge, is located in the area of Södertälje, south of Stockholm. Detailed studies of thermal cracking have been per-

Thermal Cracking in Concrete at Early Ages. Edited by R. Springenschmid. Published 1994 by E & FN Spon, 2–6 Boundary Row, London SE1 8HN, UK. ISBN: 0 419 18710 3.

formed in-situ on the massive pier shafts. From the start, the columns of the Igelsta Bridge were constructed without any specific measures with regard to controlling temperature stresses. This oversigth resulted in surface cracks. The temperature measurements in the pilot study (Hübinette, Westman, 1992) were performed on one of the massive pier shafts for the further analysis of crack risk.

2.1 Thermal stress analysis
In order to calculate and find measures against cracking, the recorded temperature from measurements made on site served as input for the stress analysis with TEMPSTRE-N.

The temperature difference between the surface layers and the core induced compressive stresses in the inner parts of the pier and tensile stresses at the surface. This proved to be in excess of the level of probable risk of thermal cracking. When the form was stripped, four days after pouring, the surface was exposed to severe temperature shock resulting in even higher tensile stresses in the surface zone, thus exceeding the limit tensile strength, see fig 4a. and 4b. Cracks were observed at the top of column and down all sides of the newly cast section. The cracks penetrated far beyond the reinforcement and therefore measures against surface cracks were considered necessary.

2.2 Different methods to prevent cracks in young concrete
In order to obtain a more accurate analysis of the concrete mixture, when using cement of type Degerhamn Std P i.e. the cement used for the Igelsta Bridge, laboratory tests were performed in order to determine the thermal- and mechanical properties:

- adiabatic and semi-adiabatic calorimetric tests
- strength development tests
- creep tests

The efficiency of different kinds of measures were studied by theoretical simulations HETT and TEMPSTRE-N:

- pouring conditions
- initial concrete temperature
- form insulation
- time for stripping the form
- cement content
- covering of the stripped pier with blanket insulation
- cooling of the concrete with embedded cooling-pipes

From the theoretical studies it appeared that the most effective way to prevent surface cracks was to cool the concrete with embedded cooling-pipes, thus reducing the temperature difference between the core of the column and the surface zone.

3 Cooling system

The cooling system in use when pouring the massive piers at the Igelsta Bridge consisted of embedded pipes mounted vertically over the full length of the columns (Larson, 1993).

See Fig. 1. The pipes were made of sheet steel. The measured air velocity was between some 4 - 8 m/s (u_{max}).

The choice of air as cooling medium instead of water was made for practical and economical reasons.

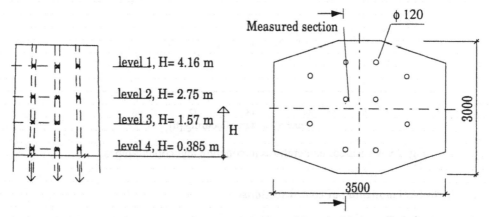

level 1, H= 4.16 m

level 2, H= 2.75 m

level 3, H= 1.57 m

level 4, H= 0.385 m

Fig. 1 Massive pier shafts at the Igelsta Bridge with embedded cooling pipes.

4 Case studies of cooled sections

The case studies at the Igelsta Bridge were made in order to confirm the predicted positive effects of air-cooling and to enable an analysis of the varying cooling effect along the pipes. The temperatures were measured at four levels, see Fig. 1. The concrete temperatures show the influence of the cooling-pipes as the temperature drops around the pipes, see Fig. 2a. The daily variation in air temperature is clearly discernible, see Fig. 2b.

In order to establish the necessary thermal properties for the concrete in the actual pour a semi-adiabatic test was performed in-situ. A sample taken from the concrete transport truck was placed in a Movable Semi-Adiabatic Equipment (MSAE) and the temperature growth in the sample was recorded.

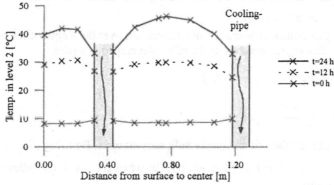

Fig. 2a. Measured temperatures at different locations in the poured section.

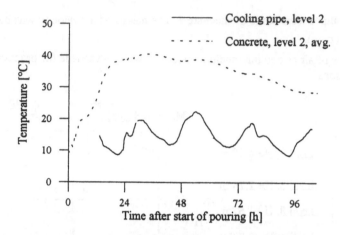

Fig. 2b. Measured temperatures in concrete and cooling-pipes.

5 Theoretical modelling of cooled sections.

Calculation of the concrete temperature development has been carried out with the aid of the computer program HETT. The results have been compared with the measured temperatures at level 2, see Fig. 2a, in which the heat transmission is almost two-dimensional. In this simulation the measured pipe temperature, $T(t)$ (°C), was used as input and the heat-transfer coefficient, h_p (W/m²°C), was adjusted until measured and calculated temperatures in level 2 correlated. The cooling effect, $\dot{q}$ (W/m), could then be expressed as;

$$\dot{q}(t) = h_p \int_0^\Omega \left(T_{w,c}(y, t) - T(t) \right) d\Omega \tag{1}$$

where $T_{w,c}(y, t)$ is the calculated pipe wall temperature, see fig 3, method 1. Since $T_{w,c}(y, t)$ varies around the wall of the cooling-pipe the temperature difference must be integrated over the periphery, Ω.

One way to estimate the change of temperature over the length of the cooling-pipe is to use the following expressions of the effect developed in a tube:

$$\dot{q}(t) = h_p \pi d \Delta x \left(T_w(t) - T(t) \right) \tag{2}$$

$$\dot{q}(t) = \rho c_p u_m \frac{\pi d^2}{4} \left(T_e(t) - T_i(t) \right) \tag{3}$$

$T_e(t)$ and $T_i(t)$ are the exit and intake bulk temperatures. By using $\Delta T = T_e(t) - T_i(t)$ and $\Delta T/\Delta x \to dT/dx$ when $\Delta x \to 0$ the following differential equation is given:

$$T' + kT = kT_w$$

$$k = \frac{4h_p}{\rho c_p u_m d} \tag{4}$$

To solve Eq.(4) T_w must be a known function of x. This function was assumed to be a second order polynomial and was defined by curve fitting to the measured temperatures at the pipe wall. Eq. (4) then has a solution:

$$T(x) = C_1 x^2 + C_2 x + C_3 + (T_0 + C_3)e^{-kx} \tag{5}$$

$$C_1 = a; \quad C_2 = b - \frac{2a}{k}; \quad C_3 = c - kC_2 = c - bk + 2a$$

T_0 is the bulk temperature at the top of the cooling-pipe ($T(0) = T_0$) and a, b, c are coefficients of the polynomial function of T_w. The cooling effect can then be calculated by inserting Eq.(5) in Eq.(2). The heat-transfer coefficient was numerically the same as in Eq.(1) and the measured temperatures at level 1 ($x = 0$ m) was used as T_0, see fig 3, method 2.

If the heat-transfer coefficient, h, is known the cooling effect and the change of temperature along the cooling-pipe can be calculated using T_w and T_0. The constant, k in Eq.(4) can be written as Eq.(6) where St is Stantons number. Stantons number can be calculated by means of (Appelqvist et. al, 1973):

$$k = \frac{4h_p}{\rho c_p u_m d} = \frac{4}{d} * St = \frac{4}{d} * 0.027 R_{e_d}^{-\frac{1}{4}} Pr^{-\frac{3}{5}} \tag{6}$$

Using Eq.(6) with $u_m = 0.82 u_{max}$ as the air velocity gives a numeric value of the constant k and the heat transfer coefficient h.

A comparison between on one hand the effect rendered with these values of k and h (method 3) and the effect according to Eq.(1) (method 1) shows that the result is somewhat high but not out of range. See Fig. 3. The differences between the results of the various methods of assessing the cooling effect may depend on various reasons. The correlation between h and u_m as derived from Eq.(6) holds true only when the flow is fully turbulent. However, the studied level of the cooling-pipe is most likely situated within the intake length and therefore in an intermediate state between laminar and turbulent flow.

Since the measured temperature at level 1 has been used as T_0 in Eq.(5) an error is introduced as the correlation between measured and bulk temperatures is not known. The properties of the cooling air included in Eq.(4) and Eq.(6) varies with air temperature and humidity. In this study the property values has been chosen for a reference temperature:

$$T_{ref} = (T_0 + \overline{T_w})/2.$$

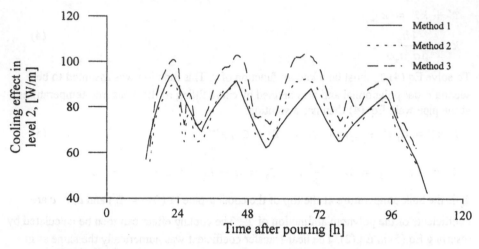

Fig. 3. Cooling effect, $\dot{q}$ (W/ms), in level 2 due to time after casting.

5.2 Stress analysis

Thermal stresses due to heat of hydration in young concrete was analysed with the aid of TEMPSTRE-N for the non-cooled and the air-cooled structure. The results are represented in Fig. 4a. and Fig. 4b, and show the calculated thermal stresses and temperatures, with the expected high tensile stresses in the surface zone for the non-cooled section. Utilization of embedded cooling-pipes reduces the temperature difference and effectively minimises tensile stresses in the surface zone.

6 Conclusions

By means of analysing tools such as TEMPSTRE-N it is today possible to calculate and optimize cooling systems with embedded cooling-pipes. The temperature variation along

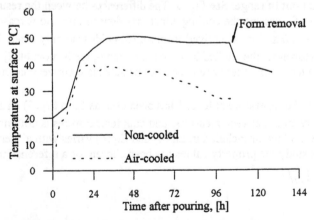

Fig. 4a. Temperatures at surface of pier shafts.

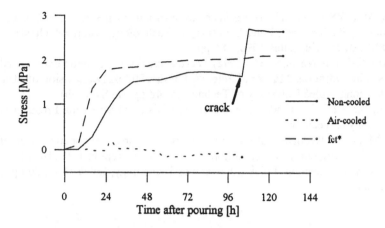

Fig. 4b. Tensile stresses at surface of pier shafts.

the cooling-pipes is an important feature when predicting cooling efficiency. In order to find this relationship, the heat-transfer conditions in the cooling-pipe may be investigated by literature studies and in laboratory tests.

A future application of the results may be to find simplified methods of designing the cooling systems. The calculations of the risk of thermal cracking may, in the design stage of a project, be carried out at different levels of accuracy depending on the available information.

Acknowledgement

This work has been supported by the Swedish Council for Building Research. Special thanks are addressed to Lars Ekroth, NCC and other personnel at the site of the Igelsta Bridge. The work was supervised by Professor Lennart Elfgren, head of the Division of Structural Engineering at Luleå University of Technology.

References

Appelqvist, B. et al. (1973) Science of fluid motion - convection heating and mass transport. Division of Applied Thermodynamics and Science of flow, Chalmers University of Technology, **Internal Report no. 73/14**, Gothenburg 1973, 66 pp, (in Swedish)

Bernander, S. Emborg, M. (1994) Risk of cracking in massive concrete structures - New developments and experiences. **Proceding from RILEM International Symposium on Avoidance of Thermal Cracking in Concrete at Early Ages**, München, Okt. 1994.

Bernander, S. Emborg, M. Daerga (1986) Thermal stresses in concrete bridge foundations placed under water. Division of Structural Engineering, Luleå University of Technology, **Technical Report 1986:27 T**, Luleå 1986, 181 pp, chap. 5. (In Swedish)

Emborg, M. (1990) Thermal stresses in concrete structures at early ages. Division of Structural Engineering , Luleå University of Technology, **Doctoral Thesis 1989:73D,** 2nd Ed, Luleå 1990, 285 pp

Hübinette, K-J. and Westman, G. (1992) Thermal cracks due to temperature gradients caused by hydration. **Master of Science Thesis 1992:021E,** Division of Structural Engineering, Luleå University of Technology, 48 pp. (in Swedish)

Jonasson, J-E. HETT - Private communication, Division of Structural Engineering, Luleå University of Technology

Larson, M. (1993) Reduction of temperature stresses in massive concrete structures using air-cooling with embedded cooling-pipes. **Technical Report 1993:10T,** Division of Structural Engineering, Luleå University of Technology, Luleå 1993, 26 pp, (in Swedish).

54 COUNTERMEASURE FOR THERMAL CRACKING OF BOX CULVERT

M. IWATA, K. SAITO, K. IKUTA and T. KAWAUCHI
East Japan Railway Company, Sendai, Japan

Abstract
Our construction office is at work on an underground railway
project whose length is about 3 km, on a continuous grade
separation project. The work on the underground railway is
under construction. But at the beginning of the
construction thermal cracking occurred at early age. So we
had to examine basic ways to control thermal cracking. As a
result, we could prevent thermal cracking in concrete at
early ages. Here we will give an outline of construction,
pre-examination and thermal records of concrete.
Keywords: C.P. Method, Cracking Index, Structural Members
(Box Culvert).

1 Introduction

Double-track box-culvert,of which side-wall is 60 ~ 80 cm,
is utilized in constructing underground railways.
Thermal crack occured on this side-wall at the early stage
constructing box-culvert. So we had to examine basic ways
to control thermal cracking.
    After the several investigations, we used 'High Range
Water Reducing Type Heat Control Agents'(HRWRTHCA), as a
countermeasure for thermal cracking at early ages. The main
elements of 'HRWRTHCA' are poly hydroxy carboxylicacid ester
and selected anion type surfactants whose effects are range
water reducing and heat controlling. On analysis, we found
that the measured temperature of concrete is 10 °C lower
than normal, when this heat control agent is used.
    This time, we constructed a box-culvert whose side wall
is 80cm in thickness and 21.5m in length. We poured the
'HRWRTHCA' into both sides of the wall. About 200㎥
concrete was poured into sides of the wall, one lift is
about 50cm. So we needed 7 hours to do the work.

2 Chemical Admixture in Use

The chemical admixture that we used is HRWRTHCA. That consists mainly of poly hydroxy carboxylicacid ester and

Thermal Cracking in Concrete at Early Ages. Edited by R. Springenschmid. Published 1994
by E & FN Spon, 2–6 Boundary Row, London SE1 8HN, UK. ISBN: 0 419 18710 3.

selected anion type surfactants, and has both the heat controll of hydration effect and the range water reducing effect.
The heat controll of hydration effect is controlling the
hydration in response to the temperature rise that a
chemical admixture has and controlls the sudden temperature
rise of concrete by the cooling effect of concrete in
process. Also, we show a temperature change and the speed
of this change in the center of a concrete test piece whose
side is 1 meter in the figure.

3 Scope of Examination

3.1 Mix Proportion Test

We considered the design and construction condition and
decreased unit cement content, Also a mix proportion test
was made to gauge experim-
entally the quantity that
made full use of the
effect of HRWRTHCA. The
mix proportion conditions
of a box-culvert on a
typical railway are shown
in the table 1.
We based upon standard mix
proportion bytest mixing
in about the following 3
ways.

Table 1 Mix proportions Condition

Slump	8 ± 2.5cm
Air Content	$4.5 \pm 1.0\%$
Nominal Strength	24MN/m³
W/C	Under 55 %

No.1   s/a(sand aggregate ratio) and unit mixing cement co-
        ntent were fixed and unit water content was changed.
No.2   s/a and W/C were fixed and unit cement content and
        unit water content were changed.
No.3   s/a,W/C,unit cement content and unit water content
        were changed.

Table 2 Result of Mixproportion Test

NO	W/C (%)	s/a (%)	Quantity of Material (kg/m³)				Admixture			Slump (cm)	Air-content (%)
			W	C	S	G	AEWA (%)	HRWA (%)	AEA (A)		
1	53.0	42.2	157	296	757	1058	0.25	—	—	7.5	4.4
2	48.3		143		778	1084	—	2.0	2.5	7.0	4.0
3	53.0				270	783	1092	—	2.0	2.5	7.0	4.0
4	55.0	44.7	145	264	829	1043	—	2.0	2.5	7.0	4.3

        ※AEWA:Air Entraining and Water Reducing Agent
        HRWA:High Range Water Reducing type Heat Contro. Agent
        AEA :Air Entraining Agent

Table 3 Result of Strength Test

NO	W/C (%)	s/a (%)	Compressive Strength / Tension Strength (MN/cm)			
			3rd DAY	5th DAY	7th DAY	28th DAY
1	53.0		17.4 / 1.51	22.6 / 2.05	27.0 / 2.21	38.0 / 2.30
2	48.3	42.2	18.9 / 1.70	31.7 / 2.21	36.8 / 2.58	48.0 / 3.06
3	53.0		16.2 / 1.32	25.6 / 2.12	29.6 / 2.36	38.7 / 2.44
4	55.0	44.7	13.6 / 1.22	22.1 / 1.98	27.4 / 2.22	39.8 / 2.41

At the same time, we made test on compression strength and
tension strength with the mix proportion. We show the result
of the mix proportion test in table 2 and the result of
the strength test in table 3. All cases we attainded strength exceeded the concrete one without HRWRTHCA (non-annexing)
on the 7th day and 28th day and judged that we had no question in strength. So we determined to use mix proportion No.4
that has least unit cement content and was considered that
was considered to have the lowerest heat of hydration.

3.2 Adiabatic Temperature Rise Test

We made an adiabatic temperature rise test before construction to forecast and analyse the effect using HRWRTCHA and to
determine the quantity of HRWRTCHA that should be ad-ded.
It was 3 way mix proportion test. One case is the quantity
of adding of mix proportion No.1 in table 2, another is mix
proportion No.4 using HRWRTCHA, and the third is the case
where the quantity of HRWRTCHA is increased by 2.5% in mix
proportion No.4. We show the adiabatic temperature rise test
in Fig. 1, and the rate of adiabatic temperature rise in
Fig. 2.

As a result, when quantity of annexing of HRWRTHCA was

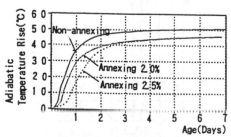

Fig. 1. Result of Adiabatic
Temperature Rise Test

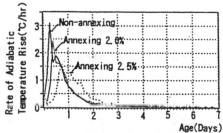

Fig. 2. Rate of Adiabatic Temperature Rise

increased by 2.5% in mix proportion No.4, adiabatic temperature rise was 4 °C lower than the case of non-annexing and
the rate of temperature rise is half. So we verified the remarkable effect of controlling heat of hydration by the test.

3.3 Prediction of Temperature Rise

On the basis of the above result, we made a two-dimensional
investigation of the case in which quantity of annexing of
HRWRTHCA increased by 2.5% in mix proportion No.4 and forecast a temperature rise in real structures. We show material
factors in the table 4 and the analytical model in Fig. 3.
The material factors were determined from the result of the
experiment. When we estimate temperature, curing temperature
has a great influence. In practice, field temperature is
forecast as about 5 °C when the construction is given. So,
we don't water particularly in the period of temperature
rise the concrete heat of hydration, but intend to carry out
heating by jet heater after the temperature drop of concrete
in the side wall and to adopt the curing method so that a
sudden temperature drop in the concrete may not happen. So,
forecasting temperature rise, we examined the environmental
temperature at curing by means of the 4 ways shown in table
5 so that the forecast reflects the examination of curing m-
ethod in practice. We show the effect of forecast of a temperature rise in Fig. 4.

Table 4 Material Factor

Specific Heat of Concrete	0.305(kcal/kg°C)
Heat Conductivity	2.2(kcal/m·hr°C)
Heat Transfer Characteristics	10(kcal/m²·hr°C)
Concrete Density	2300 (kg/m³)

Table 5 Environmental Temperature (°C)

Symbol	TEMP Rise Period	TEMP Drop Period	No.
NON	2 0	2 0	1
P1	1 5	2 0	4
P2	1 5	2 5	4
P3	2 0	2 0	4
P4	2 0	2 5	4

※Placing temperature:20°C

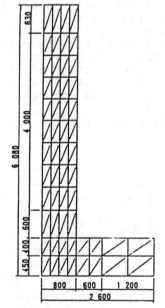

Fig. 3. Analysis Model

Consequently, when the quantity of annexing of HRWRTHCA is 2.5% in mix proportion No.4 and environmental temperature corresponds to P2 we got the result that there was an 8.7 ℃ drop in the maximum temperature and there was a 9.3 ℃ difference from the peak to the material after 7 days compared the case of non annexing.

3.4 Thermal Cracking Index

On the basis of the forecast result, we made a thermal stress analysis and calculated a thermal cracking index. The thermal cracking index is the ratio of tension strength to tension stress by a tension change in concrete. Cracking index 1 stands for a 50% probability to bring about cracking. We show the result of construction of cracking index in Fig.5. When we calculated it, the restraining block was a floor slab and the restraining coefficient was R_N : 1.0, R_{M1} :0.9, R_{M2} :3.0by the literature cited.

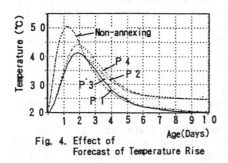

Fig. 4. Effect of Forecast of Temperature Rise

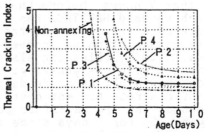

Fig. 5. Result of Thermal Cracking Index

By the calculation, we got the result that the cracking index was about 1.85 in the case of an environmental temperature P2 that had the largest cracking index and was improved about 1.00, compared with the cracking index 0.85 in the case of non annexing. By means of the above examination, we got the result that cracking could be prevented by adopting HRWRTHCA and applying a suitable curing method. So, we decided to do this construction on condition that No.4 shown in table 5 was used as mix proportion in concrete and quantity annexed with the HRWRTCHA was designed 2.5% and curing temperature could be adjusted to an environmental temperature P2.

4 Outline of Construction

We adopted a double-track box culvert with wall thickness of 0.8 meter and length of 21.5 meters. We show the configuration in Fig. 6. HRWRTHCA was used in side wall shown with oblique line in Fig.6. Meteorological data of the day and other factor are shown in table 6. As shown in Fig. 6, we meas-

Table 6 Condition of Construction

Season	Spring
Weather	Cloudy
Wind	Calm
TEMP of the Air	17~ 18 ℃
Placing TEMP	22~ 25 ℃

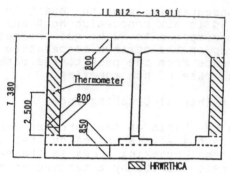

Fig. 6. Form of Box-culvert

ured the temperature change. As arrangements of construction, we placed one pump car per side wall and decided that
each lift would be about 50cm. And we poured in order. So
it took 7 hours before the work was finished. Concrete in
which we used HRWRTHCA has a little high viscosity than concrete of without it, and the flow rate is smaller. So in
this construction we increased the number of vibrators from
usual 3 to 5. Bleeding also seemed to be little more than
concrete with non-annexing. For the curing method based
upon the result of a preanalysis, we measured the temperature of concrete box and covered the box-culvert with a curing
sheet on the 2nd day after pouring when temperature rise
began. So for 5 days since the afternoon on the 3rd day
after pouring when temperature of concrete came to the peak,
we carried out the insulated curing by 2 jet heaters. After
that, we left it as it was and removed the forms on the 21th
day after pouring.

5 Result of the Construction

5.1 Temperature Change

In Fig. 7 We show changes of temperature in concrete,curing
temperature,temperature of the air and the analyzed value.
Looking at temperature change in concrete, the mesured value
is nearly equal to the analyzed value at the predicted maximum temperature. According to the analyzed temperature, it
took 1.8 days to rise to the maximum one. But the measured
temperature did not show a significant peak after pouring
and the peak value was indicated in 3days ~ 5days. Also,the
temperature change fell more siowly than the analyzed one.
We can consider various causes like a difference in pouring
and curing temperature, in the difference between the measured temperature and the analized one. So we think it is necessary to examine after this. On the other hand, we show

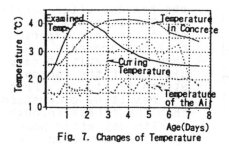

Fig. 7. Changes of Temperature

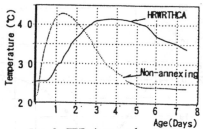

Fig. 8. TEMP changes of HRWRTHCA and Non-annexing

the temperature rise of concrete using HRWRTCHA and non-annexing concrete that was constructed at another location in autumn in Fig. 8. The wall of the box-culvert poured of non-annexing concrete was 0.7m thick, the pouring temperature was 17°C to 18°C, air temperature was 12°C, and we did not do insulated curing.
Looking at Fig. 8, in the case of using HRWRTHCA, the pouring temperature and the air temperature while reaching the maximum were about 5°C higher, but the maximum temperature of concrete were about 2°C lower than that of non-annexing concrete. Moreover, the temperature drop record shows gentle curve, comparing with the case of non-annexing concrete.
It is prdicted that the concrete's tension stress by a rapid temperature drop would be mitigated. Therefore, We can recognize that this condition is considerably beneficial in view of stress.

5.2 Examination of Cracking Index using a measured Temperature Change

As the result of the construction, cracking due to the thermal stress in mass concrete didn't occur upon removal of forms. The only one minute cracking due to the drying shrinkage occured on the center of one side-wall at the time passed four months after pouring concrete. So we calculated the thermal cracking index using the measured value based upon the temperature's record, aimed to verify cause no temperature cracking in mass concrete. As the method of calculation, we calculated the coefficient of formula of adiabatic temperature rise in a test so that calculated temperature record could approximate measured temperature one. And using the result, we calculated the temperature change, thermal stress and a cracking index. Young modulus was calculated by

$$E_o(t) = 1.5 \times 10^4 \times f'(t) \quad [kgf/cm^2] \qquad (1)$$

We multiplied it by the reduction coefficient 0.733 until

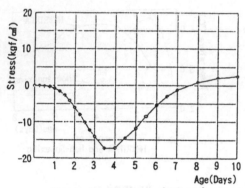

Fig. 9. Result of calculation of
Thermal Stress

the 3rd day after pouring. The strength was estimated from experimental measurements. We show the result of calculation of the thermal stress in Fig. 9. From the result, we understood that the stress through the 7th day after pouring shows values on the compressive side, and only a little tension stress occurs when we calculate through the 10th day after pouring when the temperature of concrete is about the same as the temperature of the air. Using an estimated value of tension strength which was based on experimental measurements we obtained, the result that the thermal cracking index was about 6.5. In this way, also from the analysis, the result of the construction was projected as extremely beneficial regarding stress due to thermal cracking at early ages.

6 Conclusion

As a result of the pre-examination, in case of using HRWRTHCA, the cracking index was improved about 1.00 as compared with the case of non-annexing and it was effective against cracking. Also, when the quantity of annexing of HRWRTHCA was 2.5% in mix proportion No.4, we were careful to make the construction so that the curing temperature would approach environmental temperature P2. So thermal cracking did not occur. When we used concrete with no annexing, 3 to 4 crackings per span occured. By comparison, we could thus be sure that HRWRTHCA was effective to control of thermal cracking in the box-culvert. After this, we will check further on the improvement of preexamination and construction prcedure.

7 References

M.Iwata,K.Saito,K,Syoji and H.Inokawa (1993)An Investigation
of Thermal Cracking for Box-culvert, in Proceedings of
the Japan Concrete Institute , Vol.15 No.1 pp.1161-1166

55 HIGH PERFORMANCE CONCRETE: FIELD OBSERVATIONS OF CRACKING TENDENCY AT EARLY AGE

R. KOMPEN
Norwegian Road Research Laboratory, Oslo, Norway

Abstract
Horizontal surfaces of concrete having a water/binder-ratio of 0,40 or less have shown an annoying tendency for cracking in the early plastic stage. This paper summarizes the field observations and the qualitative field experience as to the occurrence of the cracking phenomenon. Recommendations based on field trials are given for concrete mix design and for execution, on how to minimize cracking.
Keywords: Cracking, Cracking Tendency, High Performance Concrete.

1 Introduction

Since concrete with a water/binder- ratio of 0,40 came into regular use for bridges in Norway, free (unformed) surfaces, especially bridge decks, have shown an annoying tendency for cracking in the early plastic stage. This paper will summarize the field observations and the qualitative field experience from the bridge construction projects through the years 1989 - 1993, where efforts have been made to overcome the cracking problem.

2 Specification for bridge concrete

Several cases of early deterioriation of bridges in marine climate have called for improvements of the concrete quality of bridges. In 1988 the Norwegian Public Road Directorate put forward a new specification for bridge concrete:

- a mass ratio m = $w/(c+2 \cdot s)$ less or equal to 0,40.
- silica fume content (s) 2-5% by weight of Portland cement (c).
- air content 5 ± 1,5% measured in fresh concrete
- strength class C45 or higher
- consistency max 12 cm slump for superstructures, max. 16 cm for other structural members.

Due to experiences gained the slump limits are no longer valid. Slump is typically 18-20 cm, and an air content of 4,0-4,5% is normally aimed at. In several projects strength class C55 and C65 have been specified, and mass ratios down to 0,32 have been applied.

Thermal Cracking in Concrete at Early Ages. Edited by R. Springenschmid. Published 1994 by E & FN Spon, 2–6 Boundary Row, London SE1 8HN, UK. ISBN: 0 419 18710 3.

3 Types of field problems experienced

In addition to the early age cracking problem, several other and possibly interrelated field problems, have been observed in practical use of concrete with low w/b-ratios. These include the following:

1. To obtain proper workability.
2. To retain workability for a reasonable period of time.
3. Trowelling of free concrete surfaces.
4. Frequently an abnormal drop in strength when the air content exceeded a certain limit, usually 5,0-5,5% for m=0,40.
5. Testing often showed that water-impermeability was not achieved.

Compared to "ordinary concrete", low w/b-ratio concretes were very cohesive and sticky. Poker vibrators often showed poor effiency, despite high slump due to the use of superplasticizers at rather high dosages. Because the fresh concrete looks "wet", experienced concrete workers misinterpret the need for compaction.
The rapid loss of workability often experienced consists of two phenomena:

A. Ordinary slump loss as with "ordinary concrete".
B. Some kind of coagulation taking place when the concrete moves slowly or stands still due to its thixotropic character.

The extent of the field problems varied considerably from site to site and also at each site, due to both identified and unidentified parameters.

4 The cracking phenomenon

The majority of the sites using concrete with a mass ratio of 0,40 experienced cracking of free surfaces while the concrete was still in the plastic stage. The extent of cracking, both number of cracks and crack-widths, increased with reductions in the mass ratio. At a mass ratio of 0,45 only one case of plastic cracking has been observed.

The cracking occured from 15 minutes up to 6-8 hours after striking off the surface, depending on the concrete temperature, mass ratio, weather conditions and the degree of retardation. Early cracking gave generally wider cracks.
The pattern of cracking varied, mainly within three variants:

1. A large number of wide (1-2 mm) and short (10-30 cm) cracks in all directions, possibly with a main orientation at 45° with the reinforecement. This pattern is similar to ordinary plastic shrinkage cracking.
2. A small number of wide (1-3 mm) and very long (2-5 m), almost straight cracks, mainly parallel to the edges of the cast section. The cracks often coincide with discontinuities in the soffit of the slab.
3. An enormous number of fine, parallel haircracks of approx. 1,0-1,5 m length and 10-30 mm distance, mainly at 45° with the reinforcement.

An example of extreme cracking is shown in fig 1.

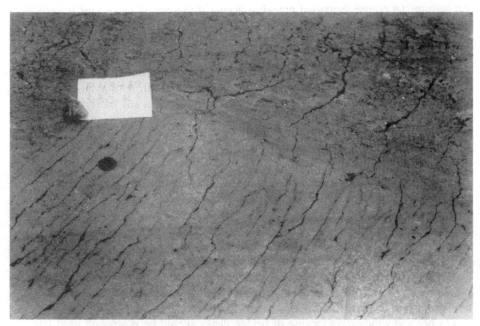

Fig. 1 An extreme case of cracking for a bridge deck.

The depth of the cracks varied, but where wide cracks were found, the depth was often 80-120 mm, that is down to and beyond the reinforcement. Drilled cores from the cracks often showed an odd shape, their width increasing with distance from the surface, see fig. 2.

It was also generally experienced that bridge decks which looked to have only modest cracking turned out to have more pronounced cracking and wider cracks when they were sandblasted.

It has to be pointed out that the cracking occured irrespective of the curing

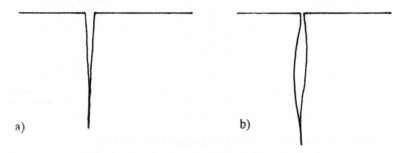

a)   b)

Fig. 2 Shape of cracks experienced from drill-cores.
    a) Ordinary plastic shrinkage cracks.
    b) Plastic cracks in low w/b-ratio concrete.

procedure. Of course increased cracking occurs if the curing procedure is poor or starts late, but neither the application of a thick layer of curing membrane nor casting in rainy weather can eliminate the cracking, if the concrete "feels for cracking".

It should also be pointed out that the cracking phenomenon has not been consistently observed, except at mass-ratios well below 0,40. At a mass-ratio of 0,40 the problem often appears and then disappears, without any easily explainable reason.

5 Conditions influencing cracking

A number of observations have been made as to which circumstances increase cracking and which measures can be taken to reduce cracking.

5.1 Weather conditions
Particularly windy weather promotes cracking, no matter if it is in combination with sunshine, overcast or rain. Dry winds are of course the worst.

High temperatures also promote cracking, but not to the same degree as wind. Sunshine also has some effect, but remarkably little compared to wind and temperature if good curing procedures are followed.

On hot and windy days casting of bridge decks should be postponed until late evening or night.

5.2 Structural design
Thick slabs have shown to have a far greater tendency for cracking than thin ones. Bridge decks having a thick core and slender wings always show more pronounced surface cracking above the thick section. Hollow core box girders with haunches also show more cracking above the haunches, see fig. 3

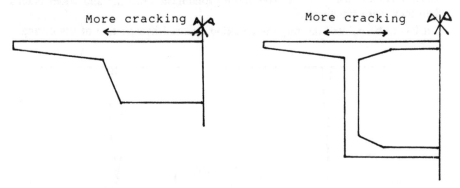

Fig.3 Areas of a bridge deck with more pronounced cracking.

There is also a marked tendency that bridge decks with a large inclination show more cracking than horizontal ones.

5.3 Casting procedure

From the "old days" there is a well established rule that deep concrete sections should be cast in layers, and that the top layer should not exceed 100-150 mm to avoid plastic settlement cracking. When the new cracking phenomenon showed up, all the "good old rules" were followed strictly. This gave no improvement, on the contrary.

Experience has shown that <u>cracking is reduced significantly if the thickness of the top layer is increased to 250-300 mm.</u> This is especially the case for thicker slabs. For thin slabs it implies that the whole depth of the slab should be cast in one operation.

Experience has also shown that the final striking off with vibratory screeds should proceed at a very low speed, to reduce cracking. Too rapid movement of the screed is expected to leave the fresh concrete surface in a state of tensile stress.

5.4 Curing procedure

Good curing procedures carried out very conscientiously are important to reduce cracking, but are no guarantee of avoiding cracking.

A curing membrane must be applied immediately after striking off. Every minute the curing membrane is delayed will increase cracking. The dosage of curing membrane should be 0,4-0,5 l/m^2 to achieve full coverage, which is well above the 0,15-0,20 l/m^2 recommended by the manufacturers. Comparative tests in the field have shown that different products have different protecting abilities, but the best ones in the field are not the ones showing best results in laboratory testing. Both wax-based and acrylic-based curing membranes are found among the most effective ones in the field.

As to the curing practice after membrane application, experiences differ. Some contractors have experienced that covering with a plastic sheet as soon as it is possible, will make the best result. It is the experience of the author that plastic sheet does not work, but that gentle sprinkling with small amounts of water (fog-curing) makes a great difference. Fog-curing should start as soon as the membrane has created a continous film or the concrete shows a loss in consistency, and should proceed until ordinary water curing can be applied.

5.5 Concrete mix design

It is a widespread opinion that cracking problems in concrete will be reduced by reducing the cement content or the cement paste content. However, field experience has shown very clearly that when facing this special early cracking problem, the opposite of that theory is true.

It has been observed that the quicker workability is lost, whatever might be the reason, the more pronounced cracking will be the result. Consequently a normal and slow loss of workability must be a primary goal for the mix design.

It has been experienced that if workability is to be retained, the water content (or the binder paste volume, if one prefers, since the w/b-ratio is fixed), has to exceed a critical limit, which is dependent on the aggregate source and the mixing procedure. This limit is quite sharp, 2-3 liters of water may decide if the workability is retained or is lost in less than 10 minutes. If the water content is below this limit, the consumption of superplasticizer will increase enormously, with no benefit as far as the cracking is concerned.

The dosage of water-reducing admixtures can be taken as a direct measure of how "stressed" the mix design is. The general experience is that dosages up to 6-8 kg/m^3 are acceptable, but that dosages above 8-10 kg/m^3 do more harm than good. This applies both for workability and cracking. Because the response of a concrete to plasticizers depends both on the cement and the aggregate characteristics, the total dosage of plasticizers does not, by itself, give an unambiguous measure of cracking tendency.

Use of constituent materials resulting in low water demand for the mix is favourable and should be aimed at. Reducing the water demand by increasing the proportion of coarse aggregate is not, however, as efficient as theory predicts. Plasticizers have a larger effect in mixes with a higher proportion of sand.

It is well known from the period where silica fume came into regular use that silica fume will promote plastic shrinkage cracking. Experience with early age cracking, however, indicates that the silica fume does not make any difference as long as the dosage is within the 5% limit, and the efficiency factor of 2,0 relative to Portland cement is applied.

From some field cases it can be inferred that long storage of the cement before use, or blowing the cement in moist weather may reduce the cracking·tendency of the concrete.

The type of water reducing admixture used is found to have a significant effect. Limiting the lignosulphonate dosage to a maximum of 3,5 kg/m^3 reduces the thixotropic tendency. Use of highly efficient superplasticizers with a long duration period also reduces the necessary dosage and the rate of workability loss.

Concrete with a slump of 10-30 mm and a mass-ratio of 0,40 seems to be free from the cracking described. There is also some indication that 230-240 mm slump concrete has far less tendency for cracking than 140-200 mm slump concrete.

6 Understanding of the cracking phenomenon on the macro level

In full scale field work a number of cracking mechanisms are acting simultaneously. The reason behind the cracking experienced for low w/b-ratio concretes is believed to be a combination of one or more of the following effects:

A) Conventional cracking mechanisms; i.e.:
    - plastic shrinkage due to external drying
    - plastic settlement taking place after compaction
    - thermal gradients created by surface cooling and/or internal heat
      of hydration
    - lateral movement on inclined surfaces
    - deformations and vibrations in the scaffolding

and a "new" phenomenon:

B)   Early volume contraction of the binder paste
   - overall volume reduction
   - differental plastic shrinkage due to internal suction of water between
     volumes of concrete.

It has been shown in two different field trials, where all the conventional cracking
mechanisms were eliminated, that the early volume contraction of the binder paste
might be large enough to create cracking.

In field work strains due to different cracking mechanisms are superimposed, and
the crack pattern might correspond to the last single mechanism making the tensile
strain capacity exeeded.

From field work it looks as if the early age cracking may be a result either from
an increase in internal volume contraction with reductions in the w/b-ratio, or
some incongruity between the rate of volume contraction and the rate of plasticity
loss or the straicapacity for the binder.

Temperature effects play undoubtedly a role in the cracking phenomenon. It
seems, however, that cooling of the surface is far more risky than temperature rise
in the core. Even more important it is to be aware of the strain due to the internal
volume contraction of the concrete when establishing criterias for allowable
temperature gradients.

7   Recommendation for reduction of plastic cracking for low w/b-ratio concretes

7.1 Concrete mix design

1. If there are possibilities for choosing constituent materials, choose materials
   having the lowest possible water demand.
2. Choose a lignosulphonate dosage of maximum 0,9% by cement weight with
   P30 cement, maximum 0,6% by HS65 cement.
3. Choose a superplasticizer product among the ones having the greatest
   efficiency and the longest duration time. In combination with air-entraining
   agents, melamine-based superplasticizers normally give the best results. If
   high silica fume dosages are used, naphtalene-based superplasticizer should
   be used to reduce the cohesiveness.
4. In hot weather and/or with long transports, use a dosage of retarder
   minimum 0,8 kg/m^3, preferably 1,0-1,5 kg/m^3, to improve workability and
   reduce thixotrophy.
5. Choose an aggregate combination giving 41-44% passing the 4 mm sieve.
6. Choose binder and water content (binder paster volume) so high that the total
   dosage of water-reducing admixtures (admixtures for retempering included)
   can be kept below approx 8 kg/m^3.
7. Be aware that a stable mixing procedure is necessary to produce a constant
   product. The mixing procedure, especially the order of entering the
   constituents into the mixer, is also decisive for the mixing efficiency and the
   resulting water demand of the mix. Cement should be wetted by water before
   coming into contact with admixtures.

8. If the fresh concrete loses workability quickly, it will also have a great tendency for cracking. The reasons for quick loss of workability might be:
- too low water content (point 6)
- cement wetted by admixture (point7)
- too high lignosulphonate dosage (point 2)
- superplasticizer with short duration time (point 3)
- high concrete temperature or aggregates with high water suction

7.2 Execution of work

1. During transport (in particular long transports) the automixer drum should be kept at a very low rotation speed, to reduce heat generation by friction in the concrete mass.
2. Upon arrival at the construction site the automixer drum should be run at maximum rotation speed for approx. 3 minutes, to eliminate coagulation. If retempering is necessary,(which should be considered normal if the transport time is 20 minutes or more),this remixing will take place automatically as a part of the retempering.
3. If the deck is thick enough to be cast in succeeding layers, the top layer should be of 25-30 cm thickness.
4. Striking off with vibratory screeds should be excecuted at a very low speed.
5. A curing membrane is to be sprayed on at a dosage of 0,4-0,5 l/m^2 immediately after striking off. For bridge decks which are to receive a waterproof membrane and an asphalt wearing course a non-wax based curing membrane is recommended.
6. As soon as the concrete has lost some of its workability , gentle water spraying (sweet water only) or preferably fog spraying should start and continue until ordinary water sprinkling can be started.
7. At temperatures below freezing point, where water cannot be supplied, the surface has to be covered by insulating sheets to avoid frost damage and/orexcssive thermal gradients.

8 Consequenses of the plastic cracking phenomenon

The plastic cracking phenomenon is regarded the most serious problem met in using low w/b-ratio concrete. There are serious worries that this phenomenon will jeopardize the quality improvements intended by the use of low w/b concretes.

By observations in the field and fullscale field trials a lot of experience has been gained on how to reduce cracking to a more acceptable level. Understanding of the mechanisms involved has, however, not reached such a level that this cracking can be completely avoided in everyday construction work. Consequenly it is strongly recommended that research should continue on the early age cracking problem, to develop both basic understanding and practical measures.

56 ESTABLISHMENT OF A NEW CRACK PREVENTION METHOD FOR DAMS BY RCD METHOD

N. SUZUKI and T. IISAKA
Department of Civil Engineering, Meijyo University, Nagoya, Japan
S. SHIRAMURA
Aichi Prefectural Office, Nagoya, Japan
A. SUGIYAMA
Department of Civil Engineering, Meijyo University, Nagoya, Japan

ABSTRACT

In the dam construction work using the RCD (Roller Compacted Dam Concrete) method, many heavy construction machines are operated being different from the conventional method, and pipe cooling becomes impossible. Also, concrete placement method is different so that new crack prevention measures become necessary. This report presents the results of temperature hysteresis analysis by the difference calculus for various cases of two-dimensional problems including concrete placement method and materials used in order to develop a crack prevention technique and also present the results of the validity based on the measured temperature and the results of the said analysis.

As a results, it was discovered that changing the amount of cement used, casting lift thickness, casting interval and others are very effective for preventing cracks and for handling the heating phenomenon. By this new method, 15 dams were already constructed and 12 dams are now under construction. Application of this method surely eliminate the occurrence of cracking.

Keywords: Casting Interval, Casting Lift Thickness, Concrete Placement, RCD Method, Structural Members (Concrete Dams).

1. Introduction

In the conventional dam construction method, concrete was casted by the block method in which joints were produced at the intervals of 15 m in dam axis direction and about 40 m intervals in the upstream and downstream direction and cracking was restricted by suppressing the temperature rise by using an artificial pipe cooling method. On the other hand, according the RCD method newly developed, the working surface for dam construction work is all made horizontal, concrete is conveyed with dump tracks and leveled with bulldozers, lateral joints are made with vibrating joint cutters, and concrete is compacted with vibrating rollers. As stated above, heavy construction machines run on the concrete surface so that pipe cooling was impossible.

Therefore, we needed new measures for preventing the cracks. As the measures, we mainly reduced the amount of cement used, lowered casting lift thickness and increased casting intervals in summer season. For trial purpose, we first reviewed and checked these measures at the Shimajigawa Dam (height of 90 m).

Thermal Cracking in Concrete at Early Ages. Edited by R. Springenschmid. Published 1994 by E & FN Spon, 2–6 Boundary Row, London SE1 8HN, UK. ISBN: 0 419 18710 3.

2. Temperature Hysteresis Analysis

The temperature hysteresis calculations for dam concrete are generally performed as one-dimensional problem by using Carlson's method. However, in the RCD method, concrete is placed without performing artificial cooling, and thus we had to first predict the temperature in concrete placement process and then to take the necessary measures against temperature. The temperature was predicted by calculations based on difference calculus for two-dimensional problem.

2.1 Calculating Equations
Next equations are the fundamental differential equations on two-dimensional heat conduction.
The calculations were analyzed by using these equations.

$$\frac{\partial u}{\partial t} = h^2 \left(\frac{\partial^2 u}{\partial x^2} + \frac{\partial^2 u}{\partial y^2} \right) + \Delta T \tag{1}$$

where u: temperature; h^2: coefficient of heat dispersion; ΔT: adiabatic temperature rise per unit time.
This is differentiated by positive type and the following difference equation is obtained.

$$U_{i,j}^{(t+\Delta t)} = U_{i,j}^{(t)} + \Delta t \cdot h^2 \left\{ \frac{U_{i-1,j}^{(t)} - 2U_{i,j}^{(t)} + U_{i+1,j}^{(t)}}{(\Delta x)^2} + \frac{U_{i,j-1}^{(t)} - 2U_{i,j}^{(t)} + U_{i,j+1}^{(t)}}{(\Delta y)^2} \right\} + \Delta T_m(i,j,t) \tag{2}$$

where, the subscripts i and j respectively express mesh points in x and y directions; Δx and Δy respectively express the mesh intervals in x and y direction; Δt expresses calculating time interval; ΔT_m expresses the adiabatic temperature rise within time Δt.
As shown in Fig. 1, the value at time $(t + \Delta t)$ is repeatedly determined by using the values 5 points at time t.

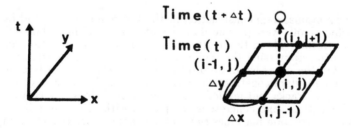

Fig.1 Mesh for Finite Difference Scheme

2.2 Assumption of calculation
The temperature hysteresis calculations were assumed as follows;

(1) Surface temperature of concrete was considered to be the same as air temperature.

(2) For the air temperature, the maximum temperature of air temperature in summer seasons for 11 years and of annual mean air temperature was used.

(3) Heating of base rock was not considered.

(4) It was assumed that the surface temperature of base rock was equal to air temperature, and the internal temperature was equal to annual mean river water temperature.

(5) Heat dispersion coefficient of base rock was the same as that of concrete.

(6) The temperature of casted concrete is equal to the air temperature on that day.

2.3 Calculating Conditions
Principal calculating conditions are as follows:

(1) Casting lift height was set to 70cm.

(2) Placement interval was 4-day cycle in summer season and 3-day cycle in other seasons, and night placement was performed when air temperature was high in summer.

(3) Placement process was performed for standard one, a process 20 days delayed and a process 30 days delayed.

(4) Amount of cement used was $C + F = 130kg/m^3$ and $160kg/m^3$ for internal concrete, $C + F = 220kg/m^3$ for external concrete and $C + F = 160kg/m^3$ and $220kg/m^3$ for rock-contact portion; mixing ratio of fly ash was 20%; also, external concrete thickness was 3 m for both the upstream and downstream portions.

(5) Adiabatic temperature rise of concrete was based on the next equations;

$$Tm = \frac{Wc \cdot Q_{28}}{C \cdot P}(1 - \exp^{-m\alpha}) \tag{3}$$

$$m = -\frac{ln(1 - Q_7/Q_{28})}{7} \tag{4}$$

Where, Tm: adiabatic temperature rise (°C) ; Wc: amount of cement used per 1 m^3 of concrete (kg/m^3); C: specific heat of concrete ($kcal/kg \cdot° C$); P: density of concrete (kg/m^3); Q_7: amount of heat generated ($kcal/kg$) in 7 days per 1 kg of cement; Q_{28}: amount of heat ($kcal/kg$) generated per 1 kg of cement in 28 days; m: constant; α: number of days.

2.4 Result of Calculations and Review
(1) Results of calculations
The temperature hysteresis analysis was made based on the conditions described previously. From the results, the maximum temperature of each case is show in Table 1.

Table 1. Analytical Results

Case Number	Casting Process	Cement Content Joint	Cement Content Inner	Class of Concrete
1	Standard	220		Joint Inner Outer
4	20 Days Delay	160	120	Joint Inner Outer
7	30 Days Delay			Joint Inner Outer
10	Standard	220	130	Joint Inner Outer
13	30 Days Delay	160		Joint Inner Outer

Case Number	Max. Temp. °C	Casting Age	Elevation (m)	Casting Temp. °C
1	35.3	Oct. 8th	213.50	15.5
	38.2	July 31st	251.40	27.3
	40.6	July 31st	251.40	27.3
4	30.9	Oct. 8th	213.50	15.5
	37.2	Aug. 5th	250.70	28.5
	37.9	Aug. 5th	250.70	28.5
7	30.9	Oct. 8th	213.50	15.5
	36.8	Aug. 4th	250.00	28.6
	37.5	Aug. 4th	250.00	28.6
10	35.3	Oct. 8th	213.50	15.5
	39.1	July 31st	251.40	27.3
	41.2	July 31st	251.40	27.3
13	30.9	Oct. 8th	213.50	15.5
	37.5	Aug. 4th	250.00	28.6
	38.0	Aug. 4th	250.00	28.6

(2) Review

(1) The maximum temperature was 41 °C for external concrete; maximum temperature during construction work by conventional method was about 40°C; and they are almost the same.

(2) Difference is about $1°C$ when the amount of $C + F$ is $120kg/m^3$ and $130kg/m^3$.

(3) In the placement process, the temperature dropped by $1.6°C$ when the process was delayed by 30 days from the standard in summer season.

(4) External concrete rose by $2°C$ in standard compared to the internal concrete and rose by $0.5°C$ in 30 days of delay.

(5) Influence of the amount of cement used for concrete at the rock-contact portion created difference of $9.2°C$ when comparing the maximum temperature between case 3 and case 9.

(6) The temperature rise due to placement start time was highest at the start of placement in July to August, followed by June and September to October.

2.5 Measured Values and Calculated Values

With respect to the temperature hysteresis of measured value and calculated value, an example of internal concrete for case 1 is shown in Fig.2. As being apparent from this Figure, the measured value and calculated value were almost the same, and the reliability of calculated value is considered to be very high. Also, the measured location for temperature was 15 m from the downstream face and near the center of dam axis at 20 m from base rock.

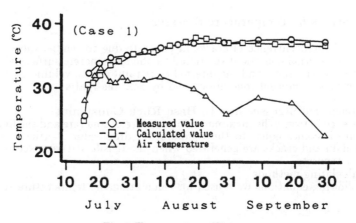

Fig.2 Temperature History

3. Final Stability Temperature

When several years has passed after the construction of a dam, the effect of heating of concrete during placement disappears, and dam temperature becomes constant due to filling of reservoir and external air temperature and the final stable temperature can be obtained. An analysis by difference calculus was performed for this final stability temperature. The results are as shown in Fig. 3. Even by the measurement thereafter, the value has been close to the results.

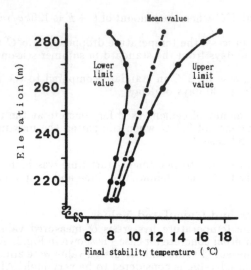

Fig. 3 Change of Final Stability Temperature

4. Calculations of Temperature Stresses

Causes of the occurrence of cracks in concrete due to temperature changes are generally the external constraint restricted by the old concrete surface with different concrete age or base rock, and the internal constraint due to volumetric change near the surface by the internal concrete caused by heat dissipation.

4.1 Temperature Stresses due to Base Rock Constraint

If difference between the maximum temperature of concrete and the final stability temperature becomes equal to the temperature drop value, tensile stresses occur inside the dam and cracks are generated when the tensile stresses exceed the tensile strength of concrete.

(1) Calculating methods

The following equations were used for calculating the temperature stresses.

$$\sigma_{max} = R \cdot E'_C \cdot \alpha(t_1 - t_2) \tag{5}$$

$$E'_C = \frac{1}{1 + 0.4\frac{E_C}{E_R}} \cdot E_C \tag{6}$$

Where, R: constraining coefficient; α: thermal expansion coefficient of concrete t_1: maximum temperature of concrete; t_2: lower limit value of final stability temperature of concrete; E'_C: virtual elastic modulus of concrete; E_C: elastic modulus of concrete; E_R: elastic modulus of foundation base rock.

(2) Results of calculations and review

Tensile stress was calculated by using equations 5 and 6 and the lower limit value of the final stability temperature of Fig. 3 and the maximum temperature in Table 1, and the results of Table 2 were obtained. Since the elastic modulus of the base rock of the Shimajigawa Dam is about 2×10^4 to 4×10^4 kgf/cm^2, a tensile stress of 21.8 kgf/cm^2 will occur at the rock-contact portion of the cases 1 and 10. On the other hand, the allowable tensile stress of the Shimajigawa Dam is about 20 kgf/cm^2, the amount of cement used at the rock-contact portion is set to 180 kg/m^3.

Table 2. Analytical Results

Case Number	Casting Process	Cement Content Joint	Cement Content Inner	Class of Concrete
1	Standard	220		Joint Inner
4	20 Days Delay	160	120	Joint Inner
7	30 Days Delay			Joint Inner
10	Standard	220		Joint Inner
13	30 Days Delay	160	130	Joint Inner

Case Number	Max. Tensile Stress (kgf/cm^2) $E_R = 1.0 \times 10^4$	$E_R = 2.0 \times 10^4$	$E_R = 4.0 \times 10^4$
1	7.0	12.8	21.8
	6.3	11.2	18.3
4	6.0	11.0	18.8
	6.1	10.8	17.6
7	6.0	11.0	18.8
	6.0	10.7	17.4
10	7.0	12.8	21.8
	6.5	11.5	18.8
13	6.0	11.0	18.8
	6.2	10.9	17.8

4.2 Temperature Stresses due to Internal Constraint

Cracks due to internal constraint occur in autumn when concrete placed in summer season has become cold. Therefore, the stress due to such a temperature difference was calculated by using finite element method assuming plane strain condition. In consequence, the tensile stress was found to be 33 kgf/cm^2 maximum. Also, a temperature gradient occurred within concrete and, if temperature distribution shape is considered to be an elastic sheet having a symmetrical parabolic shape about the center, then the temperature stress can be expressed by the following equation;

$$\sigma = -\frac{2}{3}E_C\alpha\Delta\phi \qquad (7)$$

Where, σ: stress occurred on concrete surface; α: thermal expansion coefficient of concrete; E_C elastic modulus of concrete; $\Delta\phi$: temperature difference occurred at the center and on the surface.

As shown Table 1, the temperature difference was about $20°C$ and the tensile stress calculated by equation 7 was 23 kgf/cm^2.

(1) The large tensile stress as stated above is considered to be the highest value under two-dimensional plane strain condition and, in reality, the dispersion of stresses can be considered.

(2) Stresses which cause cracks act gradually and thus the stresses are eased by creep, and cracks have not occurred in existing dams when the temperature difference is less than $20°C$. Therefore, cracks seem to hardly occur.

5. Conclusions

As a result of the analysis and reviews made as stated above, the measures against cracking for the Shimajigawa Dam by RCD method were taken as described below.

(1) Amount of cement used was reduced to a minimum; $C + F = 120$ to 130 kg/m^3 was used for internal concrete; 220 kg/m^3 for external concrete; and 180 kg/m^3 for rock-contact concrete; and mixing ratio of fly ash was set to 30%.

(2) Lift thickness of concrete placement was set to 70 cm.

(3) Placement interval was set to 4-day cycle in summer season, 3-day cycle in other seasons, and placement at night was carried out in summer time when the air temperature was high.

(4) After concrete placement, sufficient water spraying and curing was performed. Also, water spraying and curing was carried out for a long time period for the upstream and downstream surfaces of dam.

(5) Aggregates were stored in a yard where a roof cover was provided to avoid direct sun light.

We constructed the Shimajigawa Dam by the RCD method by taking the measures as stated above, and the dam was successfully constructed without the occurrence of cracks. Thereafter, 15 dams were constructed by adopting the method and measures stated above, and no cracking occurred and good-quality concrete dam were created in all the dams. In addition, even larger dams with the dam height of more than 150 m were constructed by the RCD method. In these dams, pre-cooling method with dry ice was performed for aggregates when the temperature was high in summer.

References

[1] N. Suzuki, et al.: Properties od RCD Construction and RCC Construction, Proc. of JSCE, No.406, VI-10, 1989.

[2] N. Suzuki, et al.: Technique Guide for RCD Construction Method, Sankaidoh Pub., 1989.

Author index

Acker, P. 281
Ahrens, H. 255

Bernander, S. 321, 385, 409
Bilewicz, D. 337
Bjøntegaard, Ø. 229
Bournazel, J.P. 329
Breitenbücher, R. 137, 205, 377

Dahl, P.A. 229
De Jong, R. 247
De Larrard, F. 281
De Schutter, G. 53
Dilger, W.H. 21, 313
Dobashi, Y. 305
Duddeck, H. 255

Eberhardt, M. 353
Emborg, M. 321, 385, 409

Fleischer, W. 401

Gratl, N. 63
Grenier, G. 281
Groth, P. 45, 433
Guenot, I. 103
Guerrier, F. 281
Gutsch, A. 111

Harada, H. 29
Hedlund, H. 45, 433
Hintzen, W. 145
Huckfeldt, J. 255

Iisaka, T. 79, 457
Ikuta, K. 441
Ishikawa, M. 187
Iwata, M. 441

Jonasson, J.-E. 45, 433
Justnes, H. 229

Kawauchi, T. 441
Kishi, T. 11
Koenders, E.A.B. 3

Kompen, R. 449
Koyanagi, W. 95
Kreutz, J.-St. 347
Krylov, B.A. 369
Kuiks, J.E. 273

Laplante, P. 103
Liou, D.D. 393
Lokhorst, S.J. 71, 353

Maatjes, E. 247
Maekawa, K. 11
Malinsky, E.N. 369
Mangold, M. 137, 205, 265, 361, 377
Matsui, K. 305
Matsuoka, Y. 221
Mishima, T. 171
Miyazawa, S. 221
Moranville-Regourd, M. 329
Morimoto, H. 95
Mukherjee, P.K. 197

Nagy, A. 161
Nakamura, M. 171
Nishida, N. 305

Ohshita, H. 119
Ohtani, S. 179
Okamoto, S. 221
Onken, P. 289

Paulini, P. 337
Pedersen, E.S. 63, 297
Plannerer, M. 135

Roelfstra, P.E. 37, 273
Rostásy, F.S. 111, 289

Saito, K. 441
Salet, T.A.M. 37, 273
Sato, R. 179, 313
Schillings, J.J.M. 247
Schöppel, K. 153, 213
Schrage, I. 237
Sellevold, E.J. 229

Shiramura, S. 457
Solovyanchik, A.R. 369
Springenschmid, R. 137, 153, 213, 361, 377, 401
Sugiyama, A. 79, 457
Summer, Th. 237
Suzuki, N. 457

Taerwe, L. 53
Tanabe, T. 119, 187
Tazawa, E. 221
Thelandersson, S. 161
Thielen, G. 145
Thomas,.M.D.A. 197
Tochigi, T. 29
Torrenti, J.M. 103, 281

Uehara, T. 79
Ujike, I. 313
Umehara, H. 79, 171
Ushioda, K. 305

van Breugel, K. 3, 71, 127, 353, 417

Wagenaars, W.G.L. 417
Wang, Ch. 21
Westman, G. 87

Yamada, M. 171
Yamazaki, M. 29, 425
Yoshioka, T. 179

Subject index

Additives 137
Adiabatic calorimetry 1, 3, 30, 247
Aggregates 137, 205, 377
Air cooling 433
Air entrainment 205, 377
Artificial cooling/heating 385, 409, 417, 425
Autogenous shrinkage 213, 221, 229, 329

Basements 347
Basic creep 87
Blended cement 237
Boundary heat transfer 21
Box culverts 79, 171, 441
Bridge decks 281, 401, 449
Bridges 281, 433
Bulk modulus 63

C.P. Method 171, 441
Calculation, risk of cracking 321
Calorimeter, adiabatic 3, 247
Calorimeter, semi-adiabatic 145
Casting, interval 457
Casting, lift thickness 457
Cement 137, 205, 213, 237, 377
Chemical (autogenous) shrinkage 213, 229
Chemical shrinkage 153
Clinker minerals 11
Compressive stress relaxation 95
Computer simulation 409
Concrete composition 145
Concrete placement 457
Construction of reinforced concrete 347
Contraction by hydration 119
Cooling 425
Cooling systems 255, 265, 321, 433
Crack criterion 353, 361
Crack inducer 171
Crack prediction 289
Crack, classification of 385
Crack, effects on hydration 255

Crack, spacing of 171
Crack, width of 171, 417
Cracking 111, 197, 221, 273, 337, 353, 385, 395, 401, 425
Cracking frame 137, 205, 265, 377, 401
Cracking index 441
Cracking moment 313
Cracking sensitivity 137
Cracking temperature 401
Cracking tendency 137, 205, 213, 449
Cracking, avoidance of 409
Cracking, different element geometries 53
Cracking, high performance concrete 229, 449
Cracking, reliability of analysis 305
Cracking, risk of 127, 321, 409, 433
Creep 79, 87, 221, 329
Creep, basic 87, 103
Creep prediction 71
Creep, under high tensile stresses 111
Culvert 79, 187, 441
Curing 401, 449
Curing temperature 127
Curing, artificial cooling/heating 385, 417
Curing, control 247
Curing, effect of different temperatures 369
Curing, sealed conditions 45
Curing, surface cooling 265
Curing, water shower 425

Damage due to cracks 255
Damage due to thermal stresses 255, 329
Dams 329, 457
Deformation, thermal and non-thermal 213
Diaphragm walls 393
Dynamic resonance frequency 161

Effective stress due to pore water pressure 119
Egg-shaped digester 179
Entrained air 205, 377
Equivalent thickness 53
External restraint 187
Extra reinforcement 417

Finite element methods 29, 179, 187, 297, 313, 425
Finite element transient thermal analysis 21
Fly ash 11, 377
Formwork 361
Foundations 29, 197, 255, 281, 337, 409
Fresh concrete temperature 137, 205, 377

Geometrical effects 53

Hardening concrete 21, 53, 297
Heating 409
Heat of hydration 3, 11, 21, 37, 45, 145, 153, 161, 197, 221, 289, 305, 313, 337
Heat transfer 21, 37, 305, 433
High performance concrete 87, 103, 229, 237, 449
High tensile stresses 111
Hydration, degree of 111
Hydrostatic weighing method 63
HYMOSTRUC 3, 71

Instrumentation 171, 393
Intrinsic stresses 265, 361, 401

Lightweight concrete 369
Linear creep model 111
Lock walls 377
Low heat Portland cement 29, 205

Mass concrete 29, 95, 187, 289, 305, 329, 425
Massivity 53
Material laws 289
Maturity development 385, 409
Maturity/equivalent time 103, 329

Mechanical behaviour 281, 289, 321
MEXO module 281
Micromechanics 119
Microstructure 3, 71, 127
Mix design 449
Modelling, creep 71, 79, 103
Modelling, heat of hydration 3
Modelling, hardening concrete 329
Modelling, moisture fields 45
Modelling, moisture transport 45
Modelling, multiphase 63
Modelling, relaxation 71
Modelling, temperature 45
Modelling, thermal stress 161, 289
Modelling, visco-elastic 79, 87
Modulus of elasticity 63, 145, 153, 161
Modulus of elasticity, effective 79, 313
Moisture transport modelling 37, 45
Monitoring instrumentation 393
Mould type 3
Multi-component cement minerals
Multiphase model 63

Numerical simulation 3, 53, 103, 127, 247, 255, 281, 289
Numerical simulation, finite element method 179, 187, 297, 313

Particle interaction 3
Pier shafts 433
Plastic shrinkage rig 229
Pore water pressure 119, 229
Pozzolans 11
Prediction, crack 289, 305
Prediction, creep 71
Prediction, relaxation 71
Prediction, strength 11, 247
Prediction, stress 247, 297
Prediction, temperature 11, 21, 37, 247, 297
Prestressed concrete 179
Prestress, thermal 265

Quality assurance, curing 247

RCD method 457
Reinforcement 417
Relative humidity 45, 229
Relaxation 87, 95, 145, 153, 353
Relaxation, prediction 71
Relaxation, under high tensile
    stresses 111
Resonance frequency method 161
Restraint 145, 187, 237, 289, 321,
    385, 401
Restraint stresses 153, 205, 213,
    337, 401
Rheological model 71
RILEM Round Robin Test 3

Self healing cracks 417
Self desiccation 45, 237
Semi-adiabatic calorimetry 145
Sensitivity analysis of cracking 305
Shrinkage 45
Shrinkage rig 229
Shrinkage, autogeneous 213, 221,
    329
Shrinkage, chemical 153, 213, 221,
    229
Silica fume 137, 237, 377
Silo foundation 29
Slabs 255, 337, 409
Slag 11, 197, 205
Slip detector 425
Sorption isotherms 45
Stiffness formation 63
Strength, development of 11, 145
Strength, influenced by curing
    temperatures 127
Strength, prediction of 247
Strength, the effect of slag on 197
Stress analysis 45, 49, 87, 95, 171,
    179, 273, 385
Stress fields 337
Stress level effect of relaxation 95
Stress meter 179, 187
Stress prediction 247, 297
Structural engineering experience
    347
Structural members 409
Structural members, axi-symmetrical
    179

Structural members, box culvert 79,
    171, 441
Structural members, bridge deck
    281, 401, 449
Structural members, concrete dams
    329, 457
Structural members, concrete pier
    shafts 433
Structural members, diaphragm wall
    393
Structural members, foundation
    plate 281, 337
Structural members, lock walls 377
Structural members, marine
    construction 393
Structural members, slabs 255
Structural members, tanks 247
Structural members, tunnel lining
    377
Structural members, tunnels 273,
    385, 409
Structural members, two-stage
    constructed beams 313
Superplasticized concrete 393
Superposition 111, 221, 289, 353
Superposition/incremental model
    103
Surface cooling 265
Swelling, chemical 153, 213

Tanks 247
Temperature difference limits 273
Temperature differentials 353
Temperature effects on creep 103
Temperature effects on relaxation
    95
Temperature fields 289, 337, 369
Temperature, development of 197
Temperature, prediction of 21, 37,
    247, 297
Temperature-stress testing machine
    137, 145, 205, 213, 401
Tensile strain capacity 197
Tensile stress relaxation 95
Testing equipment 137, 153, 221,
    229
TEXO module 281
Thermal distribution analysis 21

Thermal prestress 265
Thermal simulation 281
Thermal stress analysis 29, 45, 79,
    87, 95, 119, 171, 179, 273, 385
Thermal stress distribution 369
Thermal stress simulation 289
Triaxial testing 119
Triple power law 87
Tunnel lining 377
Tunnels 273, 385, 409

Viscoelastic model 87

Walls 353, 361, 409
Water migration 37, 119
Water-tight concrete 347, 417
'White troughs' 347

Zero stress temperature 361

RILEM Reports

1   Soiling and Cleaning of Building Facades
2   Corrosion of Steel in Concrete
3   Fracture Mechanics of Concrete Structures - From Theory to Applications
4   Geomembranes - Identification and Performance Testing
5   Fracture Mechanics Test Methods for Concrete
6   Recycling of Demolished Concrete and Masonry
7   Fly Ash in Concrete - Properties and Performance
8   Creep in Timber Structures
9   Disaster Planning, Structural Assessment, Demolition and Recycling
10 Application of Admixtures in Concrete
11 Interfacial Transition Zone in Concrete
12 Performance Criteria for Concrete Durability
13 Ice and Construction

RILEM Proceedings

1   Adhesion between Polymers and Concrete
2   From Materials Science to Construction Materials Engineering
3   Durability of Geotextiles
4   Demolition and Reuse of Concrete and Masonry
5   Admixtures for Concrete - Improvement of Properties
6   Analysis of Concrete Structures by Fracture Mechanics
7   Vegetable Plants and their Fibres as Building Materials
8   Mechanical Tests for Bituminous Mixes
9   Test Quality for Construction, Materials and Structures
10 Properties of Fresh Concrete
11 Testing during Concrete Construction
12 Testing of Metals for Structures
13 Fracture Processes in Concrete, Rock and Ceramics
14 Quality Control of Concrete Structures
15 High Performance Fiber Reinforced Cement Composites
16 Hydration and Setting of Cements
17 Fibre Reinforced Cement and Concrete
18 Interfaces in Cementitious Composites
19 Concrete in Hot Climates
20 Reflective Cracking in Pavements
21 Conservation of Stone and other Materials
22 Creep and Shrinkage of Concrete
23 Demolition and Reuse of Concrete and Masonry
24 Special Concretes - Workability and Mixing
25 Thermal Cracking in Concrete at Early Ages

RILEM Recommendations and Recommended Practice

RILEM Technical Recommendations for the Testing and Use of Construction
     Materials
Autoclaved Aerated Concrete - Properties, Testing and Design

RILEM

RILEM, The International Union of Testing and Research Laboratories for Materials and Structures, is an international, non-governmental technical association whose vocation is to contribute to progress in the construction sciences, techniques and industries, essentially by means of the communication it fosters between research and practice. RILEM activity therefore aims at developing the knowledge of properties of materials and performance of structures, at defining the means for their assessment in laboratory and service conditions and at unifying measurement and testing methods used with this objective.

RILEM was founded in 1947, and has a membership of over 900 in some 80 countries. It forms an institutional framework for cooperation by experts to:

- optimise and harmonise test methods for measuring properties and performance of building and civil engineering materials and structures under laboratory and service environments;
- prepare technical recommendations for testing methods;
- prepare state-of-the-art reports to identify further research needs.

RILEM members include the leading building research and testing laboratories around the world, industrial research, manufacturing and contracting interests as well as a significant number of individual members, from industry and universities. RILEM's focus is on construction materials and their use in buildings and civil engineering structures, covering all phases of the building process from manufacture to use and recycling of materials.

RILEM meets these objectives though the work of its technical committees. Symposia, workshops and seminars are organised to facilitate the exchange of information and dissemination of knowledge. RILEM's primary output are technical recommendations. RILEM also publishes the journal *Materials and Structures* which provides a further avenue for reporting the work of its committees. Details are given below. Many other publications, in the form of reports, monographs, symposia and workshop proceedings, are produced.

Details of RILEM membership may be obtained from RILEM, École Normale Supérieure, Pavillon du Crous, 61, avenue du Pdt Wilson, 94235 Cachan Cedex, France.

RILEM Reports, Proceedings and other publications are listed below. Full details may be obtained from E & F N Spon, 2-6 Boundary Row, London SE1 8HN, Tel: (0)71-865 0066, Fax: (0)71-522 9623.

Materials and Structures

RILEM's journal, *Materials and Structures*, is published by E & F N Spon on behalf of RILEM. The journal was founded in 1968, and is a leading journal of record for current research in the properties and performance of building materials and structures, standardization of test methods, and the application of research results to the structural use of materials in building and civil engineering applications.

The papers are selected by an international Editorial Committee to conform with the highest research standards. As well as submitted papers from research and industry, the Journal publishes Reports and Recommendations prepared buy RILEM Technical Committees, together with news of other RILEM activities.

Materials and Structures is published ten times a year (ISSN 0025-5432) and sample copy requests and subscription enquiries should be sent to: E & F N Spon, 2-6 Boundary Row, London SE1 8HN, Tel: (0)71-865 0066, Fax: (0)71-522 9623; or Journals Promotion Department, Chapman & Hall Inc, One Penn Plaza, 41st Floor, New York, NY 10119, USA, Tel: (212) 564 1060, Fax: (212) 564 1505.

9 780367 449308